George B. Sudworth

Nomenclature of the Arborescent Flora

of the United States - Vol. 14

George B. Sudworth

Nomenclature of the Arborescent Flora
of the United States - Vol. 14

ISBN/EAN: 9783337270728

Printed in Europe, USA, Canada, Australia, Japan

Cover: Foto ©berggeist007 / pixelio.de

More available books at **www.hansebooks.com**

Bulletin No. 14.

U. S. DEPARTMENT OF AGRICULTURE.

DIVISION OF FORESTRY.

NOMENCLATURE

OF THE

ARBORESCENT FLORA

OF THE

UNITED STATES.

BY

GEORGE B. SUDWORTH,

Dendrologist of the Division of Forestry.

Issued January 21, 1897.

PREPARED UNDER THE DIRECTION OF **B. E. FERNOW,** CHIEF OF THE DIVISION OF FORESTRY.

WASHINGTON:
GOVERNMENT PRINTING OFFICE.
1897.

LETTER OF TRANSMITTAL.

U. S. DEPARTMENT OF AGRICULTURE,
DIVISION OF FORESTRY,
Washington, D. C., June 15, 1896.

SIR: I have the honor to transmit herewith for publication a revision of the nomenclature of the arborescent flora of the United States, the result of a careful research by Mr. George B. Sudworth, the dendrologist of the Division. The object of the same is to pave the way toward establishing a uniform and as much as possible stable use of names, both scientific and vernacular, of our native trees, and thus avoid the confusion which has often arisen both in technical writings and commercial transactions from a lack of such uniform nomenclature.

Believing that both practical and scientific interests are subserved by this work, I recommend its publication.

Respectfully,

B. E. FERNOW,
Chief of Division.

Hon. J. STERLING MORTON,
Secretary of Agriculture.

INTRODUCTION.

Botanical nomenclature, just as zoological nomenclature, has been in confusion for a long time; lacking in uniformity and stability partly because of a lack of agreement as to the principles according to which the naming of plants should proceed, partly because of inaccessibility to literature when known and named plants were supposed to be new discoveries and were newly named, partly because of changes of relationship when species become subspecies or the reverse, or are referred to other or new genera. In this last respect, namely, so far as nomenclature is also an expression of classification, stability is an impossibility, since added knowledge regarding relationships requires new classification, and hence new combinations of names. Hence any attempt at a revision of the existing nomenclature can at best have for its object only to reduce the unstable condition of nomenclature to a minimum, and to bring uniformity into the manner of making new names in future, to establish principles of nomenclature. Such a revision can not be final, but must remain unstable at least in parts and for a time.

The confusion existing in the scientific names of plants is noticeable in an even greater degree in their common names, where it not infrequently happens that the same name is applied to widely differing species. The consumer may order under a given name one thing, the tradesman applies the name to another thing, and annoyance and loss is the result; as when a Western planter orders Honey Locust, meaning Gleditsia, from an Eastern nurseryman and is supplied according to New England nomenclature with Robinia. Names are neither more nor less than matters of expediency, and it is evident that both the requirements of business and of scientific order demand uniformity and stability in their application.

The need of reform in botanical nomenclature has been recognized for half a century, and attempts to establish principles upon which such a reform might be carried on have been made for the last thirty years, more notably at the botanical congresses, London (1866), Paris (1867), Genoa (1892), and Madison (1893).

Practical considerations, outside of scientific requirements, also begin to assert themselves, and with the more intensive use of vegetable products the need of a stable nomenclature is felt even on commercial reasons. The present revision of the nomenclature of the arborescent flora of the United States has grown largely out of such practical considerations, the needs of the Division of Forestry, the needs of the forester, the lumberman, the nurseryman, etc.

For this reason an attempt has also been made to codify the vernacular names in use in various regions, and it is hoped that thereby a uniform use of those selected may gradually develop. In collecting the names in use in the various localities, the Division is indebted to the working botanists in those localities, who have kindly supplied the information. There may, to be sure, have occurred omissions which in a future edition may be inserted. In choosing a vernacular name for use in future, the principle applied has been to retain and recommend the name applied in the largest number of localities, especially where the species is most common, unless it duplicate another name as frequently employed for another species, in which case the most descriptive name in common use would be substituted, or the one that most nearly translates the botanical species name. This latter principle has also been applied to such species as had no vulgar name known to us, such as many trees from Florida and other minor arborescent forms. Wherever it seemed possible, without violence to well-established usage of names, to adopt for the same genus a common name, it has been done, as Cypress for Cupressus, Cedar for Chamæcyparis, Juniper for all Juniperus. In a few instances where the name in the lumber market is well established, but differs from that applied to the plant in its native home, the former name has been so noted. To enable ready reference and pave the way for a general adoption of these names, a full synonymy grouped by States is added, so that the layman or botanist can readily determine what plant probably is really meant by the native.

To give further practical value to the publication, the nomenclature of horticultural forms has been included, as far as these may be considered well established and distinct; considerable pains have been taken in this part of the work in order to avoid in a measure the multiplicity of names with which nurserymen are loading their catalogues to their own and their customers' confusion.

The revision of the scientific nomenclature has been based upon the lines which were laid down in the code drafted by A. de Candolle and accepted by a congress of botanists held at Paris in 1867, revised by action of the Botanical Club of the American Association for the Advancement of Science. at Rochester in 1892, and at Madison in 1893. Since these codes are not readily accessible, they have been reprinted as an appendix of this bulletin, together, for comparison, with the code of the ornithologists, which has given general satisfaction.

It should, of course, be understood that as all laws made by man are not, like natural laws, infallible, but capable of varying interpretation, it becomes necessary in each individual case to have the law construed and applied by a court: a judge whose rulings will stand according to the amount of knowledge and common sense he brings to the bench. The reviser must bring to his work not only an impartial, just, and unprejudiced disposition, a clear-sighted, judicial mind, a knowledge of the laws in all their bearings, but he must have an intimate knowledge of the method of botanical description and systematics; he must have at command not only a full herbarium and the facilities for obtaining live material, but he must be thoroughly at home in the literature of botanical systematics; he must have at hand the use of a full library,[1] for in the end his work is one of literary research.

Even to those who do not feel inclined to adopt the new nomenclature the work will prove useful and welcome, on account of the full synonymy with bibliographical and historic data. In order to permit ready reference, the familiar names used in standard handbooks of botany, as Gray's Manual, Chapman's Flora of the Southern States, etc., have been printed in capital letters. Occasionally a full list of North American species, including nonarborescent species, is given in footnotes for their general interest in studying the genus.

The essential basis upon which the revision has been made is the so-called "law of priority," i. e., for species and varieties the specific or varietal name has been taken up which was first used by the author who first described the plant, and for genera the first established generic name, either alone or in combination with a type specific name. In order to avoid obscurity and uncertainty, the publication in which for the first time the binomial nomenclature was used persistently, namely, Linnæus's Species Plantarum (1st edition, 1753), has been made the starting point in accordance with an expression of the botanists of the Botanical Club of the American Association for the Advancement of Science. Objections have been made to the injustice committed in ignoring earlier names; the objectors overlook that it is not a matter of justice primarily, but of expediency, which leads to the adoption of the law of priority, and it would be inexpedient to go back to an earlier date than the one which firmly establishes our present system of notation.

The principles which have guided Mr. Sudworth in this and other matters, and his interpretations, are stated by him in his preface or elsewhere in notes.

Since this publication has for its object reform and uniformity in the use of names, and is intended to become an aid in that direction, it seems desirable to add a few words on some minor yet pertinent questions, in regard to which of late much lack of uniformity is noticeable, namely, the capitalization and hyphening of names.

Regarding scientific names, the best usage and the plan most generally adopted by zoologists seems to be to capitalize generic names invariably and to let the specific and varietal names appear in lower-case

[1] Much essential but rare literature, to be sure, is not possessed by American libraries, being accessible only in foreign libraries.

type, even when derived from patronymics. This usage, which has been adopted in this catalogue, has the advantage that the genus and species names, if used by themselves, are at once recognized as such. It is true that by this usage, namely, the decapitalization of specific names derived from patronymics, the etymological information is to a degree sacrificed, but to the real student this is so only in rare cases, because the form and character of the word will usually suggest its derivation. In citing names, to be sure. they should always be written as the author wrote.

Regarding common names. a tendency to discontinue the use of initial capital letters has lately shown itself, yet without entire consistency, for parts of names derived from patronymics are often capitalized even by those having the tendency to decapitalize otherwise.

The only reason for the nonuse of initial capital letters that we can see is the convenience of the compositor, the typewriter. or whoever prepares the manuscript. This, it appears to us, is approaching the problem from the wrong end. The convenience of the reader undoubtedly is the much more important consideration, and that requires that with the least possible effort on the part of the eye and the mind of the reader, or of the mind through the eye, the full meaning of the writer be conveyed. It is for this reason that the printed language contains such mechanical aids as capitals, hyphens, punctuation marks, quotation marks, and differences of type, which do not exist in the spoken language, because here intonation. accentuation, timing. and other means assist in conveying ideas. While, undoubtedly, from practical considerations it is desirable to reduce the mechanism for conveying ideas in printed language to the smallest amount. it would appear poor economy to save the time and mental effort of the one typesetter by requiring of the many readers however small an additional mental effort which can be avoided.

This desirable ease of apprehension, it is believed, is subserved by invariably capitalizing all names and parts[1] of composite names whenever they are used specifically as names and not as class words. In this way they are set off from the rest of the text, and the eye helps the mind of the reader to realize without effort, and immediately, that a name is intended and not description. For instance, a Drooping Juniper or a Big Tree or a Live Oak or a Remarkable Pine could not for a moment be mistaken for "a drooping juniper" or "a big tree" or "a live oak" or "a remarkable pine," the one being a specific name, the other a class description; the adjective in the one case being a part of the name, in the other simply a qualifier of a class word. The mental effort becomes more aggravated and. by the way, the artistic sense of the reader suffers more when several words are combined to

---

[1] A connecting preposition or article may be kept in lower-case type with advantage to the clearness in typography without affecting the sense or name value of the combination or the convenience of the reader, as Balm of Gilead, Lily of the Valley, etc.

make a species name, as Prince Albert's Pine or Great Oregon Fir, which the innovators would write "prince Albert's pine," "the great Oregon silver fir," and, although it may be urged that the context would readily reveal whether a pine belonging to Prince Albert is meant or a particular species, nevertheless, when all ambiguity can be avoided by so small an effort on the part of the compositor, it is worth doing. Often the convenience of the reader would not require such distinction of the name from the rest of the matter, as when we speak of the Longleaf Pine or the Incense Cedar, where no possibility of misunderstanding or need of aiding the eye exists, yet, for the sake of uniformity, it is desirable to make no exceptions and to capitalize.

As to compounding and hyphening, the greatest divergence exists, not only in regard to names, but in common language. No two dictionaries, no two grammarians, agree or are consistent; no logic or undisputed rule or law seems to prevail. Usage is therefore an unsafe guide because ever varying.

There are three ways of compounding names: By writing their parts separately side by side, or by joining them together in one word, or by hyphening them. There also seem to be three considerations upon which choice of the method may proceed, namely, grammatical considerations, logical considerations, and considerations of convenience.

We would again repeat with emphasis that written or printed language is the use of signs to convey the sense of spoken language with the least amount of mechanical work and yet with the least effort on the part of the eye and mind of the reader. Just as in spelling we gradually lose sight of the etymology, so in compounding, grammatical considerations are gradually relegated to the background, and expediency, the rapid, easy, and sure conveyance of thought to the reader, should form the guiding principle. The question whether the compound word is most easily read or most easily and correctly understood in one form or the other should decide its form.

The use of the hyphen especially should be restricted to cases where the convenience of the reader absolutely requires it. The hyphen is a means of separation rather than of joining words, which it is obvious would be more effectively done by writing them together. Conceived in this manner the hyphen means that, while according to the sense or the unit idea involved, the words should be compounded and written together, for the convenience of the eye and of the mind of the reader, we separate them by a sign. Such separation becomes desirable when either or both words are long and can not when written together be readily deciphered, or when the last letter of the first and the first letter of the second word are the same and thereby the eye is disturbed, or when some other misleading or misreading is invited by writing them together.

The use of capitals in all parts of a name combination reduces the necessity of compounding and hyphening also to a considerable degree. Where the last part of the combination is not a generic term or name but a general class word and short. compounding may advantageously be done. There are especially a number of such class words most generally used in tree names, like tree. wood. bush. berry, leaf, which are conveniently compounded. We have done so invariably (except in the synonymy). hyphening. however, when the above-stated reasons seem to make separation desirable.

For scientific names, where genus and species names are always indicated by position and where the binomial form prevents the use of more than two words, the hyphening is restricted to those cases where convenience of reading makes it desirable. Especially when the compound name is formed according to the well-established rules of the Latin language the need of a hyphen becomes less and less necessary.

To illustrate. the following type examples may serve:

Uncompounded forms.	*Compounded forms.*	*Hyphened forms.*
White Pine.	Whitewood.	Yellow-wood.
Silver Fir.	Silverleaf Poplar.	Alternate-leaf Dogwood.
She Oak.	Cherrytree.	Yate-tree.
Cherry Birch.	Hornbeam.	Dwarf Rosebay-tree.
Swamp Spanish Oak.	Oldfield Pine.	Ants-wood.
Tree Huckleberry.	Earleaf Magnolia.	Table-mountain Pine.
Poison Oak.	Poisonwood.	Wait-a-bit.

We realize that in this short rule of consulting the convenience of the reader we must leave many cases doubtful, and in the end differences of opinion must arise as to what the convenience of the reader demands. We can, of course, not enter here upon a general discussion of the intricate problem of compounding words. but we hope that by laying down a reasonable guiding principle and consistently applying it. to have paved the way to uniformity of usage.

Short names are always preferable to long names. To divest the descriptive names of all unnecessary length it seemed desirable to shorten all participle form endings, as -leaved, -fruited, -berried. -barked into -leaf. -fruit. -berry. -bark, by which also the name significance of the word is brought out more strongly, as in Earleaf Umbrella. or else to leave out these endings altogether, as Fringe Ash instead of Fringe-flowered Ash; for the same reason the possessive endings have been dropped. being unnecessary mechanical incumbrances, as in Fraser Umbrella. Where a geographical distinction like Western, American, etc.. could, when speaking or writing from or about the locality. be dropped without harm, we have placed it in brackets as indicating its superfluity under such conditions.

While the attainment of stability and uniformity of usage can not be expected either in vernacular or scientific nomenclature without some such effort as we have here made. it is hoped that friendly criticism will enable us in a future edition to improve the defects which attach naturally to such a work and to come a step nearer to the desired end.

B. E. FERNOW.

NOMENCLATURE

OF THE

ARBORESCENT FLORA

OF THE

UNITED STATES,

BY

GEORGE B. SUDWORTH,

DENDROLOGIST OF THE DIVISION OF FORESTRY.

CONTENTS.

PREFATORY REMARKS.

The following list of plants with their synonyms is intended to comprise all our arborescent flora. Popularly, size appears to be the general characteristic by which a tree is distinguished from a shrub; the line of demarcation is, however, arbitrary and often difficult to establish. In the present catalogue the definition of a tree is based on habit rather than size, and includes such woody plants as produce in nature a single trunk branching more or less above the ground.[1] Plants producing several stems from the same root stock (as in some species of Salix, Sambucus, etc.), often of large size, are accordingly excluded.

The species enumerated comprise only those found (as far as known) within the borders of the United States, numbering 492 species, together with many varieties and hybrids. A few suspected new species, chiefly *Coniferæ*, are omitted, being as yet insufficiently known. In a few instances exotic species have been introduced, when believed to be thoroughly naturalized in the United States.

SEQUENCE OF FAMILIES, GENERA AND SPECIES.

The order of sequence is based upon the recent work of A. Engler and K. Prantl, "Die natürlichen Pflanzenfamilien," with a few deviations in respect to genera in the larger families. The newer arrangement, although inconvenient to students familiar with the long-established sequence of Bentham and Hooker's Genera Plantarum, is believed to more rationally express the affinities of the natural groups under consideration. Under this arrangement some of the older families have been divided and the divisions appear as distinct families—notably Urticaceæ and Cupuliferæ. So far as known the oldest established genera are taken up, but the synonymy of genera has been omitted, except occasionally, as unimportant to the immediate purpose of this work, while that of the species has been made as full as possible.

---

[1] This is in accordance with the following definition: "Trees are woody plants, the seed of which has the inherent capacity of producing naturally, within their native limits, one main erect axis bearing a definite crown, continuing to grow for a number of years more vigorously than the lateral axes, and the lower branches dying off in time."—B. E. Fernow, in Garden and Forest, Vol. II, 410, 1888.

5

A strict chronological order is observed in the synonymy of species and varieties giving the initial citations from the oldest to the latest, including also an observance of priority in time and pagination of publications. The revision has been based upon the law of priority as recognized in the French and recent American codes. While these codes determine the general procedure, certain questions always appear which the reviser has to settle for himself, and he owes to his readers an explanation of the position taken with reference to these debatable points. These points comprise the following: The limit of the literature available for this purpose; the basis for specific, varietal, and form distinction, and differentiation of names for the same; the question of sufficiency or adequacy of description to identify the plant with the name; the ground upon which duplicated names (homonyms) in the same genus are displaced and synonyms admitted as such; the ground upon which a name is considered tenable from the bibliographical point of view, or in other words, what has been considered publication in doubtful cases; the question of correct spelling and some minor points of citation.

DENDROLOGICAL LITERATURE.

Our arborescent flora has doubtless received less critical study than almost any other class of plants, although the literature is old and extensive, the trees in their useful and ornamental aspects having always attracted the attention of explorers, travelers, and all lovers of nature.

There are few general or comprehensive works on American trees. A dozen or more authors have attempted to compass the subject, wholly or in part. The Englishman, Marshall, with his Arbustum Americanum, published in 1785, was the first, followed by the German, Wangenheim (1787), the Frenchmen, Michaux, father and son (1801–1813), the American, Nuttall (1842), Browne (1832), Emerson (1846), Gray (1850, a work never completed), Piper (1858), Cooper (1858), Engelmann (mainly 1860–1883), Vasey (1876), Curtis, Kellogg, Sargent, Ridgway, Greene, Lemmon, and other late writers.

Professor Sargent's recent Silva of North America, now nearly completed, is the only comprehensive and the most elaborate work ever issued.

Much also has been written incidentally on the arborescent flora of North America, and many species were described by Linnæus (1753), Jacquin (1760–1790), Lamarck (1783–1804), Borkhausen (1760–1800), Poiret (1810–1817), Rafinesque (1817–1838), Loudon (1838), Koch (1869–1873), and others. But much of this literature is contained in books and transactions of scientific societies long out of print or rare and difficult of access. The descriptions are, moreover, often imperfect and with too few diagnostic features, requiring a very wide and intimate acquaintance with the existing flora to safely determine their application. When original types are still preserved, doubtful descriptions may, to be sure, be verified.

The narratives of early travelers and explorers give us accounts of a few of our trees named and described more or less clearly. There is a tendency, however, among some botanists to ignore much of this literature on the ground that the descriptions are not technically drawn. Nevertheless there is no good reason why these writings should be neglected in an attempt to trace the history or to establish the earliest name of a plant, provided the sometimes loose and incomplete descriptions permit identification. For instance, we would not be justified in overlooking the fact that the Scotch explorer, David Douglas, recognized as a distinct species the California Spice Tree, although he described it only as having olive-like fruit and as producing violent sneezing, facts which enabled later botanists acquainted with his statement of the one striking character of the plant to at once place his *Laurus regia* with considerable certainty.

On the other hand, even descriptions of the present time often lack in important points, as when the describer of *Fraxinus anomala triphylla* (1896) omits to state whether it is a tree or a shrub.

A consideration of the literature of horticultural and nurserymen's varieties and other forms of our native trees is commonly excluded from writings of botanists chiefly, perhaps, because they are thought to be artificial, but a full study of the species would seem to require that all forms in nature, as well as its modifications under cultivation, should be the concern of botanists, and hence should be brought under the same general laws of nomenclature. The practitioner as well as the scientist will be benefited if a uniform nomenclature can be secured, for it will obviate many inconveniences and misunderstandings. A red-leaved tree known to be only a form of a well-known species is best designated, if necessary, in varietal or subvarietal rank, and not misleadingly as a new species. It is no less astounding than it is fallacious to see an unmistakable form of *Magnolia obovata* labeled "*Magnolia semperflorens*," based alone upon an acquired or accidental habit of flowering several times a year.

Although technically not so considered, chiefly from the lack of a date of publication, the nurseryman's attractive and expensively gotten-up trade catalogues are a species of botanical literature in which distinct cultivated forms of trees are often named, figured, or described accurately for the first time, but for lack of date the priority of the names of such plants can not always be established. It often happens also that new garden varieties or forms are introduced in trade catalogues with only a bare name, the omission of distinctive characters thus making it impossible to establish the name. A reform in this respect is recommended.

LIMITATION OF SPECIES AND VARIETIES.

The systematist naturally seeks stable limitations, but the old question of what constitutes a species, a variety, or form must always remain more or less unsettled and largely dependent upon the extent of study given a species under its various conditions. The tendency of the older and better informed and of more conservative botanists is to reduce the number of species, relegating locally and apparently specifically distinct forms to varietal rank. The common Black Maple is a very distinct species in parts of its range, but when studied over its entire range intermediate forms appear to connect it with the commoner Sugar Maple, suggesting varietal relationship. The Green Ash is equally distinct as a species, yet plants are found in the South where it passes into the Red Ash. Other examples are numerous. It may be said that one of the objects of systematic botany is to discover the natural affinities of plants, also, for general expediency, to properly designate forms displaying fairly constant and distinct features. The problem of establishing the relationship of such forms becomes most difficult, especially at the line of convergence, when the affinity to one or the other species or the right to specific rank becomes a question. This may lead to a very undesirable and unphilosophical multiplication of species, or to an unwarranted combining of fairly distinct forms. Some European systematists have sought to compass the difficulty in their species *typicæ*, *intermediæ*, *dubiæ*, etc., limitations possible, perhaps, for cultivated forms, but hardly for those in nature.

How far varietal and subvarietal distinctions may be carried for garden forms of our trees is a matter of some practical interest. The superficial distinctive features of garden forms are rendered fairly stable by the attentive and skillful horticulturist, while in nature the same plant might soon lose many of these accidental characters. The Cut-leaf Silver Maple, and the Heart-leaf Magnolia would be hard to find in nature, but are very distinct in cultivation, and must, therefore, be consigned to some category other than the synonymy of the type species. Obviously the fixed conditions under which we find such forms remaining distinct may determine the value and necessity of the segregating process.

The nomenclature of cultivated trees is deeply involved and most perplexing, but an attempt has been made in the present revision to harmonize the nomenclature of a large number of cultivated varieties and forms under each species. In a few instances it has seemed necessary to coin new names.

IDENTITY OF NAMES.

The difficulty of making sure to what plant the author intended a given name to apply—of identifying the name—has been alluded to.

In seeking to reach a just and accurate decision where the question of priority is involved, vague or incomplete descriptions have in some cases been supplemented by circumstantial evidence. Thus the fact that the habitat assigned could contain no other two-leaved pine makes the first describer's name for the Banksian Pine, *Pinus divaricata*, tenable, although the accompanying description is incomplete. The *Acer saccharum* of Marshall (Sugar Maple) is retained because all the other maples treated by Marshall are clearly enough described, by elimination, leaving little doubt but that his *A. saccharum* was applied to our Sugar Maple.

Catalogue names or others unidentifiable for lack of sufficient description of any kind (*nomina nuda*) have been replaced by the next oldest established ones, and the former cited as *nomina nuda*. A few such names have been in common use; the long-known *Quercus tinctoria* (Black Oak) of Bartram, rested chiefly upon the statement that the bark was used for a dye. Other oaks of the region are, however, known to be used for that purpose.

SYNONYMS AND HOMONYMS.

There has been a common practice of coining or maintaining the same name or term for specific, varietal, or subvarietal distinctions in the same genus, which gives rise to homonyms. The present practice has led to the necessity of considering these as synonyms and relegating all but the oldest to synonymy.

For example the three following names have been published in the genus Pinus: *Pinus sylvestris latifolia* (1858), *Pinus contorta* var. *latifolia* (1871), and *Pinus latifolia* (1889), all for different plants. The present régime of allowing but one and only the earliest term, *latifolia*, to be maintained in the genus Pinus, makes it necessary that of the above all but the one published in 1858 go into synonymy; in the case of the Arizona *Pinus latifolia* (1889), the plant was thus left without a name. This method of permitting a specific, varietal, or subvarietal term to occur only once in combination with the same generic name is the interpretation of the rule, "Once a synonym always a synonym."

Inferential homonyms arise from grouping species originally described in different genera under the same genus, by which change they appear in the same synonymy, and thus may, if the same specific terms occur, by inference fall with their specific names under the rule, "Once a synonym always a synonym." But as they have not actually been in combination with the genus name under which they are now grouped, they are not truly homonyms. Hence it seems proper not to so consider them in the selection of specific names. This method of treatment has the additional advantage of disturbing the existing nomenclature as little as possible.

Cerasus nigra Miller (1768) is a synonym of *Prunus avium* L. (1753), but Aiton described a *Prunus nigra* (1789) distinct from Linnæus's plant. By inference the later reference of *Cerasus nigra* Mill. (1768) to the genus Prunus would invalidate *Prunus nigra* Ait. (1789); but it is preferable to leave it undisturbed until either an earlier name (*Prunus nigra*) is found to exist or else until by reason of reclassification a new combination of the specific term *nigra* with the generic name *Prunus* becomes necessary.

Composite names are occasionally met with, and are those under which a writer has included two or more species or varieties which have afterward been shown to be distinct. These are perplexing and difficult to dispose of when the separation of species or varieties becomes necessary. Such a name may either serve to designate one of the separated species or varieties and also become a synonym in part of the other, or it may appear unsafe to use the name at all, when it is necessary to substitute later unique names for both.

A case of this last kind is exhibited in Linnæus's *Nyssa aquatica* (1753), which includes two distinct species, the Water Gum and the Tupelo Gum. Later, Marshall described the latter species as *Nyssa aquatica* (1785), after which Wangenheim also described it as *Nyssa uniflora* (1787).

Marshall's use or acceptance of Linnæus's name as applied to the one species is interpreted as fixing the service of that name for the Tupelo Gum, forcing Wangenheim's name of later date into synonymy. At the same time, since one name can not designate two distinct objects, Linnæus's *Nyssa aquatica*, as far as it related to the Water Gum, must appear as a synonym in part under the later *Nyssa biflora* of Walter, given to this species.

Prosopis odorata T. & F. (1845) includes the Screw Bean and Mesquite, but the latter had already been described as *P. juliflora* de C. (1825), to which, in part, *Prosopis odorata* becomes a synonym, leaving the legitimate service of *Prosopis odorata* open for the Screw Bean. To maintain Bentham's later *Prosopis pubescens* (1846) for the Screw Bean appears inconsistent, since under these conditions there is no reason why a composite name should be suppressed or reduced to synonymy except in part, unless to give way to an older name.

AUTHORSHIP AND PUBLICATION OF NAMES.

The question of authorship of a name involves the credit and the responsibility of having proposed a name with description, whether published over his own or another's signature. Publishing botanists often include in their own work names with descriptions by other botanists, in most instances the publisher carefully inserting after each the name of the author proposing them. As is well known, Torrey and Gray thus published much of Nuttall's material. In cases of this kind, although the practice has not been uniform among all botanists, the real author of the name, with the indicated source of publication, has been cited, which is believed to be the proper method, since in such cases it is not the publisher who really named the plant.

There are some doubtful cases as to when a name is to be considered properly published. The Texas Red Bud was first described and referred by Gray to *Cercis occidentalis* L. var. (1850), to which was appended as an equivalent Engelmann's manuscript name *C. reniformis*. The plant proved to be distinct, and later received another name, *C. occidentalis* var. *texensis* Watson (1878), which some authors retain to the exclusion of Engelmann's earlier name (1850). The case is clearly one in which, regardless of Gray's intentions in citing Engelmann's manuscript name, a plant simultaneously received two names, one of which being incomplete, leaves the other legitimately as the first correct name of the plant.

NAMES FOR HYBRIDS.

The earliest name applied to supposed hybrids has been preserved, whether indicating a natural unique form or a hybrid. Bartram's Oak was first described as *Quercus heterophylla*. Later, Gray determined it to be a composite form, a hybrid, and gave it a name (*Q. phellos* × *tinctoria*) indicating its possible origin. The law of priority, however, demands that the earlier name be maintained and applied so long as the form remains recognizable, regardless of its apt or inapt indication of the plant's origin.

MIXED NAMES.

A few trinomials designating garden varieties contain mixed terms—English and Latin parts. Such names, although not in accord with the desirable adoption of uniform Latin or Latinized plant names, are left unchanged, since a mixed name is none the less distinct, and may therefore serve the legitimate purpose of a name so long as needed. The retention of these names unchanged has a precedent in that of many technical names containing barbaric, specific, or varietal terms. Moreover, the substituting of a new Latin or Latinized term for an imperfect term in an already published name is an unwarranted liberty, for not even the author of a name once published is permitted to subsequently replace or radically change it.

APPROPRIATE COMMON NAMES.

Where species or varieties have not acquired common names, being known only by their technical names, an attempt has been made to supply suitable vernacular names, and, wherever possible, such names are translations of the technical name, in many instances suggesting a more or less distinctive feature of the plant, referring to a geographical limitation, or indicating the person in honor of whom the plant has been named.

EXACT CITATIONS.

So far as has been possible the names cited in synonymy are given exactly as their authors published them. The literary value of such exactness is slight, but is believed to be an essential precision, since an author is then never intentionally made to say what he did not say.

Occasionally, for the sake of brevity, a short translation of a publication's exact title is substituted. Lamarck's great work, "Encyclopédie Méthodique Botanique," is commonly cited "Dic." But for the sake of exactness, it seems preferable to cite the original title abbreviated.

CORRECTIONS IN SPELLING AND CAPITALIZATION.

The agreement of American botanists to make 1753 a date of departure for genera and species has necessitated some changes in the spelling of a few Linnæan names. The question, moreover, of how much liberty may be taken in correcting all names supposed to be erroneously formed or spelled is capable of a varied interpretation. Linnæus often used names ending in the Greek form *on* in his earlier works, and in 1753 or later changed the endings to the Latin form *um*, or vice versa. The spelling adopted in the Species Plantarum of 1753 has been preserved. Similarly in other cases of spelling altered by other authors, such as *Gleditschia* for *Gleditsia*, *Guaiacum* for *Guajacum*, *Thuia* and *Thuya* for *Thuja*, *Anona* for *Annona*, etc., the Linnæan spelling is maintained for the sake of exactness. Rafinesque and Marshall have given us a few anomalies, the correction of which is believed to be in conformity with the authors' avowed intentions. Thus the earlier *Scoria* and *Ioxylon* of Rafinesque have been changed to *Hicoria* and *Toxylon* and the *Populus deltoide* of Marshall to *P. deltoides*, the supposition being that these were typographical mistakes, in the case of Rafinesque the evidence being furnished by his later usage.

In regard to capitalization of specific or varietal names, the practice of the zoologists in uniformly using the lower-case initial letters has been followed for the sake of expediency and for securing uniformity of appearance and usage.

The writer can not help expressing the conviction in this attempt to harmonize an enormous array of names designating our trees, that the effort is contributory only to a final and fuller understanding of our forest flora. New research must constantly add to and undo much that is now thought to be stable, the standard which is believed can be reached only by the continued and combined efforts of all working botanists.

GEORGE B. SUDWORTH.

GYMNOSPERMÆ.

Family CONIFERÆ.

PINUS Linn., Spec. Pl. 1000 (1753).

Pinus strobus Linn.   **White Pine.**

SYN.—*Pinus Strobus* Linnæus, Spec. Pl. ed. 1, II, 1001 (1753).
Pinus tenuifolia Salisbury, Prodr., 399 (1796).

COMMON NAMES.

White Pine (Me., N. H., Vt., Mass., R. I., Conn., N. Y., N. J.,
Pa., Del., Va., W. Va., N. C., Ga., Ind., Ill., Wis., Mich.,
Minn., Ohio, Ont., Nebr.).
Weymouth Pine (Mass., S. C.).
Soft Pine (Pa.).
Northern Pine (S. C.).
Spruce Pine (Tenn.).

VARIETIES DISTINGUISHED IN CULTIVATION.

Pinus strobus brevifolia Loud.

SYN.—*Pinus Strobus brevifolia* Hort. ex Loudon, Arb. Frut., IV,
2280 (1838).
Pinus Strobus compressa Booth ex Loud., l. c. (1838).
Pinus Strobus nova Loddiges Cat. ed. (1836) ex Loud., l. c.
(1838).
Pinus Strobus nana Knight, Syn. Conif., 34 (1850).
Pinus compressa ex Koch, Dendrol., zw. Th. zw. Ab., 320
(1873).
Pinus Strobus pygmæa Hort. ex Beissner, Handb. Nadelh.,
291 (1891).

Pinus strobus umbraculifera Knight.

SYN.—*Pinus Strobus umbraculifera* Knight, Syn. Conif., 34 (1850).
Pinus tabuliformis Hort. ex Gordon, Pinetum, ed. 1, 240
(1858).
Pinus umbraculifera Hort. ex Koch, Dendrol., zw. Th. zw.
Ab., 320 (1873).

13

Pinus strobus minima Beissn.

SYN.—*Pinus Strobus pumila* Hort. ex Gordon, Pinetum, 2d ed., 323 (1875), not *P. pumila* Reg. (1858), nor Koch (1873).
Pinus Strobus minima Hort. ex Beissner, Handb. Conif., 55 (1887); Handb. Nadelh., 292 (1891).

Pinus strobus fastigiata Beissn.

SYN.—*Pinus Strobus fastigiata* Hort. ex Beissner, Handb. Conif., 55 (1887); Handb. Nadelh., 292 (1891).
Pinus Strobus pyramidalis Hort. ex Beissner, l. c. (1887), not *P. pyramidalis* Salisb. (1796), nor Antoine (1847), nor Gord. (1858).

Pinus strobus viridis Carr.

SYN.—*Pinus Strobus viridis* Carrière, Trait. Conif., nouv. éd., 400 (1867).

Pinus strobus gracilifolia nom. nov.

SYN.—*Pinus Strobus gracilis viridis* Hort. ex Beissner, Handb. Nadelh., 292 (1891), not *P. Strob. viridis* Carr. (1867).

Pinus strobus nivea (Knight) Carr.

SYN.—*Pinus Strobus* a *alba* Loudon, Enc. Trees, 1018 (1842), not *P. alba* Ait. (1789).
Pinus nivea Booth ex Knight, Syn. Conif., 34 (1850).
Pinus Strobus nivea Carrière, Trait. Conif., nouv. éd., 400 (1867).
Pinus Strobus argentea Hort. ex Carr., l. c. (1867).

Pinus strobus aurea Carr.

SYN.—*Pinus Strobus aurea* Hort. ex Carrière, Trait. Conif., nouv. éd., 400 (1867).

Pinus strobus variegata Carr.

SYN.—*Pinus Strobus variegata* Hort. ex Carrière, Trait. Conif., nouv. éd., 400 (1867).

Pinus strobus zebrina Beissn.

SYN.—*Pinus Strobus zebrina* Zocher ex Beissner, Handb. Nadelh., 292 (1891).

Pinus strobus prostrata [1] Hort. Kew.

SYN.—*Pinus Strobus* var. *prostrata*, ex Hand-list Conif. Roy. Gard. Kew, 101 (1896).

---

[1] So far as known this form has never been described. The name given in the Hand-list of Coniferæ of the Royal Kew Garden is unaccompanied by any description. As the name indicates, it is probably a low or prostrate form, apparently still unknown to American nurserymen.

Pinus monticola Dougl.   **Mountain White Pine.**

SYN.—*Pinus monticola* Douglas MSS. in Lambert, Desc. Gen.
Pinus, ed. 2, III, 27, t. 87 (1837).
Pinus Strobus var. *monticola* Nuttall, Sylva, III, 118 (1849).
Pinus porphyrocarpa Murray in Lawson, Pinetum Brit., I,
83, f. 1–8 (1866).
Pinus nivea Hort. ex Carrière, Trait. Conif., nouv. éd., 401
(1867), not Knight (1850).
Pinus Grozelieri Carrière, in Rev. Hort. 1869, 126 (1869).
Pinus monticola var. *minima* Lemmon, in Second Bienn. Rep.
Cal. St. Bd. For., 70 (1888).
Pinus monticola var. *digitata* Lemmon, Handb. West. Am.
Cone-b., 22 (1895).

COMMON NAMES.

White Pine (Cal., Nev., Oreg.).
Finger-cone Pine (Cal.).
Mountain Pine (Cal.).
Soft Pine (Cal.).
Little Sugar Pine (Cal.).
Mountain Weymouth Pine.
Western White Pine.

Pinus lambertiana Dougl.   **Sugar Pine.**

SYN.—*Pinus Lambertiana* Douglas, in Trans. Linn. Soc., XV, 500
(1827).
Pinus Lamberti Douglas, in Comp. Bot. Mag., II, 141
(1836)—*nomen nudum*.
Pinus Lambertiana var. *minor* Lemmon, in Second Bienn.
Rep. Cal. St. Bd. For., 70, 83 (1888).
Pinus Lambertiana var. *purpurea* Lemmon, Handb. West.
Am. Cone-b., 22 (1895).

COMMON NAMES.

Sugar Pine (Cal., Nev., Oreg.).
Big Pine.
Shade Pine (Cal.).
Great Sugar Pine.
Little Sugar Pine (var. *minor*).
Gigantic Pine (Cal. lit.).
Purple-coned Sugar Pine (var. *purpurea*.)

Pinus flexilis James.   **Limber Pine.**

SYN.—*Pinus flexilis* James, in Long's Exped., II, 27 (cited), 34,
(descr.) 35 (1823).
Pinus Lambertiana var. Hooker, Fl. Bor.-Am., II, 161 (1840).

Pinus flexilis James—Continued.

SYN.—*Pinus Lambertiana* var. *brevifolia* Endlicher, Syn. Conif.,
150 (1847).
Pinus flexilis var. *α serrulata* Engelmann, in Rothrock, Bot.
Wheeler's Rep., VI, 258 (1878).

COMMON NAMES.

White Pine (Cal., Nev., Utah, Colo., N. Mex.).
Pine (Utah, Mont.).
Bull Pine (Colo.).
Rocky Mountain White Pine (Cal.).
Rocky Mountain Pine.
Limber-twig Pine (Cal. lit.).
Western White Pine (Cal. lit.).

Pinus flexilis megalocarpa nom. nov.   **Broadcone Limber Pine.**

SYN.—*Pinus flexilis* var. *β macrocarpa* Engelmann, in Rothrock,
Bot. Wheeler's Rep., VI, 258 (1878), not *P. macrocarpa*
Lindl. (1840).

COMMON NAME.

Arizona Flexilis Pine (Cal. lit.).

Pinus albicaulis Engelm.   **White-bark Pine.**

SYN.—*Pinus flexilis* Murray, in (Rep.) Bot. Exped. Oregon, t.
(1853?),[1] not James (1823).
Pinus cembroides Newberry, in Pacif. R. R. Rep., VI, 44. 90,
f. 15 (1857), not Zucc. (1829), nor Gord. (1858).
Pinus Shasta Carrière, Trait. Conif., nouv. éd., 390 (1867).
Pinus albicaulis Engelmann, in Trans. Acad. Sci. St. Louis,
II, 209 (1868).
Pinus flexilis var. *albicaulis* Engelmann (in Gard. Chron.,
XI, 125, 1879; name cited), in Watson, Bot. Cal., II, 124
(1880).

COMMON NAMES.

White Stem Pine (Cal., Mont.).
Scrub Pine (Mont.).
Pitch Pine (Mont.).
White Bark (Oreg.).
White Bark Pine (Cal.).
Creeping Pine (Cal. lit.).
Alpine White-bark Pine (Cal. lit.).

---

[1] Presumably this paper was issued in 1852, but it appeared without any date. A
review of the paper which appeared in Hooker's Journ. Bot. and Kew Gard. Mis-
cellany (V, 315–317) in 1853 is the only evidence pointing to the possible date of the
publication.

Pinus strobiformis Engelm.   **Mexican White Pine.**

SYN.—*Pinus strobiformis* Engelmann, in Bot. Wislizenus's Rep.,
    19, 103 (1848).
    Pinus flexilis var. *reflexa* Engelm., in Wheeler's Rep., VI, 258
    (1878).
    Pinus reflexa Engelm., in Bot. Gaz., VII, 4 (1882).
    Pinus Ayacahuite var. *strobiformis* Lemmon, Handb. West.
    Am. Cone-b., 4 (1892).

COMMON NAMES.

Ayacahuite Pine.
White Pine (Ariz.).
Mexican White Pine.
Arizona White Pine.

Pinus quadrifolia Parl.   **Parry Piñon.**

SYN.—*Pinus Llaveana* Torrey, in Bot. Mex. Boundary Survey, 208,
    t. 53 (1859), not Schiede & Deppe (1838).
    PINUS PARRYANA Engelmann, in Am. Jour. Sci., 2 ser.,
    XXXIV, 332 (1862), not Gord. (1858).
    Pinus quadrifolia Parry Msc. ex Parlatore, in A. de C.,
    Prodr., XVI, sect. 2, 402 (1868).

COMMON NAMES.

Nut Pine (Cal.).
Parry's Pine (Cal.).
Parry's Nut Pine (Cal.).
Parry Nut Pine (Cal. lit.).
Piñon (Cal.).
Mexican Piñon (Cal. lit.).

Pinus cembroides Zucc.   **Mexican Piñon.**

SYN.—*Pinus cembroides* Zuccarini, in Abh. Akad. Muench., I, 392
    (1832).
    Pinus Llaveana Schiede & Deppe, in Linnæa, XII, 488 (1838).
    Pinus osteosperma Engelmann, in Bot. Wislizenus's Rep., 89
    (1848).

COMMON NAMES.

Nut Pine (Ariz., N. Mex.).
Piñon (Mex.).
Stone-seed Mexican Pinyon (lit.).
Mexican Cembra-like Pine (lit.).

Pinus edulis Engelm.   **Piñon.**

SYN.—*Pinus edulis* Engelmann, in Bot. Wislizenus's Rep., 88 (1848).
    Pinus cembroides Gordon, in Jour. Hort. Soc., London, I, 236
    (1846), not Zucc. (1829).

Pinus edulis Engelm—Continued.

SYN.—*Pinus fertilis* Roezl. ex Gordon, Pinetum, ed. 1. Suppl., 76 (1862).
Pinus monophylla var. *edulis* Jones, in Zoe, II, 251 (1891).

COMMON NAMES.

Piñon (Tex., Colo.).
Nut Pine (Tex., Colo.).
Piñon Pine (Colo.).
New Mexican Pinyon (lit.).

Pinus monophylla Torr. & Frem.   **Single-leaf Piñon.**

SYN.—*Pinus monophylla* Torrey & Fremont, in Fremont's Second Rep., 319, t. 4 (1845).
Pinus Fremontiana Endlicher, Syn. Conif., 183 (1847), in part.

COMMON NAMES.

Piñon (Cal., Ariz., Nev., Utah).
Nut Pine (Cal., Ariz., Nev., Utah).
Grey Pine (Nev.).
Nevada Nut Pine (Cal.).
Single-leaf (Cal. lit.).
Fremont's Nut Pine (Cal. lit.).

Pinus balfouriana Murr.   **Foxtail Pine.**

SYN.—*Pinus Balfouriana* "Oreg. Com." in Murray, (Rep.) Bot. Exped. Oregon, No. 618, t. 3, f. (1853?).[1]

COMMON NAMES.

Spruce Pine (Cal. lit.).
Foxtail Pine (Cal.).

Pinus aristata Engelm.   **Bristle-cone Pine.**

SYN.—*Pinus aristata* Engelmann, in Am. Journ. Sci., 2d ser., XXXIV, 331 (1862).
Pinus Balfouriana Watson, in King's Rep., V, 331 (1871), not Murray (1853?).
Pinus Balfouriana var. *aristata* Engelmann, in Wheeler's Rep., VI, 375 (1878).

COMMON NAMES.

Hickory Pine (Cal. lit.).
Bristle-cone Pine (Cal. lit.).
Foxtail Pine (Cal. lit.).

[1] See footnote p. 16.

Pinus resinosa Ait.   **Red Pine.**

SYN.—*Pinus resinosa* Solander in Aiton, Hort. Kew, ed. 1, III, 367 (1789).
   Pinus rubra Michaux f., Hist. Arb. Am., I, 46, t. 1 (1810), not Mill. (1768), nor *P. Am. rubra*. Wang. (1787), nor Lamb. (1803).
   Pinus Laricio var. *resinosa* Spach, Hist. Vég., XI, 385 (1842).

COMMON NAMES.

Red Pine (Vt., N. H., N. Y., Wis., Minn., Ont.).
Norway Pine (Me., N. H., Vt., Mass., N. Y., Wis., Mich., Minn., Ont.).
Hard Pine (Wis.).
Canadian Red Pine (Eng.).

Pinus torreyana Parry.   **Torrey Pine.**

SYN.—*Pinus Torreyana* (Parry MSS.), in Torrey, Bot. Mex. Boundary Survey, 210, t. 58, 59 (1859).
   Pinus lophosperma Lindley, in London Gard. Chronicle, 1860, 46 (1860).

COMMON NAMES.

Soledad Pine (Cal.).
Del Mar Pine (Cal.).
Lone Pine (Cal.).
Torrey Pine (Cal. lit.).
Torrey's Pine.

Pinus arizonica Engelm.   **Arizona Pine.**

SYN.—*Pinus Arizonica* Engelmann ex Rothrock, in Wheeler's Rep., VI, 260 (1878).

COMMON NAMES.

Arizona Yellow Pine (Cal.).
Arizona Pine.
Arizona 5-leaved Lumber Pine (Cal. lit.).

Pinus ponderosa Laws.   **Bull Pine.**

SYN.—*Pinus resinosa* Torrey, in Ann. Lyc. N. York, II, 249 (1828), not Ait. (1789).
   Pinus ponderosa Douglas, in Companion Bot. Mag., II, 111, 141 (1836)—*nomen nudum*.
   Pinus ponderosa Douglas in herb. ex Lawson, Man. Ag., pl. 354 (1836).
   Pinus Benthamiana Hartweg, in Journ. Hort. Soc. London, II, 189 (1847).

Pinus ponderosa Laws.—Continued.

SYN.—*Pinus brachyptera* Engelmann, in Bot. Wislizenus's Rep., 9 (1848).

Pinus macrophylla ? Torrey, in Sitgreaves's Rep., 173 (1854), not Lindl. (1839), nor Engelm. (1848).

Pinus Beardsleyi Murray, in Edinburgh New Phil. Journ., new ser., I, 286, t. 6 (1855).

Pinus Craigana Murray, l. c., 288, t. 7 (1855).

Pinus Sinclairiana Carrière, Trait. Conif., 1 éd., 355 (1855).

Pinus Engelmanni Torrey, in Pacific R. R. Rep., IV. 141 (1856), not Carr. (1854).

Pinus Parryana Gordon, Pinetum, ed. 1, 202 (1858).

Pinus Nootkatensis Manétti ex Gordon, Pinetum, ed. 1, Suppl., 67 (1862).

Pinus ponderosa var. *Benthamiana* Vasey, Cat. Forest Trees, 30; in Rep. Com. Ag., 178, 1875 (1876).

Pinus ponderosa var. *brachyptera* (Engelm.) Lemmon, in Second Bienn. Rep. Cal. St. Bd. For., 73, 98 (1888).

Pinus ponderosa var. *nigricans* Lemmon, in Bull. 7, Cal. St. Bd. For., 8 (1889).

Pinus Jeffreyi var. *ambigua* Lemmon. in Bull 7, Cal. St. Bd. For., 11 (1889).

COMMON NAMES.

Yellow Pine (Cal., Colo., Mont., Idaho, Utah, Wash., Oreg.).
Bull Pine (Cal., Wash., Utah, Idaho, Oreg.).
Big Pine (Mont.).
Long-leaved Pine (Utah, Nev.).
Red Pine.
Pitch Pine.
Southern Yellow Pine (var. *brachyptera*).
Heavy-wooded Pine (Eng.).
Western Pitch Pine.
Heavy Pine (Cal.).
Foothills Yellow Pine (*P. Benthamiana*).
Sierra Brownbark Pine (var. *nigricans*).
Montana Black Pine (var. *ambigua*) (Cal. lit.).
"Gambier Parry's Pine" (Eng. lit.).

Pinus ponderosa scopulorum Engelm.   **Rock Pine.**

SYN.—*Pinus ponderosa* var. *scopulorum* Engelmann, in Watson, Bot. Cal., II, 126 (1880).

COMMON NAMES.

Yellow Pine (Mont., Nebr.).
Bull Pine (Colo.).
Long-leaved Pine (Colo.).
Rocky Mountain Yellow Pine (lit.).

VARIETY DISTINGUISHED IN CULTIVATION.

Pinus ponderosa penduliformis [1] nom. nov.　**Weeping Bull Pine.**

SYN.—*Pinus ponderosa pendula* H. W. Sargent, in Gard. Chron., X,
236, f. 42 (1878); C. S. Sargent, in Gard. and For., I, 392,
f. 62 (1888), not *P. pendula* Ait. (1789).

Pinus apacheca [2] Lemmon.　　　　　　　　　**Apache Pine.**

SYN.—*Pinus Apacheca* Lemmon, in Erythea, II, 103, Pl. III (1894).

Pinus mayriana [3] nom. nov.　　　　　　　　　**Mayr Pine.**

SYN.—*Pinus latifolia* Sargent, in Gard. and For., II, 496, f. 135
(1889), not *P. sylvestris latifolia* Gord. (1858), nor *P. contorta* var. *latifolia* Engelm. (1871).

COMMON NAMES.

Broadleaf Pine (lit.).
Arizona Broadleaf Pine (lit.).

Pinus jeffreyi "Oreg. Com."　　　　　　　　　**Black Pine.**

SYN.—*Pinus Jeffreyi* "Oreg. Com.", in Murray, (Rep.) Bot. Exped.
Oregon, No. 731, t. 1 (1853 ?).[4]
Pinus deflexa Torrey, in Bot. Mex. Boundary Survey, 209,
t. 56, (1859), in part.
Pinus ponderosa var. *Jeffreyi* Vasey, Cat. Forest Trees, 31;
in Rep. Com. Ag. 1875, 179 (1876).
Pinus Jeffreyi var. *nigricans* Lemmon, in Second Bienn. Rep.
Cal. St. Bd. For., 74, 100 (1888), not *P. ponderosa* var.
nigricans Lem., l. c., 73, 98 (1888).
Pinus Jeffreyi var. *deflexa* (Torr.) Lemmon, l. c. (1888).
Pinus Jeffreyi var. *peninsularis* Lemmon, l. c. (1888).
Pinus Jeffreyi var. *cortex-nigra* Lemmon, Pines Pac. Slope,
7 (1888).
Pinus Jeffreyi var. (c.) *montana* Lemmon, Handb. West Am.
Cone-b., 35 (1895), not *P. montana* Mill. (1768).

---

[1] This form is supposed to have originated from seed of *P. ponderosa* collected in Oregon or California. The seed was sown at the Knap Hill nurseries in England, from which, in 1851, Mr. H. W. Sargent imported a number of plants including this weeping form. It is the only one existing at present in Eastern States with a complete weeping form, and stands in Mr. Sargent's garden at Woodenethe, in Fishkill on the Hudson, N. Y.

[2] A tree recently described and figured by Prof. J. G. Lemmon as distinct from its undoubtedly near relatives *Pinus Engelmanni* Carr., *P. ponderosa*, and *P. mayriana*. It is said to occur abundantly in the Chiricahua Mountains of southeastern Arizona. First detected in 1881, and later in 1892. I have not seen specimens of this pine, but the general appearance exhibited in the plate does not display any distinctive specific features.

[3] See prefatory remarks, p. 9.
[4] See footnote p. 16.

Pinus jeffreyi " Oreg. Com."—Continued.

SYN.—Bull Pine (Cal.).
　　Black Pine (Cal.).
　　Western Black Pine (Cal. lit.).
　　Pinos (Cal.).
　　Truckee Pine (Nev.).
　　Sapwood Pine (Cal.).
　　Jeffrey Pine (Cal. lit.).
　　Blackbark Pine (var. *nigricans* and var. *cortex-nigra*) (Cal. lit.).
　　Redbark Pine (*P. deflexa*) (Cal. lit.).
　　Peninsula Pine (var. *peninsularis*) (Cal. lit.).
　　Sierra Redbark Pine (var. *deflexa*) (Cal. lit.).
　　Peninsula Black Pine (var. *peninsularis*) (Cal. lit.).

Pinus chihuahuana Engelm.   **Chihuahua Pine.**

SYN.—*Pinus Chihuahuana* Engelmann. in Bot. Wislizenus's Rep., 103 (1848).

Chihuahua Top-Cone Pine (Cal. lit.).

Pinus contorta Loud.   **Twisted Pine.**

SYN.—*Pinus inops* Bongard, in Mém. Acad. Pétersburg, 6 sér., II, 163 (1833), not Ait. (1789).
　　Pinus contorta Douglas ex Loudon, Arb. Frut., IV, 2292, f. 2210, 2211 (1838).
　　Pinus Banksiana Lindley & Gordon. in Journ. Hort. Soc. London, V, 218 (1850). in part; not Lamb. (1803).
　　Pinus Boursieri Carrière, in Rev. Hort. 1854, 233 and f. (1854).
　　Pinus Mac-Intoshiana Hort. ex Lawson, Cat., 15 (1855).
　　Pinus muricata Bolander, in Proc. Calif. Acad., III, 227, 317, (1866), not Don (1837).
　　Pinus Bolanderi Parlatore, in de Candolle, Prodr., XVI, sect. 2, 379 (1868).
　　Pinus Saskatchawensis Hook. ! mss. ex Parlatore, in A. de C., l. c., 381 (1868).
　　Pinus Tamarac Murray, in Gard. Chron., 191 (1869).
　　Pinus contorta var. *Bolanderi* Lemmon, in Erythea, II, 176 (1894), not Vasey (1875).
　　Pinus contorta var. *Hendersoni* Lemmon, l. c. (1894).

Scrub Pine (Cal., Nev., Idaho, Mont., Oreg.).
Pine (Utah).

Pinus contorta Loud.—Continued.

SYN.—Knotty Pine (Mont.).
Twisted Pine (Mont., Nev., Idaho).
Tamarack (Cal.).
Sand Pine (Oreg.).
North Coast Scrub Pine (Cal. lit.).
Bolander's Pine (var. *Bolanderi*).
Henderson's Pine (var. *Hendersoni*).

Pinus murrayana "Oreg. Com."   **Lodgepole Pine.**

SYN.—*Pinus Murrayana* "Oreg. Com.", in Murray, (Rep.) Bot.
Exped. Oregon, No. 740, t. 3, f. 2 (1853?).
Pinus inops var. *Bentham*, Pl. Hartweg, 337 (1857).
Pinus contorta Newberry, in Pacif. R. R. Rep., VI, 34, 90,
t. 5, f. 11 (1857), not Dougl. ex Loud. (1838).
Pinus contorta var. *latifolia* Engelmann, in King's Rep., V,
331 (1871).
Pinus contorta var. *Bolanderi* Vasey, Cat. Forest Trees, 29;
Rept. Com. Agr. 1875, 177 (1876), not *P. Bolanderi* Parl.
(1868).
Pinus contorta var. *Murrayana* Engelmann, in Watson,
Bot. Cat., II, 126 (1880).
Pinus Murrayana var. *Sargentii* Mayr, Wald. Nordam. Holz.,
349 (1890).

COMMON NAMES.

Tamarack (Wyo., Utah, Mont., Cal.).
Prickly Pine (Utah).
White Pine (Mont.).
Black Pine (Wyo.).
Lodgepole Pine (Wyo., Mont., Idaho).
Spruce Pine (Colo., Idaho, Mont.).
Tamarack Pine (Cal.).
Murray Pine (Cal. lit.).

Pinus sabiniana Dougl.   **Sabine Pine.**

SYN.—*Pinus Sabiniana* Douglas, in Trans. Linn. Soc., XVI, 749
(1833).
Pinus Sabinii Douglas, in Companion Bot. Mag., II, 150
(1836)—*nomen nudum.*

COMMON NAMES.

Sabine's Pine (Cal. lit.).
Gray-leaf Pine (Cal.).

Pinus coulteri Lamb.   **Coulter Pine.**

SYN.—*Pinus Coulteri* Lambert (Mss.) in Don, in Trans. Linn.
Soc., XVII, 440 (1837).

Pinus coulteri Lamb.—Continued.

Syn.—*Pinus macrocarpa* Lindley, in Bot. Reg. Misc., XXVI, 61, (1840).
Pinus Sabina Coulteri Loudon, Encycl. Pl. 985, f. 1839–1841 (1841).
Pinus Sabina Coulteri vera, Loud., l. c. (1841).
Pinus Sabiniana Macrocarpa Hort. ex Gordon, Pinetum, ed. 1, 201 (1858).
Pinus Sabina var. Hort. ex Carrière, Trait. Conif. 1 éd., 336 (1855).
Pinus Sabiniana major Manetti ex Gordon, Pinetum, ed. 1, Suppl., 66 (1862).

COMMON NAMES.

Coulter's Pine (Cal.).
Nut Pine (Cal., Idaho).
Bigcone Pine (Cal.).
Largeconed Pine (Eng. lit.).

Pinus radiata[1] Don.   **Monterey Pine.**

Syn.—*Pinus Californiana* Loiseleur, in Nouv. Duham., V, 243 (1812).—?
Pinus adunca Bos cex Desfontaines, Tabl. ed. 2, 247 (1815).—?
Pinus radiata Don, in Trans. Linn. Soc., XVII, 422 (1836).
Pinus tuberculata Don, l. c., post *P. radiata* (1836).
PINUS INSIGNIS Douglas in herb. ex Loudon, Arb. Frut., IV, 2265, f. 2170–2172 (1838).
Pinus Montereyensis Rauch ex Gordon, Pinetum, ed. 1, 197 (1858).
Pinus rigida Hooker & Arnott, in Bot. Beechey's Voyage, 160 (1841), not Mill. (1768).
Pinus Sinclairii Hooker & Arnott, l. c., 392, 393, t. 93 (1841), in part.
Pinus insignis macrocarpa Hartweg, in Journ. Hort. Soc. Lond., III, 226 (1846), not *P. macrocarpa* Lindl. (1840).
Pinus insignis Dougl. var. fide Engelmann ex Watson, in Proc. Am. Acad. Sci., XI, 119 (1876).
Pinus insignis var. binnata Engelm., in Watson, Bot. Cal., II, 128—adv. sheets, 1879—(1880).
Pinus insignis var. (a) radiata (Don) Lemmon, in Second Bienn. Rep. Cal. St. Bd. For., 76, 114 (1888).
Pinus insignis var. (b) levigata Lemmon, l. c. (1888).
Pinus insignis var. sub-lœvis Lemmon, Pines Pac. Slope, 10 (1888).

---

[1] See Garden and Forest, V, 64 (1892); Rep. Sec. Agric. 1892, 328 (July, 1893); Erythea, I, 224 (Nov., 1893).

Pinus radiata Don—Continued.

SYN.—*Pinus radiata* var. (*a*) *tuberculata* Lemmon, Handb. West
Am. Cone-b., 41 (1895).
Pinus radiata var. (*b*) *binata* Lemmon, l. c., 42 (1895).

COMMON NAMES.

Monterey Pine (Cal.).
Spreading-cone Pine (Cal. lit.) (var. *radiata*).
Nearly-smooth-cone Pine (Cal. lit.) (var. *levigata* and *sub-
lœvis*).
Remarkable Pine (Cal. lit.).
Small-coned Monterey Pine (var. *tuberculata*) (Cal. lit.).
Two-leaved Insular Pine (var. *binata*).

Pinus attenuata Lemmon.   **Knobcone Pine**.

SYN.—*Pinus Californica* Hartweg, in Journ. Hort. Soc. London, II,
189 (1847), not *P. Californiana* Loisel. (1812).
PINUS TUBERCULATA Gordon, in Journ. Hort. Soc. London,
IV, 218 and f. (1849), not Don (1836).
Pinus attenuata Lemmon, in Mining and Scientif. Press,
Jan. 16; in Gard. and For., V, 65 (1892).

COMMON NAMES.

Knobcone Pine (Oreg., Idaho, Cal.).
Prickly-cone Pine (Idaho).
Sun-loving Pine (Cal. lit.).
Sunny-slope Pine (Cal. lit.).
Narrow-cone Pine (Cal. lit.).
Tuberculated Coned Pine (Eng. lit.).

Pinus tæda Linn.   **Loblolly Pine**.

SYN.—*Pinus Tœda* Linnæus, Spec. Pl. ed. 1, II, 1000, excl. habitat
"*Canadœ paludosis*" (1753).
Pinus Tœda α tenuifolia Aiton, Hort. Kew, ed. 1, III, 36ラ
(1789).
Pinus tada Rafinesque, Flor. Ludovic., 162 (1817)—*nomen
nudum*.

COMMON NAMES.

Loblolly Pine (Del., Va., N. C., S. C., Ga., Ala., Fla., Miss.,
La., Tex., Ark.).
Oldfield Pine (Del., Va., N. C., S. C., Ga., Ala., Fla., Miss.,
La., Tex., Ark.).
Torch Pine (Eng. lit.).
Shortleaf Pine (La.).
Rosemary Pine (Va., N. C.).
Slash Pine (Va., N. C., in part).

Pinus tæda Linn.—Continued.

SYN.—Longschat Pine (Del.).
Longshucks (Md., Va.).
Black Slash Pine (S. C.).
Frankincense Pine (lit.).
Shortleaf Pine (Va., N. C., S. C.).
Bull Pine (Texas and Gulf region).
Virginia Pine.
Sap Pine (Va., N. C.).
Meadow Pine (Fla.).
Cornstalk Pine (Va.).
Black Pine (Va.).
Foxtail Pine (Va., Md.).
Indian Pine (Va., N. C.).
Spruce Pine (Va., in part).
Bastard Pine (Va., N. C.).
Yellow Pine (north Ala., N. C.).
Swamp Pine (Va., N. C.).
Longstraw Pine (Va., N. C., in part).

Pinus rigida Mill.   **Pitch Pine.**

SYN.—*Pinus rigida* Miller, Gard. Dict. ed. 8, No. 10 (1768).
Pinus Tæda var. *rigida* Aiton, Hort. Kew, ed. 1, III, 368
(1789).
Pinus Tæda var. *α* Poiret, in Lamarck, Enc. Méth. Bot., V,
340 (1804).
Pinus Fraseri Loddiges, Cat., 50 (1836), not Pursh (1814).
Pinus Loddigesii Loudon, Arb. Frut. IV, 2269 (1838).
Pinus rigida var. *lutea*[1] Kellerman ex Bot. Gaz., XVII, 280
(1892), not *P. lutea* Walt. (1788), nor Gord. (1858).

COMMON NAMES.

Pitch Pine (Vt., N. H., Mass., R. I., Conn., N. Y., N. J.,
Pa., Del., W. Va., N. C., S. C., Ga., Ohio, Ont., Md.,
Eng.).
Long-leaved Pine (Del.).

---

[1] The author distinguishes this form from the species by the thinner, scarcely furrowed, reddish-yellow bark, and by the deeper yellow, more durable and more distinctly marked heartwood. I have not seen specimens of this tree, but from the characters cited believe it to be one of the many forms of this variable species. The superficial appearance of the wood, size, and definition of the annual rings, thickness, and other characters of the bark vary greatly under different soil and moisture conditions and without essential change in the specific characters of the flowers, cones, or foliage. The great adaptability of the Pitch Pine to a wide range of soil conditions—from pure sand to a rich loam—makes it possible to point out many superficially distinct forms, which agree, however, in the essential features of the species.

Pinus rigida Mill.—Continued.

SYN.—Longschat Pine (Del.).
    Hard Pine (Mass.).
    Yellow Pine (Pa.).
    Black Pine (N. C.).
    Black Norway Pine (N. Y.).
    Rigid Pine (Eng. lit.).
    Sap Pine (lit.).

Pinus serotina Michx. .   **Pond Pine.**

SYN.—*Pinus serotina* Michaux, Fl. Bor.-Am., II, 205 (1803).
    Pinus Tæda β alopecuroidea Aiton, Hort. Kew., ed. 2, V, 317
      (1813).
    Pinus rigida var. *serotina* Loudon, Encycl. Pl. ed. 1, 979,
      f. 1824–1827 (1829).
    Pinus alopecuroides Hort. ex Gordon, Pinetum, ed. 1, 209
      (1858).
             COMMON NAMES.

Pond Pine (N. C., S. C., Fla., Miss., La.).
Loblolly Pine (N. C., Fla.).

Pinus virginiana Mill.   **Scrub Pine.**

SYN.—*Pinus Virginiana* Miller, Gard. Dict., ed. 8, No. 9 (1768).
    PINUS INOPS Solander ex Aiton, Hort. Kew, ed. 1, III, 367
      (1789).
    Pinus Tæda var. *Virginiana* Poiret, in Lamarck, Enc. Méth.
      Bot., V, 340 (1804).
    Pinus turbinata Bosc ex. Loudon, Enc. Trees, 975 (1842).
    Pinus ruthenica Hort. ex Carrière, Trait. Conif. nouv. éd., 471
      (1867).
             COMMON NAMES.

Jersey Pine (N. J., Pa., Del., N. C., S. C.).
Scrub Pine (R. I., N. Y., Pa., Del., N. C., S. C., Ohio).
Short Shucks (Md., Va.).
Shortschat Pine (Del.).
Spruce Pine (N. J., N. C.).
Shortleaved Pine (N. C.).
Cedar Pine (N. C.).
River Pine (N. C.). ·
Nigger Pine (Tenn.).
New Jersey Pine (lit.).

Pinus clausa (Engelm.) Sargent.   **Sand Pine.**

SYN.—*Pinus clausa* Chapman ex Vasey, in Gard. Month., XVIII,
    151 (1876)—*nomen nudum.*

Pinus clausa (Englem.) Sargent—Continued.

Syn.—*Pinus inops* var. *clausa* Engelmann, in Bot. Gaz., II, 125 (1877); in Trans. Acad. Sci., St. Louis, IV, 183 (1880).
Pinus clausa Sargent in Tenth Census of the United States, IX (Cat. For. Trees N. A.), 199 (1884).

COMMON NAMES.

Sand Pine (Fla.).
Oldfield Pine (Fla.).
Florida Spruce Pine (Ala.).
Scrub Pine (Fla.).
Spruce Pine (Fla.).
Upland Spruce Pine (Fla.).

Pinus pungens Michx. f.   **Table-mountain Pine.**

Syn.—*Pinus pungens* Lambert ex Michaux f., Hist. Arb. Am., I, 61, t. 5 (1812).

COMMON NAMES.

Table-mountain Pine (Pa., Del., S. C., Md.).
Southern Mountain Pine (Tenn.).
Prickly Pine (N. C.).

Pinus muricata Don.   **California Swamp Pine.**

Syn.—*Pinus muricata* Don, in Trans. Linn. Soc., XVII, 441 (1837).
Pinus Edgariana Hartweg, in Journ. Hort. Soc. London, III, 217 (1848).
Pinus inops var. Bentham, Pl. Hartweg, 337 (1857).
Pinus contorta Bolander, in Proc. Calif. Acad., III, 227 (1866), not Loud. (1838).
Pinus muricata var. *Anthoni* Lemmon, Handb. West. Am. Cone-b., 10 (1892).

COMMON NAMES.

Swamp Pine (Cal.).
Dwarf Marine Pine (Cal.).
Prickle-cone Pine (Cal.).
Bishop's Pine (Cal. and Eng. lit.).
Anthony's Prickle-Cone Pine (var. *Anthoni*).
Obispo Pine (Cal.).

Pinus echinata Mill.   **Shortleaf Pine.**

Syn.—*Pinus echinata* Miller, Gard. Dict., ed. 8, No. 12 (1768).
Pinus Virginiana var. *echinata* Du Roi, Harbk., II, 38 (1772).
Pinus Tæda γ *variabilis* Aiton, Hort. Kew, ed. 1, III, 368 (1789).

Pinus echinata Mill.—Continued.

SYN.—PINUS MITIS Michaux, Fl. Bor.-Am., II, 204 (1803).
Pinus variabilis Lambert, Pinus, ed. 1, I, 22, t. 15 (1803).
Pinus Royleana Jamieson ex Lindley, in Journ. Hort. Soc.,
IX, 52 (1855).
Pinus lutea Loddiges ex Gordon, Pinetum, ed. 1, 170 (1858),
not Walter (1788).
Pinus Roylei Lindley ex Gord., l. c. (1858).
Pinus intermedia Fischer ex Gordon, Pinetum, ed. 1, 170 (1858),
not Du Roi (1772).
Pinus rigida Porcher, Resources S. States, 504 (1863), not
Mill. (1768).

COMMON NAMES.

Yellow Pine (N. Y., N. J., Pa., Del., Va., N. C., Ala., Miss.,
La., Ark., Mo., Ill., Ind., Kans. (scarce), Ohio).
Shortleaved Pine (N. C., S. C., Ga., Ala., Miss., Fla., La., Tex.,
Ark.).
Spruce Pine (Del., Miss., Ark.).
Bull Pine (Va.).
Shortschat Pine (Del.).
Pitch Pine (Mo.).
Poor Pine (Fla.).
Shortleaved Yellow Pine.
Yellow Pine (N. C., Va.; Eng. lit.).
Virginia Yellow Pine (Va., in part).
North Carolina Yellow Pine (N. C. and Va., in part).
North Carolina Pine (N. C. and Va., in part).
Carolina Pine (N. C. and Va., in part).
Slash Pine (N. C., Va., in part).
Oldfield Pine (Ala., Miss.).

Pinus glabra Walt. .   **Spruce Pine.**

SYN.—*Pinus glabra* Walter, Fl. Caroliniana, 237 (1788).
Pinus mitis β ? paupera Wood, Cl. Book, 660 (1869).

COMMON NAMES.

Spruce Pine (S. C., Ala., Fla.).
Cedar Pine (Miss.).
White Pine (Fla.).
Walter's Pine (S. C.).
Lowland Spruce Pine (Fla.).
Poor Pine (Fla.).

Pinus divaricata (Ait.) Gord.   **Jack Pine.**

SYN.—*Pinus sylvestris δ divaricata* Aiton, Hort. Kew., ed. 1, III,
366 (1789).

Pinus divaricata (Ait.) Gord.—Continued.

SYN.—PINUS BANKSIANA Lambert, Pinus, ed. 1, I, 7, t. 3 (1803).
Pinus Hudsonica Poiret, in Lamarck, Enc. Méth. Bot., V, 339 (1804).
Pinus rupestris Michaux f., Hist. Arb. Am., I, 49, t. 2 (1810).
Pinus Banksii Douglas, in Companion Bot. Mag., II, 152 (1836)—*nomen nudum*.
Pinus divaricata Hort. ex Gordon, Pinetum, ed. 1, 163, (1858); (Aiton) Sudworth, in Bull. Torr. Bot. Club, XX, 44 (1893).

COMMON NAMES.

Scrub Pine (Me., Vt., N. Y., Wis., Mich., Minn., Ont.).
Gray Pine (Vt., Minn., Ont.).
Jack Pine (Mich., Minn., Canada).
Princes Pine (Ont.).
Black Jack Pine (Wis).
Black Pine (Minn.).
Cypress (Quebec to Hudson Bay).
Canada Horn-cone Pine (Cal. lit.).
Chek Pine.
Sir Joseph Bank's Pine (Eng.).
"Juniper" (Canada).

Pinus palustris Mill.   **Longleaf Pine.**

SYN.—*Pinus palustris* Miller, Gard. Dict., ed. 8, No. 14 (1768).
Pinus lutea Walter, Fl. Caroliniana, 237 (1788).
Pinus australis Michaux f., Hist. Arb. Am., I., 64, t. 6 (1803).
Pinus serotina Hort. Cf. Bon Jard. 976 (1837), ex Antoine, Conif., 23 (1840–1847), not Michx. (1803).
Pinus Palmiensis Fr. Gard. ex Gordon, Pinetum, ed. 1, Suppl. 63 (1862).
Pinus Palmieri Manetti ex Gord., l. c. (1862).

COMMON NAMES.

Longleaved Pine (Del., N. C., S. C., Ga., Ala., Fla., Miss., La., Tex.).
Southern Pine (N. C., Ala., Miss., La.).
Yellow Pine (Del., N. C., S. C., Ala., Fla., La., Tex.).
Turpentine Pine (N. C.).
Rosemary Pine (N. C.).
Brown Pine (Tenn.).
Hard Pine (Ala., Miss., La.).
Georgia Pine (Del.).
Fat Pine (Southern States).

Pinus palustris Mill.—Continued.

SYN.—Southern Yellow Pine (general).
Southern Hard Pine (general).
Southern Heart Pine (general).
Southern Pitch Pine (general).
Heart Pine (N. C. and South Atlantic region).
Pitch Pine (Atlantic region).
Longleaved Yellow Pine (Atlantic region).
Longleaved Pitch Pine (Atlantic region).
Longstraw Pine (Atlantic region).
North Carolina Pitch Pine (Va., N. C.).
Georgia Yellow Pine (Atlantic region).
Georgia Pine (general).
Georgia Heart Pine (general).
Georgia Longleaved Pine (Atlantic region).
Georgia Pitch Pine (Atlantic region).
Florida Yellow Pine (Atlantic region).
Florida Pine (Atlantic region).
Florida Longleaved Pine (Atlantic region).
Texas Yellow Pine (Atlantic region).
Texas Longleaved Pine (Atlantic region).

Pinus heterophylla (Ell.) Sudworth.   **Cuban Pine.**

SYN.—*Pinus Tœda* var. *heterophylla* Elliott, Sk. Bot. S. C. Ga., II,
636 (1824).
PINUS CUBENSIS Grisebach, in Mem. Am. Acad., VIII, pt. 2,
530 (1863), not Hort. ex Gord. (1858).
Pinus Cubensis var. *terthrocarpa* Wright in Grisebach, Cat.
Pl. Cuben., 217 (1866).
Pinus Elliottii Engelmann ex Vasey, Cat. Forest Trees, 30;
in Rep. Com. Ag., 1875, 178 (1876).
Pinus Elliottii Engelm., in Trans. Acad. Sci., St. Louis, IV,
186, t. 1, 2, 3 (1879).
Pinus heterophylla (Ell.) Sudworth, in Bull. Torr. Bot. Club;
XX, 45 (1893).

COMMON NAMES.

Slash Pine (Ala., Miss., Ga., Fla.).
Swamp Pine (Fla., Miss., Ala., in part).
Bastard Pine (Ala. lumbermen, Fla.).
Meadow Pine (Cal., Fla., eastern Miss., in part).
Pitch Pine (Fla.).
She Pitch Pine (Ga.).
She Pine (Ga., Fla.).
Spruce Pine (southern Ala.)

LARIX Andanson, Fam. Pl. II, 480 (1763).

Larix laricina (Du Roi) Koch.   **Tamarack.**

SYN.—*Pinus laricina* Du Roi, Obs. Bot., 49 (1771).

Pinus intermedia Du Roi, Harbk. Baumz., II, 115 (1772).

Pinus Larix rubra Marshall, Arb. Am., 103 (1785).

Pinus-Larix alba Marsh., l. c., 104 (1785).

Pinus-Larix nigra Marsh., l. c. (1785).

Pinus pendula Solander in Aiton, Hort. Kew, ed. 1, III, 369 (1789).

LARIX AMERICANA Michaux, Fl. Bor.-Am., II, 203 (1803).

Pinus microcarpa Lambert, Pinus, ed. 1, 1, 56, t. 37 (1803).

Abies pendula Poiret, in Lamarck, Enc. Méth. Bot., VI, 514 (1804).

Larix tenuifolia Salisbury, in Trans. Linn. Soc., VIII, 314 (1807).

Larix pendula Salisbury, l. c. (1807).

Larix microcarpa Desfontaines, Hist. Arb., II, 597 (1809).

Pinus americana Steudel, Nom. Bot., ed. 1, 621 (1821), not DuRoi (1771).

Larix intermedia Loddiges, Cat., 50 (ed. 1836)—nomen nudum.

Larix Americana rubra Loudon, Arb. Frut., IV, 2400 (1838).

Larix Americana var. *pendula* Loud., l. c. (1838).

Larix Americana var. *prolifera* Loud., l. c., 2401 (1838).

Abies microcarpa Lindley & Gordon, in Journ. Hort. Soc., V, 213 (1850).

Larix Fraseri Curtis ex Gordon, Pinetum, ed. 1, 129 (1858).

Larix decidua var. *Americana* Henkel & Hochstetter, Nadel-hölz., 133 (1865).

Larix laricina Koch, Dendrol., zw. Th. zw. Ab., 263 (1873).

Larix laricina var. *microcarpa*, Lemmon, in Third Bienn. Rep. Cal. St. Bd. For. 108 (1890).

Larix laricina var *pendula* Lemmon, l. c. (1890).

COMMON NAMES.

Larch (Vt., Mass., R. I., Conn., N. Y., N. J., Pa., Del., Wis., Minn., Ohio, Ont.).

Tamarack (Me., N. H., Vt., Mass., R. I., N. Y., N. J., Pa., Del., Ind., Ill., Wis., Mich., Minn., Nebr., Ohio, Ont.).

Hackmatack (Me., N. H., Mass., R. I., Del., Ill., Minn., Ont.).

American Larch (Vt., Del., Nebr., Wis. nurserymen).

Juniper (Me., N. Bruns. to Hudson Bay).

Black Larch (Minn.).

Epinette Rouge (Quebec).

Ka-nch-tens = "The leaves fall" (Indians, N. Y.).

Red Larch (Mich.).

Hacmack (lit.).

33

Larix occidentalis Nutt.   **Western Larch.**

SYN.—*Pinus Larix* Douglas, in Companion Bot. Mag., II, 109 (1836)—*nomen nudum;* not Linn. (1753).
Larix occidentalis Nuttall, Sylva, III, 143, t. 120 (1849).
Larix Americana var. *brevifolia* Carrière, Trait. Conif., nouv. éd., 357 (1867).
Pinus Nuttallii Parlatore in de Candolle, Prodr., XVI, sect. 2, 412 (1868).

COMMON NAMES.

Tamarack (Oreg.).
Hackmatack.
Larch (Idaho, Wash., etc.).
Red American Larch.
Western Tamarack.
Great Western Larch (Cal. lit.).
Western Larch (Eng.).

Larix lyallii Parl.   **Lyall Larch.**

SYN.—*Larix Lyallii* Parlatore, Enum. Sem. Hort. Reg. Mus. Flor., 259, 1863 (1863).
Pinus Lyallii Parlatore in de Candolle, Prodr., XVI, sect. 2, 412 (1868).

COMMON NAMES.

Tamarack (Idaho, Wash., Oreg.).
Larch (Idaho, Wash., Oreg.).
Mountain Larch.
Lyall's Larch (lit.).
Woolly Larch (Cal. lit.).

PICEA Link, in Abh. Akad. Berl., 179 (1827).

Picea mariana (Mill.) B. S. P.   **Black Spruce.**

SYN.—*Abies Mariana* Miller, Gard. Dict., ed. 8, No. 5 (1768).
Pinus Mariana Du Roi, Obs. Bot., 38 (1771).
Pinus-Abies canadensis Marshall, Arb. Am., 103 (1768).
Pinus nigra Aiton, Hort. Kew, ed. 1, III, 370 (1789).
Abies nigra Du Roi, Harbk. Baumz., ed. Pott, II, 182 (1800).
PICEA NIGRA Link, Handb., II, 478 (1831).
Abies denticulata Michaux, Fl. Bor.-Am., II, 206 (1803).
Pinus denticulata Steudel, Nom. Bot., ed. sec., II, 337 (1841).
Pinus Marylandica Hort. ex Antoine, Conif., 88 (1840–1847).
Abies alba Chapman, Fl. S. States, ed. 1, 435 (1860), not Michx. (1803), nor Mill. (1768).
Abies arctica Murray, in. Journ. Bot., V, 253 (1867), not Cunningham (1858).

18158—No. 14——3

Picea mariana (Mill.) B. S. P.—Continued.

Syn.—*Abies Novæ-Angliæ* Koch. Dendrol., zw. Th. zw. Ab., 240 (1873).

Abies Americana Koch, l. c., 241 (1873).

Pinus Americana nigra Hort. ex Beissner. Handb. Conif., 58 (1887).

Picea Mariana (Mill.) B. S. P., in Prelim. Cat. Anth. Pter. N. Y., 71 (1888).

Picea nigra Mariana Hort. ex Beissn., Handb. Nadelh., 336 (1891).

Abies nigra Mariana Hort. ex Beissn., l. c.. (1891).

Abies Marylandica Hort. ex Sargent, in Tenth Cens. U. S. IX (Cat. For. Trees N. A.), 203 (1884).

COMMON NAMES.

Black Spruce (N. H., Vt., Mass., R. I., N. Y., Pa., W. Va., N. C., S. C., Wis., Mich., Minn., Ont., Eng.).

Double Spruce (Me., Vt., Minn.).

Blue Spruce (Wis.).

Spruce (Vt.).

White Spruce (W. Va.).

Yew Pine (W. Va.).

Juniper (N. C.).

Spruce Pine (W. Va., Pa.).

He Balsam (Del., N. C.).

Epinette Jaune (Quebec).

VARIETIES DISTINGUISHED IN CULTIVATION.

Picea mariana doumetti (Carr.) Beissn.

Syn.—*Picea nigra Doumetti* Carrière, Trait. Conif., 1. éd. 242 (1855).

Abies nigra Doumetti Hort. ex. Beissner, Handb. Conif., 58 (1887).

Picea Mariana Doumetti Hort. ex Beissn., l. c. (1887).

Abies Mariana Doumetti Hort. ex. Beissn., l. c. (1887).

Abies Doumetti Hort. ex Beissn., l. c. (1887).

Picea mariana pumila (Carr.) nom. nov.

Syn.—*Abies nigra pumila* Hort. ex Carrière, Trait. Conif., 1, éd. 242 (1855).

Picea nigra fastigiata Carr., l. c. (1855).

Abies nigra fastigiata Hort. ex Gordon, Pinetum, ed. 1, 8 (1858).

Picea mariana humilis nom. nov.

Syn—*Picea nigra nana* Hort. ex Beissner. Handb. Conif., 58 (1887), not *P. alba nana* Carr. (1855).

Picea mariana humilis nom. nov.—Continued.

SYN.—*Picea Mariana nana* Hort. ex Beissn., l. c. (1887), not *P. alba nana* Carr. (1855).

Abies nigra nana Hort. ex Beissn., l. c. (1887).

Abies Mariana nana Hort. ex Beissn., l. c. (1887).

Picea mariana argenteo-variegata (Beissn.) nom. nov.

SYN.—*Picea nigra argenteo-variegata* Hesse ex Beissner, Handb. Nadelh., 337 (1891).

Abies nigra argenteo-variegata Hort. ex Beissn., l. c. (1891).

Picea mariana aurescens nom. nov.

SYN.—*Picea nigra aurea* Hesse ex Beissner, Handb. Nadelh., 337 (1891), not *P. excelsa aurea* Carr. (1855).

Abies nigra aurea Hort. ex Beissn., l. c. (1891), not *A. excelsa aurea* Hort. ex Carr. (1867).

Picea mariana albescens nom. nov.

SYN.—*Picea nigra glauca* Carrière, Trait. Conif., 1, éd. 242 (1855), not *Pin. glauca* Moench. (1785).

Abies nigra glauca Hort. ex Carr., l. c. (1855).

Picea rubra (Poir.) Diet. **Red Spruce.**

SYN.—*Pinus Americana rubra* Wangenheim, Beitr. Holz., 75, t. 16, f. 80 (1787), not *P. rubra* Mill. (1768).

Pinus Americana Gaertner, Fruct., II, 60, t. 91, f. 1. (1791), not Du Roi (1771).

Pinus rubra Lambert, Pinus, ed. 1, I, 48, t. 28 (1803), not Mill. (1768).

Abies rubra Poiret, in Lamarck, Enc. Méth. Bot., VI, 520 (1804).

Abies minuta Poiret in Lam., l. c., 523 (1804).

Abies pectinata Poiret in Lam., l. c., not de C. (1805).

Abies nigra var. *rubra* Michaux f., Hist. Arb. Am., I, 123 (1810).

Picea rubra Dieterich, Fl. Berl., II, 785 (1824).

Abies rubra var. *arctica* Lindley & Gordon, in Journ. Hort. Soc. Lond., V, 211 (1850).

Abies arctica Cunningham ex Gordon, Pinetum, ed. 1, 11 (1858).

Picea nigra var. *rubra* Engelmann, in London Gard. Chron. 1879, 334 (1879).

Abies Americana rubra Hort. ex Beissner, Handb. Nadelh., 338 (1891).

COMMON NAMES.

Red Spruce.

North American Red Spruce (foreign lit.).

VARIETIES DISTINGUISHED IN CULTIVATION.

Picea rubra cœrulea (Loud.) Forbes.

SYN.—*Abies* (n.) *rubra* 2 *cœrulea* Loudon, Arb. Frut.. IV, 2316 (1838).
Abies cœrulea Booth ex Loud., l. c. (1838).
Picea rubra cœrulea Forbes. Pine. Wob., 99 (1839).
Abies cœrulescens Hort. ex Koch, Dendrol., zw. Th. zw. Ab.,
242 (1873).

Picea rubra pendula Carr.

SYN.—*Picea rubra pendula* Carrière, Trait. Conif., nouv. éd.. 323,
(1867).

Picea rubra gracilis (Knight) Carr.

SYN.—*Abies rubra gracilis* Knight, Syn. Conif., 37 (1850).
Picea rubra gracilis Carrière, Trait. Conif.. nouv. éd.. 323
(1867).

Picea canadensis (Mill.) B. S. P.   **White Spruce.**

SYN.—*Abies Canadensis* Miller, Gard. Dict., ed. 8, No. 4 (1768).
Pinus Canadensis Du Roi, Obs. Bot.. 38 (1771), not Linn.
(1753).
Pinus laxa Ehrhart, Beitr., III, 24 (1788).
Pinus alba Aiton, Hort. Kew., ed. 1, III, 371 (1789).
Pinus tetragona Moench, Meth., 364 (1794).
Abies alba Michaux. Fl. Bor.-Am., II, 207 (1803). not Mill.
(1768).
Abies currifolia Salisbury, in Trans. Linn. Soc., VIII, 315
(1807).
PICEA ALBA Link, Handb., II, 478 (1831).
Picea rubra var. *violacea* Endlicher, Syn. Conif., 114 (1847).
Picea nigra var. *glauca* Carrière, Trait. Conif., éd. 1, 242
(1855).
Pinus virescens Neilreich, Fl. Nied. Nachtr., 63 (1866).
Abies arctica Murray, in Seemann, in Journ. Bot.. V, 253, t.
69, f. 1, 8–13 (1867).
Picea Tschugatskoyæ Hort. ex Carrière. Trait. Conif., nouv.
éd., 319 (1867).
Abies laxa Koch, Dendrol.. zw. Th. zw. Ab., 243 (1873).
Abies alba var. *arctica* Parlatore, in de Candolle. Prodr.,
XVI, sect. 2, 414 (1868).
Abies virescens Hinterh. ex Nyman, Conspect., 673 (1878).
Picea Canadensis (Mill.) B. S. P.. Prelim. Cat. Anth. Pter., 71
(1888).
Picea laxa Sargent, in Gard. & For.. II, 496 (1888).

Picea canadensis (Mill.) B. S. P.—Continued.

Syn.—*Picea alba Dakotaii* (auth. ?), ex Ann. Rep. Nebr. Hort. Soc. 1889, 68 (1889).

Pinus Americana alba Hort. ex Beissner, Handb. Conif., 59 (1887).

Abies Americana alba Hort. ex Beissn., Handb. Nadelh., 340 (1891).

COMMON NAMES.

White Spruce (Vt., N. H., Mass., R. I., N. Y. Wis., Mich., Minn., Ont.).

Single Spruce (Me., Vt., Minn.).

Black Spruce (Pa. (Meehan)).

Skunk Spruce (Wis., Me., Ont.).

Cat Spruce (Me.).

Spruce (Vt.).

Pine (Hudson Bay).

Double Spruce (Vt.).

VARIETIES DISTINGUISHED IN CULTIVATION.

Picea canadensis glauca (Moench) nom. nov.

Syn.—*Pinus glauca* Moench, Verzeich. Baum. Weissn., 73 (1785).

Abies rubra violacea Loudon, Arb. Frut., IV, 2316 (1838).

Abies cærulea Forbes, Pinet. Wob., 99 (1839), not Booth ex Loud. (1838).

Picea cærulea Link, in Linnæa, XV, 522 (1841), not *P. rubra cærulea* Forb. (1839).

Pinus rubra β violacea Endlicher, Syn. Conif., 114 (1847).

Abies glauca Hort. ex Carrière, Trait. Conif., éd. 1, 238 (1855).

Abies alba glauca Plumbly ex Gordon, Pinetum, ed. 1, 3 (1858).

Abies alba argentea Hort. ex Gord., l. c. (1858).

Abies alba cærulea Hort. ex. Carrière, Trait. Conif., nouv. éd., 320 (1867).

Picea alba cærulea Carr., l. c. (1867), not *P. rubra cærulea* Forb. (1839).

Abies Americana cærulea Beissner, Handb. Conif., 59 (1887).

Picea alba glauca Hort. ex Beissn., l. c. (1887); Handb. Nadelh., 341 (1891), not *P. nigra glauca* Carr. (1855).

Picea alba·argentea Hort. ex Beissn., l. c. (1887).

Picea glauca Hort. ex Beissn., l. c. (1887).

Picea canadensis acutissima (Beissn.) nom. nov.

Syn.—*Picea alba acutissima* Hort. ex Beissner, Handb. Nadelh., 342 (1891).

Picea acutissima Hort. ex. Beissn., l. c. (1891).

Abies acutissima Hort. ex. Beissn., l. c. (1891).

Picea canadensis compressa (Beissn.) nom. nov.

SYN.—*Picea alba compressa* Hort. ex Beissner, Handb. Nadelh., 342 (1891).

Picea canadensis nana (Loud.) nom. nov.

SYN.—*Abies alba nana* Loudon, Enc. Trees, 1030 (1842).
Picea alba nana Carrière, Trait. Conif., 1 éd., 239 (1855).
Abies alba prostrata Hort. ex. Carr., l. c. (1855).

Picea canadensis nana glaucifolia nom. nov.

SYN.—*Picea alba nana glauca* Hort. ex Beissner, Handb. Nadelh., 342 (1891), not *P. nigra glauca* Carr. (1855), nor *P. alba glauca* Beissn. (1887).
` *Abies alba nana glauca* Hort. ex Beissn., l. c. (1891), not *A. glauca* Carr. (1855).

Picea canadensis echinoformis (Carr.) nom. nov.

SYN.—*Picea alba echinoformis* Carrière, Trait. Conif., 1 éd., 239 (1855).
Abies alba echinoformis Hort. ex Carr., l. c. (1855).

Picea canadensis compacta gracilis (Beissn.) nom. nov.

SYN.—*Picea alba compacta gracilis* Breinig ex Beissner, Handb. Nadelh., 343 (1891).
Picea alba compacta pyramidalis Smith. ex Beissn., l. c. (1891).
Abies alba compacta pyramidalis Hort. ex Beissn., l. c. (1891).

Picea canadensis compressiformis nom. nov.

SYN.—*Picea alba fastigiata* Carrière, Trait. Conif., nouv. éd., 321 (1867), not *P. nigra fastigiata* Carr. (1855).

Picea canadensis nutans nom. nov.

SYN.—*Picea alba pendula* Hort. ex Carrière, Trait. Conif., nouv. éd., 321 (1867), not *P. excelsa pendula* Carr. (1855).
Abies alba pendula Hort. ex Carr., l. c. (1867).

Picea canadensis aurea (Beissn.) nom. nov.

SYN.—*Picea alba aurea* Beissner, Handb. Conif., 59 (1887).
Abies alba aurea Beissn., l. c. (1887).

Picea engelmanni Engelm.   **Engelmann Spruce.**

SYN.—*Abies nigra* Engelmann, in Am. Journ. Sci., 2 ser., XXXIII, 330 (1838), not Du Roi (1800), nor Poiret in Lam. (1804).
Abies alba? Torrey, in Fremont's Rep., 97 (1845).
Abies Engelmanni Parry, in Trans. Acad. Sci. St. Louis, II, 122 (1863)—*nomen nudum*.

Picea engelmanni Engelm.—Continued.

SYN.—*Picea Engelmanni* Parry sub Abiete ex Engelmann, in Trans.
Acad. Sci. St. Louis, II, 212 (1863).
Pinus commutata Parlatore, in de Candolle, Prodr., XVI,
sect. 2, 417 (1868).
Abies commutata Gordon, Pinetum, 2d ed., 5 (1875).
Picea Engelmanni var. *Franciscana* Lemmon, Handb. West.
Am. Cone-b., 51 (1895).

COMMON NAMES.

Engelmann's Spruce (Utah).
Balsam (Utah).
White Spruce (Oreg., Colo., Utah, Idaho).
White Pine (Idaho).
Mountain Spruce (Mont.).
Arizona Spruce (Cal. lit.).

VARIETIES DISTINGUISHED IN CULTIVATION.

Picea engelmanni griseifolia nom. nov.

SYN.—*Abies Engelmanni glauca* Veitch, Man. Conif., 69 (1881), not
A. glauca Carr. (1855).
Picea Engelmanni glauca Beissner, Handb. Conif., 60 (1887);
Handb. Nadelh. 345 (1891), not *P. nigra glauca* Carr.
(1855).

Picea engelmanni argyrophylla nom. nov.

SYN.—*Picea Engelmanni argentea* Hort. ex Beissner, Handb. Nadelh.,
345 (1891).

Picea engelmanni minutifolia nom. nov.

SYN.—*Picea Engelmanni microphylla* Hesse ex Beissner, Handb.
Nadelh., 345 (1891), not *P. microphylla* Carr. (1867).
Abies Engelmanni microphylla Hort. ex Beissn., l. c. (1891),
not *A. microphylla* Raf. (1832).

Picea pungens Engelm.   **Blue Spruce**.

SYN.—*Abies Menziesii* Engelmann, in Am. Journ. Sci., 2 ser.,
XXXIV, 330 (1838), not Lindl. (1833).
Picea Menziesii Engelmann, in Trans. Acad. Sci. St. Louis, II,
214 (1863), not Carr. (1855).
Picea pungens Englemann, in London Gard. Chron. 1879,
334 (1879.)
Abies Menziesii Parryana Andre, in Ill. Hort., XXIII, 198
(1876).
Abies Engelmanni glauca Veitch, Manual Conif., 69 (1881).

Picea pungens Engelm.—Continued.

Syn.—*Picea Pungence,* ex Ann. Rep. Nebr. Hort. Soc. 1889, 67 (1889).

Picea commutata Hort. ex Beissner. Handb. Nadelh., 346 (1891).

Abies Parryana Hort. ex Beissn., l. c. (1891).

Picea Parryana Hort. ex Parry. in Gard. Chron., II. 725 (1883).

COMMON NAMES.

Parry's Spruce (Utah).
Blue Spruce (Colo., Oreg., N. Mex.).
Spruce (Mont.).
Balsam (Colo., Mont., Utah).
White Spruce (Utah, Mont., Colo.).
Silver Spruce (Colo.).
Colorado Blue Spruce (Colo.).
Prickly Spruce (lit.).

VARIETIES DISTINGUISHED IN CULTIVATION.

Picea pungens glaucescens nom. nov.

Syn.—*Picea pungens glauca* Beissner, Handb. Conif., 60 (1887); Handb. Nadelh., 347 (1891), not *P. alba glauca* Beissn., l. c. (1887) ante, nor *P. canadensis glauca* (Moench) Sudw.=*Pinus glauca* Moench (1785).

Picea Parryana glauca Beissn., l. c. (1887).

Picea pungens cyanea nom. nov.

Syn.—*Picea pungens cærulea* Hort. ex Beissner, Handb. Nadelh., 347 (1891), not *P. cærulea* Forbes (1839).

Picea Parryana cærulea Hort. ex Beissn., l. c. (1891).

Picea pungens argentea Beissn.

Syn.—*Picea pungens argentea* Beissner. Handb. Conif., 60 (1887); Handb. Nadelh., 347 (1891).

Picea Parryana argentea Beissn., l. c. (1887).

Picea Menziesi argentea Hort. ex Beissn., Handb. Nadelh., 347 (1891), not *P. Menziesii* Carr. (1855).

Picea Engelmanni glauca Hort. ex Beissn., l. c. (1891), not Beissn. (1887).

Abies Engelmanni glauca Hort. ex Beissn., l. c. (1891), not Beissn. (1887).

Picea pungens glauca pendens nom. nov.

Syn.—*Picea pungens glauca pendula* Koster u. Cie ex Beissner, Handb. Nadelh., 348 (1891), not *P. excelsa pendula* Carr. (1855).

Picea sitchensis (Bong.) Trautv. & Mayer.   **Sitka Spruce.**

SYN.—*Pinus Sitchensis* Bongard, in Mém. Acad. Pétersb. sér. 6, II, 164 (1831).
  Abies trigona Rafinesque, in Atlant. Journ., 119 (1832).—?
  Abies falcata Raf., l. c., 120 (1832).—?
  Abies Menziesii Lindley, in Penn. Cycl., I, 32 (1833).
  Pinus Menziesii var. *crispa* Antoine, Conif., 85, t. 35, f. 2 (1840-1847).
  Pinus Jezoënsis Antoine, l. c., 97, t. 37, f. 1 (1840-1847).
  Picea Sitchensis Trautvetter & Mayer, Fl. Ochot., 87 (1847).
  Abies Sitchensis Lindley & Gordon, in Journ. Hort. Soc. London, V, 212 (1850).
  Picea Menziesii Carrière, Man. Pl., IV, 339 (1854).
  Picea Jezoësis Carrière, Trait. Conif., 1 éd., 255 (1855).
  Pinus Menziesii Douglas Mss. in Lambert, Pinus, 1 ed., III, 161, t. 71 (1837).
  Sequoia Rafinesquei Carrière, Trait. Conif., nouv. éd., 213 (1867).
  Abies Merkiana Fisch. ex Parlatore, in de Candolle, Prodr., XVI, sect. 2, 418 (1868).
  Picea Sitkœnsis Mayr, Wald. Nordam., 338 (1890).

COMMON NAMES.

Tideland Spruce (Cal., Oreg., Wash.).
Menzies Spruce.
Western Spruce.
Great Tideland Spruce (Cal. lit.).

Picea breweriana Watson.   **Weeping Spruce.**

SYN.—*Picea Breweriana* Watson, in Proc. Am. Acad. Sci., XX, 378 (1885).

COMMON NAMES.

Weeping Spruce.
Brewer's Spruce.
Siskiyou Spruce (Germ. lit.).

TSUGA Carr., Trait. Conif., 185 (1855).

Tsuga canadensis (Linn.) Carr.   **Hemlock.**

SYN.—*Pinus Canadensis* Linnæus, Spec. Pl., ed. sec., II, 1412 (1763).
  Abies Americana Miller, Gard. Dict. ed. 8, No. 6 (1768).
  Pinus-Abies americana Marshall, Arb. Am., 103 (1785).
  Abies Canadensis Michaux, Fl. Bor.-Am., II, 206 (1803), not Mill. (1768).
  Abies taxifolia Rafinesque, New Fl. and Bot. 1st pt., 38 (1836), not Poir. (1804).

Tsuga canadensis (Linn.) Carr.—Continued.

> SYN.—*Abies taxifolia* var. *patula* Raf., l. c., 39 (1836).
> *Picea Canadensis* Link, in Linnæa, XV, 524 (1841).
> *Tsuga Canadensis* Carrière, Trait. Conif., 1 éd., 189 (1855).

COMMON NAMES.

Hemlock (Me., N. H., Vt., Mass., R. I., Conn., N. Y., N. J., Pa., Del., Va., N. C., S. C.. Ky.. Wis.. Mich., Minn., Nebr., Ohio, Ont.).
Hemlock Spruce (Vt., R. I., N. Y., Pa.. N. J., W. Va., N. C., S. C.; England, cult.).
. Spruce (Pa., W. Va.).
Spruce Pine (Pa., Del., Va., N. C.).
Oh-neh-tah="Greens on the stick" (N. Y., Indians).
Canadian Hemlock (lit.).
New England Hemlock (lit.).

VARIETIES DISTINGUISHED IN CULTIVATION.

Tsuga canadensis pumila nom. nov.

> SYN.—*Abies canadensis nana* Hort. ex Carrière, Man. Pl., IV, 334 (1854), not *A. Tsuga nana* Sieb. et Zucc. (1844).
> *Tsuga canadensis nana* Carr., Trait Conif., 1 éd., 190 (1855), not *T. Sieb. nana*, Carr., l. c. (1855) ante!

Tsuga canadensis compacta minima nom. nov.

> SYN.—*Tsuga canadensis compacta nana* Hort. ex Beissner, Handb. Nadelh., 402 (1891), not *T. Sieb. nana* Carr. (1855).
> *Abies canadensis compacta nana* Hort. ex Beissn., l. c., not *A. can. nana* Carr. (1854).

Tsuga canadensis globosa Beissn.

> SYN.—*Tsuga canadensis globosa* Beissner, Handb. Conif., 65 (1887); Handb. Nadelh., 402 (1891).
> *Abies canadensis globosa* Beissn., l. c. (1887) and (1891).
> *Tsuga canadensis globularis* Hort. ex Beissn., l. c. (1891).
> *Abies canadensis globularis* Hort. ex Beissn., l. c. (1891).

Tsuga canadensis gracilis Carr.

> SYN.—*Abies Canadensis gracilis* Waterer ex Gordon, Pinetum, ed. 1, Suppl., 9 (1862), not *A. rubra gracilis* Knight (1850).
> *Tsuga Canadensis gracilis* Carrière, Trait. Conif., nouv. éd., 249 (1867).

Tsuga canadensis milfordensis Nichol.

> SYN.—*Tsuga canadensis milfordensis* Nicholson, Dic. Gard., IV, 101 (1889).

43

Tsuga canadensis erecta nom. nov.

SYN.—*Tsuga canadensis fastigiata* Hort. ex Beissner, Handb. Nadelh., 402 (1891), not *T. Dougl. fastigiata* Carr (1855).
Abies canadensis fastigiata Hort. ex Beissn., l. c. (1891), not *A. Dougl. fastigiata* Knight (1850).

Tsuga canadensis columnaris Beissn.

SYN.—*Tsuga canadensis columnaris* Bolle ex Beissner, Handb. Nadelh., 402 (1891).

Tsuga canadensis macrophylla Beissn.

SYN.—*Tsuga canadensis macrophylla* Hort. ex Beissner, Handb Nadelh., 402 (1891).
Abies canadensis macrophylla Hort. ex Beissn., l. c. (1891).

Tsuga canadensis paucifolia nom. nov.

SYN.—*Tsuga canadensis sparsifolia* Beissner, Handb. Nadelh., 402 (1891), not *T. Dougl. sparsifolia* Carr. (1861).

Tsuga canadensis microphylla Beissn.

SYN.—*Tsuga canadensis microphylla* Beissner, Handb. Conif., 65 (1887); Handb. Nadelh., 403 (1891).
Abies canadensis var. *microphylla* Lindley ex Hoopes, Book Everg., 188 (1868), not *A. microphylla* Raf. (1832).

Tsuga canadensis parvifolia (Veitch) Beissn.

SYN.—*Abies canadensis parvifolia* Veitch, Man. Conif., 115 (1881).
Tsuga canadensis parvifolia Beissner, Handb. Conif., 65 (1887); Handb. Nadelh., 403 (1891).

Tsuga canadensis pendula Beissn.

SYN.—*Tsuga canadensis pendula* Beissner, Handb. Conif., 65 (1887); Handb. Nadelh., 403 (1891).
Abies canadensis pendula Beissn., l. c. (1887) and (1891), not *A. excel. pendula* Loud. (1842).

Tsuga canadensis aurea Beissn.

SYN.—*Tsuga canadensis aurea* Beissner, Handb. Conif., 65 (1887); Handb. Nadelh., 403 (1891).
Abies canadensis aurea Beissn., l. c. (1887) and (1891), not *A. excelsa aurea* Hort. ex Carr. (1867).

Tsuga canadensis albo-spica (Gord.) Beissn.

SYN.—*Abies canadensis alba-spica* Barron ex Gordon, Pinetum, 2d ed., 421 (1875).
Tsuga canadensis albo-spica Beissner, Handb. Conif., 65 (1887); Handb. Nadelh., 403 (1891).

Tsuga canadensis albo-spica (Gord.) Beissn.—Continued.

Syn.—*Tsuga canadensis alba-spica* Nicholson, Dic. Gard., IV, 101 (1889).

Tsuga canadensis argenteo-variegata nom. nov.

Syn.—*Tsuga canadensis fol. argent. variegata* Hort. ex Beissner, Handb. Nadelh., 403 (1891).
Abies canadensis fol. argent. variegata Hort. ex Beissn., l. c. (1891).

Tsuga canadensis argentifolia nom. nov.

Syn.—*Tsuga canadensis* var. *argentea*,[1] ex Hand-list Conif. Roy. Gard. Kew, 63. (1896), not *T. pattoniana argentea* Beissn. (1891).

Tsuga caroliniana Engelm.   **Carolina Hemlock.**

Syn.—*Tsuga caroliniana* Engelmann, in Bot. Gaz.. VI, 223 (1881).
Abies Caroliniana Chapman, Fl. S. States, 2d ed., Suppl., 650 (1887).

COMMON NAMES.

Hemlock (N. C., S. C.).
Southern Hemlock (lit.).

Tsuga mertensiana (Bong.) Carr.   **Western Hemlock.**

Syn.—*Pinus Mertensiana* Bongard, Observat. Vég. Sitka. 45 (1831).
Pinus Canadensis Bong., in Mém. Acad. Pétersburg, 6 sér., II. 163 (1832), not Linn. (1763).
Abies heterophylla Rafinesque, in Atlant. Journ.. 119 (1832).
Abies microphylla Raf., l. c. (1832).
Abies Mertensiana Lindley & Gordon, in Journ. Hort. Soc. London. V, 211 (1850).
Picea Mertensiana Hort. ex Gordon, Pinetum. ed. 1. 18 (1858).
Abies? Canadensis Cooper, in Smithsonian Rep. 1858, 262 (1859), not Mill. (1768), nor Desf. (1809).
Abies Bridgei Kellogg. in Proc. Calif. Acad., II, 8 (1863).
Abies Albertiana Murray, in Proc. Hort. Soc. London, III, 149 (1863).
Abies taxifolia Jeffrey ex Gordon. Pinetum, ed. 1, 18 (1858).
Abies Williamsonii Bridges ex Gordon. Pinetum. ed. 1, Suppl., 12 (1862), not Newberry (1857).
Tsuga Mertensiana Carrière. Trait. Conif.. nouv. éd., 250 (1867).

---

[1] So far as known this variety has never been described. The name given to the plant in the Kew Garden list indicates that the foliage is glaucous or silvery, but no characters are given. It appears not to be distinguished in America.

Tsuga mertensiana (Bong.) Carr.—Continued.

SYN.—*Pinus Pattoniana* McNab, in Proc. Roy. Irish Acad., 2 ser.,
II, 211, 212, t. 23, f. 2 (1876), not Parl. (1868).
Abies Pattonii McNab, in Journ. Linn. Soc., XIX, 208 (1882).
Tsuga Canadensis var. *Mertensiana* Newberry ex Zabel, in
Forst. Blät., IX, 209 (1885).

COMMON NAMES.

Hemlock Spruce (Cal.).
Western Hemlock (Cal.).
Hemlock (Oreg., Idaho, Wash.).
Western Hemlock Spruce (lit.).
California Hemlock Spruce (Eng.).
Western Hemlock Fir (Eng.).
Prince Albert's Fir (Eng.).
Alaska Pine (Northwestern lumbermen).

VARIETY DISTINGUISHED IN CULTIVATION.

Tsuga mertensiana latifolia nom. nov.

SYN.—*Tsuga Mertensiana macrophylla* Beissner, Handb. Nadelh.,
404 (1891), not *T. canadensis macrophylla* Hort. ex Beissner, l. c. (1891) ante!
Tsuga canadensis macrophylla Hort. Beissn., l. c. (1891).
Abies canadensis macrophylla Hort. Beissn., l. c., not Beissn.,
l. c. (1891) ante!

Tsuga pattoniana (Jeffr.) Engelm.   **Alpine Hemlock.**

SYN.—*Abies Pattoniana* Jeffrey ex Murray, in (Rep.) Bot. Exped.
Oregon, No. 430, t. 4, f. (1853).[1]
Abies Hookeriana Murray, in Edinburgh New Phil. Journ.,
new ser., I, 289, t. 9, f. 11–17 (1855).
Picea Californica Carrière, Trait. Conif., 1 éd., 261 (1855).—?
Abies Williamsonii Newberry, in Williamson, in Pacific R. R.
Rep., VI, 53, 90, t. 7, f. 19 (1857).
Abies Pattonii Gordon, Pinetum, ed. 1, 10 (1858).
Pinus Pattoniana Parlatore, in de Candolle, Prodr., XVI,
sect. 2, 429 (1868).
Tsuga Pattoniana Engelmann, in Watson, Bot. Calif., II, 121,
(1880).
Hesperopeuce Pattoniana Lemmon, in Third Bienn. Rep. Cal.
Bd. For., 126 (1890).

COMMON NAMES.

Williamson's Spruce (Cal.).
Weeping Spruce (Cal.).

---

[1] See footnote, p. 16.

Tsuga pattoniana (Jeffr.) Engelm.—Continued.

SYN.—Alpine Spruce (Cal.).
Hemlock Spruce (Cal.).
Patton's Spruce.
Alpine Western Spruce.

Tsuga pattoniana hookeriana (Carr.) Lemmon. **Hooker Hemlock.**

SYN.—*Tsuga Hookeriana* Carrière, Trait. Conif., nouv. éd., 252 (1867).
Tsuga Pattoniana var. *Hookeriana* Lemmon, Handb. West.
Am. Cone-b.. 54 (1895).

VARIETY DISTINGUISHED IN CULTIVATION.

Tsuga pattoniana argentea Beissn.

SYN.—*Tsuga Pattoniana argentea* Beissner. Handb. Nadelh., 410
(1891).

PSEUDOTSUGA Carr., Trait. Conif., nouv. éd.. 256 (1867).

Pseudotsuga taxifolia (Poir.) Britton. **Douglas Spruce.**

SYN.—*Pinus taxifolia* Lambert, Pinus, ed. 1, 51, t. 33 (1803), not
Salisb. (1796).
Abies taxifolia Poiret, in Lamarck, Enc. Méth. Bot., VI,
523 (1804).
Abies mucronata Rafinesque, in Atlant. Journ., 120 (1832).
Abies mucronata var. *palustris* Raf., l. c. (1832).
Abies Douglasii Lindley, in Penn. Cycl.. I, 32 (1833).
Pinus Douglasii Sabine, Mss. in Lambert, Pinus. ed. 2, III,
163, t. 90 (1837).
Abies Douglasii var. *taxifolia* Loudon, Arb. Frut., IV, 2319,
f. 2231 (1838).
Abies californica Hort. ex Steudel, Nom. Bot., ed. sec., I, 1
(1840).
Picea Douglasii Link, in Linnæa, XV, 524 (1841).
Pinus Douglasii var. *brevibracteata* Antoine, Conif., 84, t. 33.
f. 4 (1847).
Tsuga Douglasii Carrière, Trait. Conif., 1 éd.. 192 (1855).
Tsuga Douglasii var. *taxifolia* Carr., l. c. (1855).
Abies Drummondi Hort. ex Gordon, Pinetum, ed. 1, 16 (1858).
Abies obliquata Rafinesque ex Gordon, Pinetum, ed. 1, Suppl.,
10 (1862).
Abies obliqua Bongard ex Gordon, l. c. (1862).
PSEUDOTSUGA DOUGLASII Carrière, Trait. Conif., nouv.
éd.. 256 (1867).
Pseudotsuga Douglasii taxifolia Carr., l. c., 258 (1867).
Picea mucronata Carr., l. c.. 312 (1867).
Pseudotsuga Douglasii denudata Carr., l. c., 792 (1867).

Pseudotsuga taxifolia (Lam.) Britton—Continued.

SYN.—*Pseudotsuga Lindleyana* Carr., in Rev. Hort., 152 (1868).
Pseudotsuga taxifolia Britton, in Trans. N. Y. Acad. Sci.,
VIII, 74 (1889).
Tsuga taxifolia Kuntze, Revis. Gen. Pl., II, 802 (1891).
Pseudotsuga mucronata (Raf.) Sudworth, in Contr. U. S. Nat.
Herb., III, No. 4, 266 (1895).
Pseudotsuga taxifolia var. *suberosa* Lemmon, Handb. West.
Am. Cone-b., 57 (1895).

COMMON NAMES.

Red Fir (Oreg., Wash., Idaho, Utah, Mont., Colo.).
Douglas Spruce (Cal., Colo., Mont.).
Douglas Fir (Utah, Oreg., Colo.).
Yellow Fir (Oreg., Mont., Idaho, Wash.).
Spruce (Mont.).
Fir (Mont.).
Oregon Pine (Cal., Wash., Oreg.).
Red Pine (Utah, Idaho, Colo.).
Puget Sound Pine (Wash.).
Douglas Tree.
Cork-barked Douglas Spruce (var. *suberosa* Cal. lit.).

VARIETIES DISTINGUISHED IN CULTIVATION.

Pseudotsuga taxifolia pendula[1] (Beissn.) nom. nov.
Weeping Douglas Spruce.

SYN.—*Pinus Douglasi pendula* Parlatore, in de Candolle, Prodr.,
XVI, sect. 2, 430 (1868), not *P. pendula* Soland. in Ait.
(1789).
Abies taxifolia pendula Masse ex Neumann, in Flor. des
Serres, VIII, 186 (1853), not *A. pendula* Lindl. & Gord.
(1850).
Abies Douglasi pendula Gord., Pinetum, 2d ed., 27, (1875).
Pseudotsuga Douglasi pendula Engelmann ex Beissner,
Handb. Conif., 66 (1887); Handb. Nadelh., 417 (1891).
Tsuga Douglasi pendula Hort. ex. Beissn., l. c. (1887) and
(1891).

Pseudotsuga taxifolia pendula cærulea nom. nov.

SYN.—*Pseudotsuga Douglasi glauca pendula* Smith ex Beissner,
Handb. Nadelh., 418 (1891), not *P. Dougl. pendula* Beissn.
(1887).
Tsuga Douglasi glauca pendula Hort. ex Beissn., l. c. (1891).

---

[1] This form is not uncommon in nature.

Pseudotsuga taxifolia pendula cærulea nom. nov.—Continued.

Syn.—*Abies Doulgasi glauca pendula* Hort. ex Beissn., l. c. (1891).
Pinus Douglasi glauca pendula Hort. ex Beissn., l. c. (1891).

Pseudotsuga taxifolia glauca (Beissn.), nom. nov.

Syn.—*Pseudotsuga Douglasi glauca* Beissner, Handb. Conif., 66
(1887); Handb. Nadelh., 419 (1891).
Tsuga Douglasi glauca Beissn., l. c. (1891).
Abies Douglasi glauca Beissn., l. c. (1891).
Pinus Douglasi glauca Hort. ex Beissner, Handb. Nadelh.,
419 (1891).

Pseudotsuga taxifolia fastigiata (Knight) nom. nov.

Syn.—*Abies Douglasi fastigiata* Knight, Syn. Conif., 37 (1850).
Tsuga Douglasii fastigiata Carrière, Trait. Conif., 1 éd. 193
(1855).
Tsuga Douglasii sparsifolia Carr., in Rev. Hort. 1861, 243
(1861).
Pseudotsuga Douglasi fastigiata Carr., Trait. Conif., nouv. éd.,
257 (1867).
Pseudotsuga Douglasi stricta Carr., l. c., 258 (1867).

Pseudotsuga taxifolia standishiana (Gord.) nom. nov.

Syn.—*Abies Douglasi Standishiana* Gordon, Pinetum. ed. 1, Suppl.,
10 (1862).
Pseudotsuga Douglasi Standishi Hort. ex Beissner, Handb.
Nadelh.. 418 (1891).
Pseudotsuga Douglasi Standishiana Hort. ex Beissn., l. c.
(1891).

Pseudotsuga taxifolia dumosa (Carr.) nom. nov.

Syn.—*Pseudotsuga Douglasii dumosa* Carrière, Trait. Conif., nouv.
éd., 258 (1867).
Pseudotsuga Douglasi monstrosa Hort. ex Carr., l. c. (1867).
Abies Douglasi monstrosa Hort. ex Beissner, Handb. Conif.,
66 (1887).
Tsuga Douglasi monstrosa Beissn., l. c. (1887).

Pseudotsuga taxifolia compacta (Beissn.) nom. nov.

Syn.—*Pseudotsuga Douglasi compacta* Beissner, Handb. Conif., 66
(1887).
Tsuga Douglasi compacta Beissn., l. c. (1887).
Abies Douglasi compacta Beissn., l. c. (1887).

Pseudotsuga taxifolia elegans (Beissn.) nom. nov.

Syn.—*Pseudotsuga Douglasi elegans* Hort. ex Beissner, Handb.
Nadelh., 419 (1891).

Pseudotsuga taxifolia elegans (Beissn.) nom. nov.—Continued.

SYN.—*Tsuga Douglasi elegans* Hort. ex Beissn., l. c. (1891).
Abies Douglasi elegans Hort. ex Beissn., l. c. (1891), not *A.
excelsa elegans* Hort. ex Knight (1850).

Pseudotsuga taxifolia argentea (Beissn.) nom. nov.

SYN.—*Pseudotsuga Douglasi argentea* Koster ex Beissner, Handb.
Nadelh., 419 (1891).

Pseudotsuga taxifolia argentea densa nom. nov.

SYN.—*Pseudotsuga Douglasi argentea compacta* Haus. ex Beissner,
Handb. Nadelh., 419 (1891), not *P. Dougl. compacta* Beissn.
(1887).

Pseudotsuga taxifolia variegata (McDon.) nom. nov.

SYN.—*Abies Douglasii variegata* McDonald, in Gard. Chron., 1583
(1871).

Pseudotsuga taxifolia stairi (Beissn.) nom. nov.

SYN.—*Pseudotsuga Douglasi Stairi* Hort. ex Beissner, Handb.
Nadelh., 420 (1891).
Tsuga Douglasi Stairi Hort. ex Beissn., l. c. (1891).
Abies Douglasi Stairi Hort. ex Beissn., l. c. (1891).

Pseudotsuga taxifolia brevifolia[1] (Hort. Kew) nom. nov.

SYN.—*Pseudotsuga Douglasii* var. *brevifolia*, ex Hand-list Conif.
Roy. Gard. Kew, 87 (1896).

Pseudotsuga taxifolia revoluta[2] (Hort. Kew) nom. nov.

SYN.—*Pseudotsuga Douglasii* var. *revoluta*, ex Hand-list Conif. Roy.
Gard. Kew, 87 (1896).

Pseudotsuga macrocarpa[3] (Torr.) Mayr.   **Bigcone Spruce.**

SYN.—*Abies Douglasii* var. *macrocarpa* Torrey, in Ives's Rep., 28
(1861).
Abies macrocarpa Vasey, in Gard. Monthly, 1876, 22 (1876).

---

[1] No description is given of this variety in the Kew Garden list, nor so far as known
has it ever been characterized. The name indicates that it is the short-leaved form
not uncommon in nature, especially at high elevations. It is probably not cultivated by American nurserymen.

[2] This variety appears never to have been described, the Kew Garden name (l. c.)
being unaccompanied by characters. It is distinguished mainly by the strongly
revolute margins of its foliage.

[3] Prof. J. G. Lemmon is doubtless under the impression that this combination was
first made by himself (in Third Bienn. Rep. Cal. St. Bd. For., 134, Aug., 1890). Dr.
Mayr's publication (l. c.) of the same combination in a work appearing likewise in
1890 would make it difficult to decide which author is responsible but for the fact
that Dr. Mayr's preface bears the date "März, 1889," and Professor Lemmon's preface
"August 30, 1890," indicating the possibility that Dr. Mayr at least formed the combination first, if he did not actually and formally publish it first.

50

Pseudotsuga macrocarpa (Torr.) Mayr—Continued.

SYN.—*Pseudotsuga Douglasii* var. *macrocarpa* Engelmann, in Watson, Bot. Cal., II, 120 (1880).
Pseudotsuga macrocarpa[1] Mayr, Wald. Nordam., 278 (März, 1890); Lemmon, in Third Bienn. Rep. Cal. St. Bd. For., 134 (Aug., 1890).

COMMON NAMES.

Spruce (Cal.).
Hemlock (Cal.).
Bigcone Spruce (lit.).
Bigcone Douglas Spruce (lit.).

ABIES Jussieu, Gen., 414 (1789).

Abies fraseri (Pursh) Lindl.　　　　　　　**Fraser Fir.**

SYN.—*Pinus Fraseri* Pursh, Fl. Am. Sept., II, 639 (1814).
Abies balsamea β *Fraseri* Nuttall, Genera, II, 223 (1818).
Pinus balsamea var. *Fraseri* Torrey, Compend. Fl. N. States, 359 (1826).
Abies humilis La Pylaye, in Mém. Soc. Linn. Par., IV, 437 (1826).—?
Abies Fraseri Lindley, in Penn. Cycl., I, 30 (1833).
Picea Fraseri Loudon, Arb. Frut., IV, 2340, f. 2243, 2244 (1838).
Abies balsamea β *Fraseri* Spach., Hist. Vég., XI, 422 (1842).

COMMON NAMES.

Balsam (N. C., S. C.).
Balsam Fir (N. C., S. C.).
Double Fir Balsam (Tenn.).
Double Spruce (N. C.).
She Balsam (N. C.).
She Balsam Fir (N. C.).
Mountain Balsam (N. C.).
Healing Balsam.

Abies balsamea (Linn.) Mill.　　　　　　　**Balsam Fir.**

SYN.—*Pinus balsamea* Linnæus, Spec. Pl., ed. 1, II, 1002 (1753).
Abies balsamea Miller, Gard. Dic., ed. 8, No. 3 (1768).
Pinus-Abies Balsamea Marshall, Arb. Am., 102 (1785).
Pinus taxifolia Salisbury, Prodr., 399 (1796).
Abies balsamifera Michaux, Fl. Bor.-Am., II, 207 (1803).

[1] See footnote p. 49.

Abies balsamea (Linn.) Mill.—Continued.

SYN.—*Picea balsamea* Loudon, Arb. Frut., IV, 2339, f. 2240, 2241 (1838).
Picea aromatica Carrière, Trait. Conif., nouv. éd., 310 (1867).
Picea Fraseri Emerson, Trees & Shr. Mass., ed. 1, 88 (1878).

COMMON NAMES.

Balsam Fir (N. H., Vt., Mass., R. I., N. Y., Pa., W. Va., Wis., Mich., Minn., Nebr., Ohio, Ont.; Eng. cult.).
Balsam (Vt., N. H., N. Y.).
Canada Balsam (N. C.).
Balm of Gilead (Del.).
Balm of Gilead Fir (N. Y., Pa.).
Blister Pine (W. Va.).
Fir Pine (W. Va.).
Fir Tree (Vt.).
Single Spruce (N. Bruns. to Hudson Bay).
Silver Pine (Hudson Bay).
Sapin (Quebec).
Cho-koh-tung = "Blisters" (N. Y. Indians).

VARIETIES DISTINGUISHED IN CULTIVATION.

Abies balsamea hudsonia (Knight) Veitch.

SYN.—*Picea Fraseri Hudsonia* Knight, Syn. Conif., 39 (1850).
Abies Hudsonia Bosc ex Carrière, Trait. Conif., 1 éd., 200 (1855).
Abies Fraseri var. *Hudsoni* Carr., l. c. (1855).
Abies Fraseri Hudsoni Carr., l. c. (1855).
Abies minor Duhamel ex Gordon, Pinetum, ed. 1, 143 (1858), not *Gilbert* (1792).
Picea Hudsonia Hort. ex Gordon, Pinetum, ed. 1, 148 (1858).
Abies Hudsonica Carrière, Trait. Conif., nouv. éd., 271 (1867).
Abies balsamea Hudsonia Veitch, Man. Conif., 88 (1881).
Picea hudsonica Hort. ex Beissner, Handb. Conif., 70 (1887); Handb. Nadelh., 465 (1891).
Abies balsamea hudsonica Beissn., l. c. (1887) and (1891).

Abies balsamea brachylepis Willk.

SNY.—*Abies balsamea* var. *brachylepis* Willkomm, in Delect. Bot. Dorpat. (1868).

Abies balsamea longifolia (Loud.) Endl.

SYN.—*Picea balsamea longifolia* Booth ex Loudon, Enc. Trees, 1044 (1842).

Abies balsamea longifolia (Loud.) Endl.—Continued.

SYN.—*Abies balsamea longifolia* Endlicher, Syn. Conif., 103 (1847).
Abies invalensis Hort. ex Carrière, Trait. Conif., nouv. éd.,
293 (1867).

Abies balsamea cærulea Carr.

SYN.—*Abies balsamea cærulea* Carrière, Trait. Conif., nouv. éd.,
294 (1867).
Abies excelsa Fraseri Hort. ex Carr., l. c. (1867).

Abies balsamea hemispháerica nom. nov.

SYN.—*Abies balsamea nana* Carrière, Trait. Conif., nouv. éd., 294
(1867), not *A. Tsuga nana* Sieb. et Zucc. (1844).
Abies balsamea tenuifolia Hort. ex Carrière, l. c. (1867), not
A. excelsa tenuifolia Loud. (1842).
Abies balsamea globosa Hort. ex Beissner, Handb. Nadelh.,
465 (1891), not *A. can. globosa* Beissn. (1887).

Abies balsamea prostrata (Knight) Carr.

SYN.—*Picea balsamea prostrata* Knight, Syn. Conif., 39 (1850).
Abies balsamea prostrata Hort. ex Carrière, Trait. Conif., 1.
éd., 218 (1855).

Abies balsamea paucifolia nom. nov.

SYN.—*Abies balsamea denudata* Carrière, Trait. Conif., nouv. éd.,
294 (1867), not *A. excelsa denudata* Gord. (1862).

Abies balsamea nudicaulis Carr.

SYN.—*Abies balsamea nudicaulis* Carrière, Trait. Conif., nouv. éd.,
294 (1867).

Abies balsamea versicolor nom. nov.

SYN.—*Abies balsamea variegata* Carrière, Trait. Conif., 1 éd., 218
(1855), not *A. pectinata variegata* Forb. (1839).

Abies balsamea argentifolia nom. nov.

SYN.—*Abies balsamea argentea* Beissner, Handb. Conif., 70 (1887);
Handb. Nadelh., 466 (1891), not *A. alba argentea* Hort. ex
Carr. (1867).

Abies lasiocarpa (Hook.) Nutt.   **Alpine Fir**.

SYN.—*Pinus (Abies) lasiocarpa* Hooker, Fl. Bor.-Am., II, 163 (1842),
not Hort.
Abies lasiocarpa Nuttall, Sylva, III, ed. 1, 138 (1849).
Abies grandis Engelmann, in Am. Journ. Sci., 2 ser., XXXIV,
330 (1862), not Lindl. (1833).

Abies lasiocarpa (Hook.) Nutt.—Continued.

SYN.—*Abies subalpina* Engelmann ex Ward, in Am. Nat., X, 555
(1876).
Picea amabilis Gordon, Pinetum, ed. 1, 154 (1858), in part.
Abies bifolia Murray, in Proc. Hort. Soc. London, III, 318,
f. 51–56 (1863).
Abies subalpina var. *fallax* Engelmann, in Trans. Acad. Sci.,
St. Louis, III, 597 (1878).
Pinus amabilis Parlatore, in de Candolle, Prodr., XVI, sect.
2, 426 (1868), in part.
Picea bifolia Murray, in Lond. Gard. Chron., III, f. 97 (1875).

COMMON NAMES.

Sub-Alpine Fir (Utah).
Balsam (N. Mex., Colo., Utah, Idaho, Oreg.).
White Fir (Idaho, Mont.).
White Balsam (Cal.).
Oregon Balsam Tree (Cal.).
Pumpkin Tree (Cal.).
Alpine Fir (Cal.).
Mountain Balsam (mts. of Utah and Idaho).
Down-cone Fir (lit.).
Downy-cone Sub-Alpine Fir (Cal. lit.).

VARIETY DISTINGUISHED IN CULTIVATION.

Abies lasiocarpa cærulescens (Beissn.) nom. nov.

SYN.—*Abies subalpina cœrulescens* Fröbel ex Beissner, Handb.
Nadelh., 467 (1891).

Abies grandis Lindl.   **Great Silver Fir.**

SYN.—*Pinus grandis* Douglas, (Mss. in Herb. Lond. Hort. Soc.,
1830) in Companion Bot. Mag., II, 147 (1836)—*nomen
nudum*.
Abies aromatica Rafinesque, in Atlant. Journ., 119 (1832).—?
Abies grandis Lindley, in Penn. Cycl., I, 30 (1833).
Picea grandis Loudon, Arb. Frut., IV, 2341, f. 2245, 2246
(1838).
Abies amabilis Murray, in Proc. Hort. Soc. London, III, 310,
f. 3–9 (1863), not Forb. (1839).
Abies Gordoniana Carrière, Trait. Conif., nouv. éd., 298
(1867).
Abies grandis oregona Beissner, Handb. Conif., 71 (1887).
Abies oregona Hort. ex Beissn., Handb. Nadelh., 476 (1891).

Abies grandis Lindl.—Continued.

COMMON NAMES.

SYN.—White Fir (Cal., Oreg., Idaho).
Silver Fir (Mont., Idaho).
Yellow Fir (Mont., Idaho).
Oregon White Fir (Cal.).
Western White Fir.
Grand or Oregon White Fir (Cal. lit.).
Great California Fir (lit.).

VARIETIES DISTINGUISHED IN CULTIVATION.

Abies grandis aurifolia nom nov.

SYN.—*Abies grandis aurea* Hesse ex Beissn., Handb. Nadelh., 479 (1891), not *A. nigra aurea* Carr (1867).

Abies grandis crassa nom nov.

SYN.—*Abies grandis compacta* Hesse ex Beissner, Handb. Nadelh., 479 (1891), not *A. Dougl. compacta* Beissn. (1887), nor *A. concolor violacea compacta* Beissn. (1891).

Abies concolor (Gord.) Parry.   **White Fir.**

SYN.—*Abies concolor* Lindley & Gordon, in Journ. Hort. Soc. London, V, 210, No. 15 (1850)—*nomen nudum.*
Pinus lasiocarpa Hooker, ex Oreg. Com. in (Rep.) Bot. Exped. Oregon, No. 393, t. 4, f. (1853?)[1], not Hook. (1842).
Abies balsamea Bigelow, in Pacif. R. R. Rep., IV, 18 (1856), not Mill. (1768).—?
Picea grandis Newberry, in Pacif. R. R. Rep., VI, 46, f. 16, Pl. VI (as *P. grandies*) (1857), in part; not Loud. (1838).
Pinus grandis Newberry, l. c., 46 (1857), in part.
Picea concolor Gordon, Pinetum, ed. 1, 155 (1858).
Pinus concolor Engelmann ex Gordon, l. c. (1858).
Picea Parsonii Hort. Am. ex Gordon, Pinetum, ed. 1, Suppl., 52 (1862).
Abies grandis Carrière, Trait. Conif., nouv. éd., 296 (1867), in part; not Lindl. (1833).
Pinus grandis Parlatore, in de Candolle Prodr., XVI, sect. 2, 427 (1868), in part.
Abies Parsonii Hort. ex Sargent, in Tenth Cens. U. S., IX, (Cat. For. Trees N. A.), 212 (1884).
Abies concolor Parry, in Am. Nat., IX, 204 (1875).

---

[1] See footnote p. 16.

Abies concolor (Gord.) Parry—Continued.

SYN.—*Abies Parsoniana* Hort. (Barron, Cat. 1860) ex Masters, in Gard. Chron., XIII, 648 (1880).

COMMON NAMES.

White Fir (Cal., Idaho, Utah, Colo.).
Balsam Fir (Cal., Idaho, Colo.).
Silver Fir (Cal.).
Balsam (Cal.).
White Balsam (Utah).
Bastard Pine (Utah).
Balsam Tree (Idaho).
Black Gum (Utah).
California White Fir (Cal.).
Colorado White Fir (Cal. lit.).
Concolor Silver Fir (Eng. lit.).

VARIETIES DISTINGUISHED IN CULTIVATION.

Abies concolor pendens (Beissn.) nom. nov.

SYN —*Abies concolor* var. *lasiocarpa pendula* Hort. ex Beissner, Handb. Nadelh., 475 (1891), not *A. lasiocarpa* Nutt. (1849), nor *A. pectinata pendula* Hort. Carr. (1855).

Abies concolor varia nom. nov.

SYN.—*Abies concolor* var. *lasiocarpa variegata* Hort. ex Beissner, Handb. Nadelh., 475 (1891), not *A. pectinata variegata* Forb. (1839).

Abies concolor purpurea nom. nov.

SYN.—*Picea concolor* var. *violacea* Murray, in London Gard. Chron., 464, f. 94, 95 (1875).
Abies concolor violacea Hort. ex Beissner, Handb. Conif., 72 (1887), not *A. rubra violacea* Loud. (1838).
Picea concolor violacea Roezl., in Gard. Chron., XII, 1879, 684 (1879).

Abies concolor purpurea compressa nom. nov.

SYN.—*Abies concolor violacea compacta* Hort. ex Beissner, Handb. Nadelh., 476 (1891), not *A. Dougl. compacta* Beissn. (1887).

Abies concolor angustata nom. nov.

SYN.—*Abies concolor fastigiata* Chargueraud, in Rev. Hort., 1889, 428 (1889), not *A. Dougl. fastigiata* Knight (1850).

Abies concolor lowiana (Murr.) Lemmon.   **Pale-leaf White Fir.**

Syn.—*Abies Lowiana* Murray, Syn. Var. Conif., 27 (1850).
　　Abies lasiocarpa Lindley & Gordon, in Journ. Hort. Soc., V,
　　　210 (1850), not Nutt. (1849).
　　Picea Lowiana Gordon, Pinetum, ed. 1, Suppl.. 53 (1862).
　　Picea lasiocarpa Hort. ex Gordon, l. c. (1862).
　　Abies grandis var. *Lowiana* Hoopes. Book of Everg.. 212
　　　(1868).
　　Picea Parsonii Fowler. in Gard.Chron.. 394 (1872). not Gord.
　　　(1862).
　　Picea Lowii Fowler, l. c. (1872).
　　Pinus Lowiana McNab, in Proc. Roy. Irish Acad.. 2 ser.. II,
　　　680, t. 46. f. 5 (1876).
　　Abies concolor var. *lasiocarpa* Beissner, Handb. Conif., 71
　　　(1887). not *A. lasiocarpa* Nutt. (1849).
　　Picea Parsoniana Barron ex Beissn.. l. c. (1887).
　　Abies amabilis Hort. ex Beissn.. Handb. Nadelh., 473 (1891),
　　　not Forb. (1839).
　　Abies californica vera Hort. ex Beissn., l. c. (1891).
　　Abies grandis var. *pallida* Masters ex Lemmon. West. Am.
　　　Cone-b.. 14 (1892).
　　Abies concolor var. *Lowiana* Lemmon. Handb. West. Am.
　　　Cone-b., 64, (1895).

COMMON NAMES.

Pale-leaved White Fir (Cal. lit.).
California White Fir.

Abies venusta (Dougl.) Koch.   **Bristle-cone Fir.**

Syn.—*Pinus venusta* Douglas, in Companion Bot. Mag.. II, 152
　　　(1836).
　　Pinus bracteata D. Don. in Trans. Linn. Soc., XVII, 443
　　　(1836).
　　Picea bracteata Loudon, Arb. Frut., IV, 2348, f. 2256 (1838).
　　Abies bracteata Hooker & Arnott, in Bot. Beechey's Voyage.
　　　394 (1841).
　　Abies venusta Koch, Dendrol., zw. Th. zw. Ab.. 210 (1873).

COMMON NAMES.

Fringed Spruce (Idaho).
Bristle-cone Fir (Cal.).
Santa Lucia Fir (Eng. lit.).

Abies amabilis (Loud.) Forb.   **Amabilis Fir.**

Syn.—*Pinus amabilis* Douglas, (Mss. in Herb. Lond. Hort. Soc.,
　　　1830) in Companion Bot. Mag., II, 93 (1836)—*nomen
　　　nudum*.

Abis amabilis (Loud.) Forb.—Continued.

SYN.—*Pinus grandis* Lambert, Pinus, ed. 1, III, t. 26 (1837), not Dougl. (1836).

Picea amabilis Loudon, Arb. Frut., IV, 2342, f. 2247, 2248 (1838).

Abies amabilis Forbes, Pinetum Wob., 125, t. 44 (1839).

Abies grandis Murray, in Proc. Hort. Soc. London, III, 308, f. 18–21 (1863), not Lindl. (1833).

Abies grandis var. *densiflora* Engelmann, in Trans. Acad. Sci., St. Louis, III, 599 (1878).

COMMON NAMES.

Red Fir (Mont.).
Red Silver Fir (Western mts.).
Fir (Cal.).
Lovely Red Fir (Cal. lit.).
Lovely Fir (Cal. lit.).
Amabilis, or Lovely Fir (Cal. lit.).
"Larch" (Oreg. lumbermen).

Abies nobilis Lindl.   **Noble Fir.**

SYN.—*Pinus nobilis* Douglas, (Mss. in Herb. Lond. Hort. Soc. 1830) in Companion Bot. Mag., II, 93, 147 (1836)—*nomen, nudum*.

Abies nobilis Lindley, in Penn. Cycl., I, 30 (1833).

Picea nobilis Loudon, Arb. Frut., IV, 2342, f. 2249, 2250, (1838).

Picea amabilis Hort. ex Carrière, Trait. Conif. nouv. éd., 269 (1867), not Gord. (1858).

Pseudotsuga nobilis McNab, in Proc. Roy. Irish Acad., ser. 2, II, sub t., 49 (1877).

Abies magnifica Engelmann, in Watson, Bot. Calif., II, 119 (1890), in part.

Abies nobilis oregona Beissner, Handb. Nadelh., 485 (1891), not *A. grandis oregona* Beissn. (1887).

COMMON NAMES.

Red Fir (Cal., Oreg.).
"Larch" (Oreg. lumbermen).
Noble Fir (Oreg.).
Bigtree (Indians, Idaho).
Feather-cone Red Fir (Cal. lit.).
Noble or Bracted Red Fir (Cal. lit.).
Tuck Tuck (Pacific Indians).

VARIETIES DISTINGUISHED IN CULTIVATION.

Abies nobilis glaucifolia nom. nov.

SYN.—*Abies nobilis glauca* Hort. ex Carrière, Trait. Conif., nouv.
éd., 269 (1867), not *A. glauca* Carr. (1855).
Abies nobilis argentea Hort. ex Beissner, Handb. Conif., 71
(1887), not *A. alba argentea* Hort. ex Carr. (1867).

Abies nobilis robustifolia nom. nov.

SYN.—*Abies nobilis robusta* Hort. ex Beissner, Handb. Nadelh., 488
(1891), not Carr. (1867).

Abies magnifica Murr.   Shasta Fir.

SYN.—*Abies campylocarpa* Murray, in Trans. Bot. Soc. Edinburgh,
VI, 370 (1860).—?
Abies magnifica Murray, in Proc. Royal Soc. London, III,
318, f. 25–33 (1863).
Abies s. g. *Picea magnifica* Murray, l. c., 319 (1863).
Pinus amabilis Parlatore, in de Candolle. Prodr.. XVI, sect.
2, 426 (1868), in part.
Picea magnifica Hort. ex Carrière. Trait. Conif.. nouv. éd.,
269 (1867).
Abies amabilis Hort. ex Carr.. l. c. (1867), not Forb. (1839).
Picea amabilis magnifica Hort. ex Carr.. l. c. (1867).
Abies nobilis robusta Veitch ex Carr.. l. c. (1867).
Picea amabilis robusta Hort. ex Carr., l. c. (1867).
Pinus magnifica McNab, in Proc. Roy. Irish Acad., ser. 2,
II, 700 (1876), not Roezl (1875).
Pseudotsuga magnifica McNab. l. c., sub t., 49 (1877).
Abies nobilis Engelmann. in Watson. Bot. Calif.. II. 119
(1880), in part.
Pseudotsuga magnifica McNab, in Proc. Royal Irish Acad.,
2d ser., II, 700, t. 49, f. 30, 30a (1887).
Abies nobilis var. *magnifica* Masters. in Journ. Linn. Soc.
Bot., XXII, 189 (1887).
Abies magnifica var. *Shastensis* Lemmon. in Third Bienn.
Rep. Cal. St. Bd. For., 145 (1890).—?

COMMON NAMES.

Red Fir (Cal.).
California Red-bark Fir (Cal.).
Magnificent Fir (Cal. lit.).
California Red Fir (Cal. lit.).
Shasta Red Fir (var. *Shastensis*).
Shasta Fir (Cal. lit.).
Golden Fir (Cal. lit.).

VARIETIES DISTINGUISHED IN CULTIVATION.

Abies magnifica cyanea nom. nov.

SYN.—*Abies magnifica glauca* Hort. ex Beissner, Haudb. Nadelh. 484 (1891), not *A. glauca* Hort. ex Carr. (1855).

Abies magnifica xanthocarpa[1] Lemmon.   **Yellow-fruit Fir.**

SYN.—*Abies nobilis* var. *robusta* Masters, in Gard. Chron., XXIV, 657, f. 147 (1885), not *A. nobilis robusta* Veitch ex Carr., (1867).

Abies magnifica var. *Xanthocarpa* Lemmon, in Third Bienn. Rep. Cal. St. Bd. For., 145 (1890).

Abies shastensis Lemmon, ex Hand-list Conif. Roy. Gard. Kew, 83 (1896), not *A. magnifica Shastensis* Lemmon (1890).

COMMON NAME.

Yellow-fruited Fir (Cal. lit.).

TAXODIUM Rich., in Ann. Mus. Par., XVI, 298 (1810).

Taxodium distichum (Linn.) Rich.   **Bald Cypress.**

SYN.—*Cupressus disticha* Linnæus, Spec. Pl., ed. 1, II, 1003 (1753).

Cupressus læta Salisbury, Prodr., 397 (1796).

Taxodium distichum Richard, in Ann. Mus. Par., XVI, 298 (1810).

Schubertia disticha Mirbel, in Mém. Mus. Par., XIII, 75 (1825).

Cupressus Americana Catesby ex Endlicher, Syn. Conif., 68 (1847), not Trautv. (1846).

Cupressus imbricata Nuttall ex Gordon, Pinetum, ed. 1, 307 (1858).

Cuprespinnata disticha Nelson Pinac., 61 (1866).

Cuprespinnata sinensis Nels., l. c. (1866).

COMMON NAMES.

Bald Cypress (Del., N. C., S. C., Ala., la., Fla., Tex., Ark., Mo., Ill., Ind.).

White Cypress (N. C., S. C., Fla., Miss.).

Black Cypress (N. C., S. C., Tex., Ala.).

Red Cypress (Ga., Miss., Tex., La.).

Swamp Cypress (La.).

Cypress (Del., N. C., S. C., Fla., Miss., Ky., Mo., Ill.).

Deciduous Cypress (Del., Ill., Tex.).

Southern Cypress (Ala.)

---

[1] Mr. Lemmon (l. c.) describes this variety as distinct in nature. The golden color of the immature cones is, however, lost at maturity.

VARIETIES DISTINGUISHED IN CULTIVATION.

Taxodium distichum pendulum (Forb.) Carr.

SYN.—*Taxodium sinense pendulum* Forbes, Pinet. Wob., 180 (1839).
Taxodium distichum sinense pendulum Loudon. Enc. Trees, 1078 (1842).
Glyptostrobus pendulus Endlicher, Syn. Conif., 71 (1847).
Taxodium distichum pendulum Carrière, Trait. Conif., 1 éd., 145 (1855).
Taxodium sinense Noisette ex Carrière, l. c., 152 (1855).
Taxodium distichum sinense Loudon ex Gordon, Pinetum, ed. 1, 309 (1858).

Taxodium distichum pendulum elegans Beissn.

SYN.—*Taxodium distichum pendulum elegans* Hort. ex Beissner, Handb. Nadelh., 152 (1891).

Taxodium distichum pendulum novum Beissn.

SYN.—*Taxodium distichum pendulum novum* Smith ex Beissner, Handb. Nadelh., 152 (1891).

Taxodium distichum patens (Ait.) Endl.

SYN.—*Cupressus disticha α patens* Aiton, Hort. Kew., ed. 2, V, 323 (1813).
Cupressus disticha β nutans Ait., l. c. (1813).
Taxodium distichum var. *nutans* Endlicher, Syn. Conif., 68 (1847).
Taxodium distichum var. *patens* Endl., l. c. (1847).

Taxodium distichum denudatum Carr.

SYN.—*Taxodium distichum denudatum* Carrière, Trait. Conif., 1 éd., 145 (1855).
Taxodium denudatum Hort. ex Carr., l. c., nouv. éd., 182 (1867).

Taxodium distichum imbricaria (Nutt.) nom. nov.

SYN.—*Cupressus disticha β imbricaria* Nuttall, Genera, II, 224 (1818).
Taxodium adscendens Brongniart, in Ann. Sci. Nat., 1 sér., XXX, 128 (1833).
Schubertia disticha imbricaria Spach, Hist. Vég., XI, 349 (1842).
Taxodium distichum fastigiatum Knight, Syn. Conif., 21 (1850).
Cupressus disticha fastigiata Hort. ex Carrière, Trait. Conif., nouv. éd., 181 (1867).

Taxodium distichum knighti Carr.

SYN.—*Taxodium distichum Knighti* Carrière, Trait. Conif., nouv.
éd., 183 (1867).
Taxodium distichum pyramidale Hort. Angl. ex Beissner,
Handb. Conif., 43 (1887); Handb. Nadelh., 153 (1891).

Taxodium distichum pyramidatum Carr.

SYN.—*Taxodium distichum pyramidatum* Carrière, in Rev. Hort.
1859, 65 (1859).
Taxodium pyramidatum Hort. ex Carr., Trait. Conif., nouv.
éd., 184 (1867).

Taxodium distichum microphyllum (Brong.) Henk. & Hochst.

SYN.—*Taxodium microphyllum* Brongniart, in Ann. Sc. Nat., XXX,
1 sér., 182 (1833).
Schubertia disticha microphylla Spach, Hist. Vég., XI, 350
(1842).
Taxodium distichum microphyllum Henkel & Hochstetter,
Nadelhölz, 261 (1865).

Taxodium distichum intermedium Carr.

SYN.—*Taxodium distichum intermedium* Carrière, in Rev. Hort.
1859, 63 (1859).

Taxodium distichum nanum Carr.

SYN.—*Taxodium distichum nanum* Hort. ex Carrière, Trait. Conif.,
1 éd., 145 (1855).

Taxodium distichum nigrum Gord.

SYN.—*Taxodium distichum nigrum* Hort. ex Gordon, Pinetum, ed. 1,
305 (1858).

SEQUOIA Endlicher, Syn. Conif., 147 (1847).

Sequoia washingtoniana (Winsl.) nom. nov.   **Bigtree.**

SYN.— *Wellingtonia gigantea* Lindley, in London Gard. Chron., 1853,
819, 823 (1853).
SEQUOIA GIGANTEA Decaisne, in Bull. Bot. Soc. France, I,
70 (1854), not Endl. (1847).
Taxodium Washingtonianum Winslow, in Calif. Farmer,
Sept , 1854 (1854).
Washingtonia Californica Winslow, l. c. (1854).
Sequoia Wellingtonia Seeman, in Bonplandia, III, 27 (1855).
Taxodium giganteum Kellogg & Behr, in Proc. Calif. Acad.
Sci., I, 53 (1855); reprint ed. 2, 51 (1873).
Gigantabies Wellingtoniana Nelson, Pinac., 79 (1866).

Sequoia washingtoniana (Winsl.) nom. nov.—Continued.

SYN.—Sequoia (Cal.).
Bigtree (Cal.).
Giant Sequoia (Cal.).
Mammoth-tree (Cal., and in Eng. cult.).

VARIETIES DISTINGUISHED IN CULTIVATION.

Sequoia washingtoniana pendula (Beissn.) nom. nov.

SYN.—*Sequoia gigantea pendula* Hort. ex Beissner, Handb. Nadelh., 164 (1891).
Wellingtonia gigantea pendula Hort. ex Beissn., l. c. (1891).

Sequoia washingtoniana glauca (Gord.) nom. nov.

SYN.—*Sequoia gigantea glauca* Hort. ex Gordon, Pinetum, 2d ed., 381 (1875).
Wellingtonia gigantea glauca Hort. ex Beissner, Handb. Nadelh., 165 (1891).

Sequoia washingtonia aurea (Beissn.) nom. nov.

SYN.—*Sequoia gigantea aurea* Hort. ex Beissner, Handb. Nadelh., 165 (1891).
Sequoia gigantea lutea Hort. ex Beissn., l. c. (1891).
Wellingtonia gigantea aurea Hort. ex Beissn., l. c. (1891).
Wellingtonia gigantea lutea Hort. ex Beissn., l. c. (1891).

Sequoia washingtoniana argentea (Beissn.) nom. nov.

SYN.—*Sequoia gigantea argentea* Hort. ex Beissner, Handb. Nadelh., 165 (1891).
Wellingtonia gigantea argentea Hort. ex Beissn., l. c. (1891).

Sequoia washingtoniana variegata (Gord.) nom. nov.

SYN.—*Wellingtonia gigantea variegata* Hort. ex Gordon, Pinetum, 2d ed., 416 (1875).
Sequoia gigantea variegata Hort. ex Beissner, Handb. Nadelh., 165 (1891).

Sequoia washingtoniana holmsi (Beissn.) nom. nov.

SYN.—*Sequoia gigantea Holmsi* Smith ex Beissner, Handb. Nadelh., 165 (1891).
Wellingtonia gigantea Holmsi Smith ex Beissn., l. c. (1891).

Sequoia washingtoniana glaucescens pyramido-compacta nom. nov.

SYN.—*Sequoia glauca pyramidalis compacta* Otin ex Chargueraud, in Rev. Hort., 1889, 476 (1889).
Sequoia gigantea glauca pyramidalis compacta Hort. ex Beissner, Handb. Nadelh., 165 (1891).

Sequoia washingtoniana pygmæa (Beissn) nom. nov.

SYN.—*Sequoia gigantea pygmæa* Hort. ex Beissner, Handb. Nadelh.,
165 (1891).
Wellingtonia gigantea pygmæa Hort. ex Beissn., l. c. (1891).

Sequoia sempervirens (Lamb.) Endl.   **Redwood.**

SYN.—*Taxodium sempervirens* Lambert, Pinus, 114 (1803).
Abies trigona Rafinesque, in Atlant. Journ., 119 (1832).
Taxodium sp. Douglas, in Companion Bot. Mag., II, 150
(1836).
Abies religiosa Hooker & Arnott, in Bot. Beechey's Voyage,
160 (1841).
Schubertia sempervirens Spach, Hist. Vég., XI, 353 (1842).
Sequoia gigantea Endlicher, Syn. Conif., 198 (1847).
Sequoia sempervirens Endl., l. c., 198 (1847).
Sequoia religiosa Presl, Epimel. Bot., 357 (1849).
Gigantabies taxifolia Nelson, Pinac., 78 (1866).

COMMON NAMES.

Redwood (Cal., and Am. lit.).
Sequoia (Cal.).
Coast Redwood (Cal.).
California Redwood (Eng. lit.).

VARIETIES DISTINGUISHED IN CULTIVATION.

Sequoia sempervirens gracilis Carr.

SYN.—*Sequoia sempervirens gracilis* Carrière, Trait. Conif., nouv.
éd., 211 (1867).

Sequoia sempervirens taxifolia Carr.

SYN.—*Sequoia sempervirens taxifolia* Hort. ex Carrière, Trait.
Conif., nouv. éd., 211 (1867).

Sequoia sempervirens adpressa Carr.

SYN.—*Sequoia sempervirens adpressa* Carrière, Trait. Conif., nouv.
éd., 211 (1867).
Sequoia pyramidata Hort. ex Carr., l. c. (1867).

Sequoia sempervirens picta nom. nov.

SYN.—*Sequoia sempervirens variegata* Carrière, in Rev. Hort. 1890,
330 (1890), not *S. wash. variegata* (Gord. 1875).

Sequoia sempervirens albo-spica (Gord.) Beissn.

SYN.—*Taxodium sempervirens albo spica* Hort. ex Gordon, Pinetum,
2d ed., 381 (1875).
Sequoia sempervirens albo spica Veitch, Man. Conif., 212
(1881).

Sequoia sempervirens albo-spica (Gord.) Beissn.—Continued.

Syn.—*Sequoia sempervirens albo-spicata* Hort. ex Beissner, Handb.
Nadelh., 159 (1891).
Taxodium sempervirens albo-spica Beissn.. Handb. Conif., 43
(1887); Handb. Nadelh, 159 (1891).

Sequoia sempervirens glauca Gord.

Syn.—*Taxodium sempervirens glauca* Hort. ex Gordon, Pinetum,
2d ed., 381 (1875).
Taxodium sempervirens glaucum Hort. ex Gord.. l. c. (1875).

LIBOCEDRUS Endlicher, Syn. Conif., 42 (1847).

Libocedrus decurrens Torr.   **Incense Cedar.**

Syn.—*Libocedrus decurrens* Torrey, in Smithsonian Contrib., VI
(Plantæ Fremont.), 7, pl. 3 (1853).
Thuya Craigiana Murray, in (Rep.) Bot. Exped. Oregon, No.
750, t. 5 (1853?).[1]
. *Thuya gigantea* Carrière, in Rev. Hort. 1854, 224, f. 12–14
(1854), in part; not Nutt. (1834).
Heyderia decurrens Koch, Dendrol.. zw. Th. zw. Ab., 179
(1873).

COMMON NAMES.

White Cedar (Cal., Oreg.).
Cedar (Cal., Oreg.).
Incense Cedar (Cal., Oreg.).
Post Cedar (Cal., Nev., Idaho).
Juniper (Nev.).
Bastard Cedar (Cal., Wash.).
Red Cedar (Idaho).
California Post Cedar (Cal. lit.).

VARIETIES DISTINGUISHED IN CULTIVATION.

Libocedrus decurrens depressa Gord.

Syn.—*Libocedrus decurrens depressa* Schott ex Gordon. Pinetum,
2d ed.. 426 (1875).

Libocedrus decurrens columnaris Beissn.

Syn.—*Libocedrus decurrens columnaris* Hort. ex Beissner, Handb.
Nadelh.. 30 (1891).

Libocedrus decurrens compacta Beissn.

Syn.—*Libocedrus decurrens compacta* Hort. ex Beissner, Handb.
Nadelh.. 30 (1891).

---

[1] See footnote. p. 16.

Libocedrus decurrens glauca Beissn.

SYN.—*Libocedrus decurrens glauca* Hort. ex Beissner, Handb.
Nadelh., 30 (1891).
Thuya Craigiana glauca Hort. ex Beissner, l. c. (1891).

THUJA[1] Linn., Spec. Pl., 1002 (1753).

Thuja occidentalis Linn.   (Atlantic) **Arborvitæ.**

SYN.—*Thuja occidentalis* Linnæus, Spec. Pl., ed. 1, 1002 (1753).
Thuya odorata Marshall, Arb. Am., 152 (1785).
Thuya obtusa Moench, Meth., 691 (1794).
Cupressus Arbor-Vitæ Targioni-Tozzetti, Obs. Bot., III, 71
(1808).
Retinospora ericoides Zuccarini ex Gordon, Pinetum, Suppl.,
91 (1862), in part.
Biota ericoides Hort. ex Carrière, Trait. Conif., nouv. éd.,
141 (1867).
Juniperus ericoides Hort. ex Masters, in Journ. Linn. Soc.,
XVIII, 495 (1881).

COMMON NAMES.

Arborvitæ (Me., Vt., Mass., R. I., Conn., N. Y., N. J., Pa.,
Del., Va., W. Va., Ind., Ill., Wis., Mich., Minn., Ohio,
Ont.).
White Cedar (Me., N. H., Vt., R. I., Mass., N. Y., N. J., Va.,
N. C., Wis., Mich., Minn., Ont.).
Cedar (Me., Vt., N. Y.).
American Arbor Vitæ (N. Y. and in cult. Eng.).
Oo-soo-ha-tah=Feather-leaf (Indians).
Vitæ (Del.).
Atlantic Red Cedar (Cal. lit.).

VARIETIES DISTINGUISHED IN CULTIVATION.

Thuja occidentalis ellwangeriana (Gord.) Beissn.

SYN.—*Retinospora Ellwangeriana* Barry ex Gordon, Pinetum, 2d
ed., 362 (1875).
Thuja Ellwangeriana Hort. ex Gordon, l. c. (1875).
Thuya occidentalis Ellwangeriana Beissner, Handb. Conif.,
27 (1887).
Thuya occidentalis Tom Thumb Hort. ex Beissner, l. c. (1887).

---

[1]Linnæus wrote *Thuya* in the Systema (ed. 1, 1735), but *Thuja* in the Hortus
Cliffortianus (449, 1737), also in the various editions of his Genera Plantarum and in
the Species Plantarum.

18158—No. 14——5

Thuja occidentalis spaethi Beissn.

Syn.— *Thuya occidentalis Spœthi* Smith ex Beissner, Handb. Nadelh., 39 (1891).
Thuya occidentalis Ohlendorffi Hort. ex Beissn., l. c. (1891).
Thuya tetragona Hort. ex Beissn., l. c. (1891), not Endl. (1847).

Thuja occidentalis wareana Gord.

Syn.—*Thuia Wareana* Hort. ex Carrière, Trait. Conif., 1 éd., 104 (1855).
Thuja Caucasica Hort. ex Gordon, Pinetum, ed. 1, Suppl., 103 (1862).
Thuja Sibirica Hort. ex Gord., l. c., 104 (1862).
Thuja Occidentalis Wareana Knight ex Gord., l. c. (1862).
Thuia occidentalis robusta Carrière, Trait. Conif., nouv. éd., 109 (1867).
Thuya plicata Wareana Hort. ex Beissner, Handb. Nadelh., 40 (1891).

Thuja occidentalis wareana lutescens Beissn.

Syn.—*Thuya occidentalis Wareana lutescens* Hesse in Hort. ex Beissner, Handb. Nadelh., 40 (1891).

Thuja occidentalis wareana globosa Beissn.

Syn.—*Thuya occidentalis Wareana globosa* Hort. ex Beissner, Handb. Nadelh., 40 (1891).

Thuja occidentalis densa Gord.

Syn.—*Thuja Occidentalis densa* Gordon ex Gordon, Pinetum, ed. 1, Suppl., 103 (1862).
Thuja compacta Standish ex Gord., l. c. (1862).

Thuja occidentalis walthamensis Gord.

Syn.—*Thuja Occidentalis Walthamensis* Paul ex Gordon, Pinetum, 2d ed., 406 (1875).

Thuja occidentalis fastigiata Beissn.

Syn.—*Thuya occidentalis fastigiata* Hort. ex Beissner, Handb. Nadelh., 40 (1891).
Thuya occidentalis pyramidalis Hort. ex Beissn., l. c. (1891).
Thuya occidentalis stricta Hort. ex Beissn., l. c. (1891).
Thuya occidentalis columnaris Hort. ex Beissn., l. c. (1891).

Thuja occidentalis fastigiata nova Beissn.

Syn.—*Thuya occidentalis fastigiata nova* Hort. ex Beissner, Handb. Nadelh., 41 (1891).

Thuja occidentalis l'haveana Beissn.

SYN.—*Thuya occidentalis l'Haveana* Hort. ex Beissner, Handb. Nadelh., 40 (1891).

Thuja occidentalis rosenthali Beissn.

SYN.—*Thuya occidentalis Rosenthali* Ohlendorff ex Beissner, Handb. Nadelh., 41 (1891).

Thuja occidentalis viridis Beissn.

SYN.—*Thuya occidentalis viridis* Hort. ex Beissner, Handb. Nadelh., 41 (1891).
Thuya occidentalis erecta viridis Hort. ex Beissn., l. c. (1891).
Thuya occidentalis atrovirens[1] Hort. ex Beissn., l. c. (1891).

Thuja occidentalis theodonensis Beissn

SYN.—*Thuya occidentalis Theodonensis* Hort. ex Beissner, Handb. Nadelh., 41 (1891).
Thuya occidentalis magnifica Hort. ex Beissn., l. c. (1891).

Thuja occidentalis tatarica Beissn.

SYN.—*Thuya occidentalis tatarica* Hort. ex Beissner, Handb. Nadelh., 41 (1891).
Biota tatarica Hort. ex Beissn., l. c. (1891), not Lindl. & Gord. (1850).

Thuja occidentalis riversi Beissn.

SYN.—*Thuya occidentalis Riversi* Hort. ex Beissner, Handb. Nadelh., 41 (1891).
Thuya occidentalis sp. Rivers in Hort. ex Beissn., l. c. (1891).
Thuya sp. Rivers in Hort. ex Beissn., l. c. (1891).

Thuja occidentalis vervæneana Gord.

SYN.—*Thuya occidentalis Verræneana* Hort. ex Gordon, Pinetum, ed. 1, Suppl., 103 (1862).
Thuja Verræneana Van-Geert ex Gord., l. c. (1862).
Thuya occidentalis aurescens Hort. ex Beissner, Handb. Nadelh., 41 (1891).

Thuja occidentalis lutea Veitch.

SYN.—*Thuia occidentalis lutea* Veitch, Man. Conif., 262 (1881).
Thuya occidentalis lutea Hort. ex. Beissner, Handb. Nadelh., 41 (1891).

---

[1] This variety is often cultivated as a form distinct from *T. occid. viridis*, but the distinction is one of very slight degree in color.

Thuja occidentalis lutea humilis nom. nov.

SYN.—*Thuya occidentalis lutea nana* Hort. ex Beissner, Handb. Nadelh., 41 (1891), not *Th. Sphær. nana* Hort. ex Gord. (1858), nor *Th. occid. recurva nana* Hort. ex Carr. (1867).

Thuja occidentalis aurea Gord.

SYN.—*Thuja Occidentalis aurea* Maxwell in Hort. ex Gordon, Pinetum, 2d ed., 431 (1875).
Thuya plicata aurea Hort. ex Beissner, Handb. Nadelh., 42 (1891).

Thuja occidentalis varia nom. nov.

SYN.—*Thuia occidentalis variegata* Veitch, Man. Conif., 262 (1881), not *Th. variegata* Carr. (1855).

Thuja occidentalis pendula Gord.

SYN.—*Thuya occidentalis pendula* Gordon, Pinetum, ed. 1, Suppl., 103 (1862).

Thuja occidentalis pendula glaucescens nom. nov.

SYN.—*Thuya occidentalis pendula glauca* Hort. ex Beissner, Handb. Nadelh., 42 (1891), not *Th. Craig. glauca* Hort. ex Beissn., l. c. (1891), ante!

Thuja occidentalis reflexa Carr.

SYN.—*Thuia occidentalis reflexa* Hort. ex Carrière, Trait. Conif., nouv. éd., 110 (1867).

Thuja occidentalis bodmeri Beissn.

SYN.—*Thuya occidentalis Bodmeri* Hort. ex Beissner, Handb. Nadelh., 42 (1891).
Thuya Bodmeri Hort. ex Beissn., l. c. (1891).

Thuja occidentalis athrotaxoides Beissn.

SYN.—*Thuya occidentalis athrotaxoides* Hort. ex Beissner, Handb. Nadelh., 42 (1891).

Thuja occidentalis recurvata Beissn.

SYN.—*Thuya occidentalis recurvata* Hort. ex Beissner, Handb. Nadelh., 42 (1891).
Thuya recurvata Hort. ex Beissn., l. c. (1891).

Thuja occidentalis recurvata argenteo-variegata Beissn.

SYN.—*Thuya occidentalis recurvata argenteo-variegata* Hort. ex Beissner, Handb. Nadelh., 42 (1891).

69

Thuja occidentalis recurva pusilla nom. nov.

SYN.—*Thuia occidentalis recurva nana* Hort. ex Carrière, Trait.
Conif., nouv. éd., 111 (1867), not *Th. nana* Carr (1855), nor
Th. sphaer. nana Gord. (1858), nor Gord. (1862).
Thuia recurva nana Hort. ex Carr., l. c. (1867).

Thuja occidentalis denudata Beissn.

SYN.—*Thuya occidentalis denudata* Hort. ex Beissner, Handb.
Nadelh., 43 (1891).

Thuja occidentalis asplenifolia Carr.

SYN.—*Thuia occidentalis asplenifolia* Hort. ex Carrière, Trait.
Conif., 1 éd., 104 (1855).
Thuia asplenifolia Hort. ex Carr., Trait. Conif., nouv. éd.,
106 (1867).

Thuja occidentalis gracilis Gord.

SYN.—*Thuja Occidentalis gracilis* Scott ex Gordon, Pinetum, 2d
ed., 431 (1875).

Thuja occidentalis filicoides Beissn.

SYN.—*Thuya occidentalis filicoides* Hort. ex Beissner, Handb.
Nadelh., 43 (1891).

Thuja occidentalis cristata Gord.

SYN.—*Thuya Occidentalis cristata* Cripps ex Gordon, Pinetum, 2d
ed., 404 (1875).
Thuya cristata Hort. ex Beissner, Handb. Nadelh., 43 (1891).

Thuja occidentalis boothi Beissn.

SYN.—*Thuya occidentalis Boothi* Hort. ex Beissner, Handb. Nadelh.,
43 (1891).

Thuja occidentalis globosa Gord.

SYN.—*Thuja Occidentalis globosa* Hort. ex Gordon, Pinetum, 2d ed.,
405 (1875).
Thuya occidentalis globosa compacta Hort. ex Beissner,
Handb. Nadelh., 43 (1891), not *Th. occ. compacta* Beissn.
(1887).
Thuya occidentalis globosa viridis Hort. ex Beissn., l. c.
(1891), not *Th. occ. viridis* Beissn., l. c. (1891), ante!
Thuya globosa Hort. ex Beissn., l. c. (1891).

Thuja occidentalis globularis Beissn.

SYN.—*Thuya occidentalis globularis* Lamb. & Reiter. ex Beissner,
Handb. Nadelh., 43 (1891).

Thuja occidentalis hoveyi Gord.

Syn.—*Thuja Occidentalis Hoveyi* Hort. ex Gordon, Pinetum, 2d ed., 405 (1875).
Thuja Hoveyi Hort. ex Gord., l. c. (1875).

Thuja occidentalis spihlmanni Beissn.

Syn.—*Thuya occidentalis Spihlmanni* Smith ex Beissner, Handb. Nadelh., 43 (1891).

Thuja occidentalis frœbeli Beissn.

Syn.—*Thuya occidentalis Frœbeli* Hort. ex Beissner, Handb. Nadelh., 43 (1891).

Thuja occidentalis parva nom. nov.

Syn.—*Thuya occidentalis pumila* Hort. ex Beissner, Handb. Nadelh., 44 (1891), not *Th. gig. pumila* Gord. (1875).

Thuja occidentalis albo-variegata Beissn.

Syn.—*Thuya occidentalis albo-variegata* Hort. ex Beissner, Handb. Nadelh., 44 (1891).

Thuja occidentalis aureo-variegata Beissn.

Syn.—*Thuya occidentalis aureo-variegata* Hort. ex Beissner, Handb. Nadelh., 44 (1891).

Thuja occidentalis argentea Gord.

Syn.—*Thuja Occidentalis argentea* Gordon, Pinetum, 2d ed., 404 (1875).

Thuja occidentalis alba Gord.

Syn.—*Thuja Occidentalis alba* Masters ex Gordon, Pinetum, 2d ed., 431 (1875).
Thuya occidentalis albo-spica Hort. ex Beissner, Handb. Nadelh., 44 (1891).
Thuya occidentalis Victoria Hort. ex Beissn., l. c. (1891).

Thuja occidentalis little-gem Beissn.

Syn.—*Thuya occidentalis Little Gem* Hort. ex Beissner, Handb. Nadelh., 44 (1891).

Thuja occidentalis silver-queen Beissn.

Syn.—*Thuya occidentalis Silver Queen* Hort. ex Beissner, Handb. Nadelh., 44 (1891).

Thuja plicata Don.   **Pacific Arborvitæ.**

Syn.—*Thuya plicata* Don. Hort. Cantab., ed. 6, 249 (1811).
THUYA GIGANTEA Nuttall, in Journ. Phila. Acad., VII, 52 (1834).

71

Thuja plicata Don—Continued.

SYN.—*Thuya Menziesii* Douglas Mss., Hook., in Herb. Delessert
ex Carrière, Trait. Conif., 1 éd., 106 (1855).
Thuia Lobbii Hort. ex Gordon, Pinetum, ed. 1, 323 (1858).
Thuja Lobbiana Hort. ex Gord., l. c. (1858).
Thuya occidentalis var. *plicata* Loudon ex Gord., l. c., 325
(1858).
Thuja Douglasii Nutt. ex Parlatore, in de Candolle, Prodr.,
XVI, sect. 2, 457 (1868).
Thuya craigiana Hort., ex Hand-list Conif. Roy. Gard. Kew,
49 (1896), not Murr. (1853 ?).
Thuya gigantea var. *plicata* Don, ex Hand-list, l. c. (1896).

COMMON NAMES.

Red Cedar (Idaho, Oreg., Wash.).
Canoe Cedar (Oreg., Wash.).
Arbor Vitæ (Cal.).
Shinglewood (Idaho).
Gigantic Cedar (Cal.).
Cedar (Oreg.).
Gigantic Red Cedar (Cal. lit.).
Western Cedar.
Gigantic or Pacific Red Cedar (Cal. lit.).
Lobb's Arbor Vitæ (in cult. Eng.).
Pacific Red Cedar (Cal. lit.).

VARIETIES DISTINGUISHED IN CULTIVATION.

Thuja plicata gracillima (Beissn.) nom. nov.

SYN.— *Thuya gigantea gracillima* Hort. ex Beissner, Handb. Conif.,
30 (1887).
Thuya gigantea gracilis Hort. ex Beissn., Handb. Nadelh., 48
(1891), not *T. occid. gracilis* Gord. (1875).
Thuya Lobbi gracilis Hort. ex Beissn., l. c. (1891).

Thuja plicata atrovirens (Gord.) nom. nov.

SYN.—*Thuya gigantea atrovirens* Hort. ex Gordon, Pinetum, 2d ed.,
430 (1875).
Thuya Lobbi atrovirens R. Smith ex Gord., l. c. (1875).

Thuja plicata aurescens (Beissn.) nom. nov.

SYN.—*Thuya gigantea aurescens* Beissner, Handb. Conif., 30 (1887).
Thuya gigantea semperaurea Hort. ex Beissn., Handb.
Nadelh., 49 (1891).
Thuya Lobbi semperaurea Hort. ex Beissn., l. c. (1891).
Thuya gigantea lutescens Hort. ex Beissn., l. c. (1891).

Thuja plicata aurescens (Beissn.) nom. nov.—Continued.

SYN.—*Thuya Lobbi lutescens* Hort. ex Beissn., l. c. (1891).
 Thuya gigantea var. *plicata lutea*, ex Hand-list Conif. Roy.
 Gard. Kew, 49 (1896)—*nomen nudum* (as to rank).

Thuja plicata argenteo-versicolor nom. nov.

SYN.—*Thuya plicata argenteo-rariegata* Hort. ex Beissner, Handb.
 Nadelh., 46 (1891); not Beissn., l. c., 42 (1891).

Thuja plicata flava nom. nov.

SYN.—*Thuya plicata aurea* Beissner, Handb. Conif., 30 (1887), not
 Th. aurea Hort. ex Carr. (1855).
 Thuya gigantea aurea, Hort. ex Beissner, Handb. Nadelh.,
 49 (1891).
 Thuya Lobbi aurea Hort. ex Beissn., l. c. (1891).

Thuja plicata variegata Carr.

SYN.—*Thuia plicata variegata* Carrière, Trait. Conif., 1 éd., 102
 (1855).
 Thuja plicata aureo-variegata Hort. ex Beissner, Handb.
 Nadelh., 46 (1891), not *Th. occ. aur.-variegata*, Beissn., l. c.
 (1891), ante!
 Thuya gigantea aureo-rariegata Hort. ex Beissn., l. c., 49 (1891).
 Thuya Lobbi aureo-rariegata Hort. ex. Beissn., l. c. (1891).
 Thuja gigantea rariegata Hort. ex Gordon, Pinetum, 2d ed.,
 403 (1875).
 Thuja Lobbii rariegata Hort. ex. Gord., l. c. (1875).

Thuja plicata compacta (Carr.) Beissn.

SYN.—*Thuya Occidentalis compacta* Carrière, Trait. Conif., 1 éd.,
 104 (1855).
 Thuya occidentalis nana Carr., l. c., nouv. éd., 109 (1867).
 Thuya plicata compacta Hort. ex Beissner, Handb. Nadelh.,
 45 (1891).
 Thuya gigantea compacta, ex Hand-list Conif. Roy. Gard.
 Kew, 49 (1896).

Thuja plicata llaveana Gord.

SYN.—*Thuja dumosa* Gordon, Pinetum, ed. 1, Suppl., 102 (1862).
 Thuya Occidentalis dumosa Hort. ex Gord., l. c. (1862).
 Thuja occidentalis nana Hort. ex Gord., l. c. (1862).
 Thuja minor Paul ex Gord., l. c. (1862).
 Thuja nana Hort. ex Gord., l. c. (1862).
 Thuja plicata Llaveana Hort. ex Gord., l. c. (1862).
 Thuya plicata dumosa Hort. ex Gord., l. c. (1862).
 Thuja Antarctica Hort. ex Gord., l. c. (1862).

Thuja plicata llaveana Gord.—Continued.

SYN.—*Thuja pigmæa* Hort. ex Gord., Pinetum, 2d ed., 401 (1875).
Thuja prostrata Hort. ex Gord., l. c. (1875).
Thuja recurva nana Hort. ex Gord., l. c. (1875).
Biota prostrata Hort. ex Gord., l. c. (1875).

Thuja plicata minima Gord.

SYN.—*Thuja plicata minima* Smith ex Gordon, Pinetum, 2d ed., 408 (1875).
Thuya plicata pygmæa Beissner, Handb. Conif., 30 (1887), not *Thuia pygmæa* Hort. ex Carr. (1867).

Thuja plicata erecta (Gord.) nom. nov.

SYN.—*Thuja gigantea erecta* Smith ex Gordon, Pinetum, 2d ed., 403 (1875).
Thuja Lobbii erecta Hort. ex Gord., l. c. (1875).

Thuja plicata pumila (Gord.) nom. nov.

SYN.—*Thuja gigantea pumila* Hort. ex Gordon, Pinetum, 2d ed., 431 (1875).
Thuja Lobbii pumila Smith ex Gord., l. c. (1875).

Thuja plicata penduliformis[1] nom. nov.

SYN.—*Thuya gigantea* var. *pendula,*[2] ex Hand-list Conif. Roy. Gard. Kew, 49 (1896), not *Th. occid. pendula* Gord. (1862).

Thuja plicata cristatiformis[3] nom. nov.

SYN.—*Thuya gigantea* var. *plicata cristata,*[1] ex Hand-list Conif. Roy. Gard. Kew, 49 (1896), not *Th. occid. cristata* Cripps ex Gord. (1875).

CUPRESSUS Linn., Spec. Pl., 1002 (1753).

Cupressus macrocarpa Hartweg.   **Monterey Cypress.**

SYN.—*Cupressus macrocarpa* Hartweg, in Journ. Hort. Soc. London, II, 187 (1847).
Cupressus torulosa Lindley & Paxton, in Flow. Gard., I, 167, f. 105 (1850), not Don (1825).
Cupressus Hartwegii Carrière, in Rev. Hort. 1855, 233 (1855).

---

[1] A variety distinguished by its pendulous branchlets, and so far as known undescribed.
[2] Published without description.
[3] A variety distinguished chiefly by its short, recurved branchlets, often resembling a cock's comb.

Cupressus macrocarpa Hartweg—Continued.

Syn.—*Cupressus Reinwardti* Hort. ex Gordon, Pinetum, ed. 1, Suppl., 25 (1862).

COMMON NAME.

Monterey Cypress (Cal.).

VARIETIES DISTINGUISHED IN CULTIVATION.

Cupressus macrocarpa angulata Lemmon.

Syn.—*Cupressus macrocarpa* var. *fastigiata* Knight, Syn. Conif., 20 (1850), not *C. fastigiata* de C. (1805).
Cupressus Hartwegii var. *fastigiata* Carrière, Trait. Conif., nouv. éd., 169 (1867).
Cupressus macrocarpa var. *angulata* Lemmon, in Third Bienn. Rep. Cal. St. Bd. For., 181 (1890).

Cupressus macrocarpa lambertiana (Carr.) Masters.

Syn.—*Cupressus Lambertiana* Carrière, Trait. Conif., 1 éd., 124 (1855).
Cupressus macrocarpa var. *Lambertiana* Masters, in Journ. Hort. Soc., 343 (1896).

Cupressus macrocarpa crippsii Masters.

Syn.—*Cupressus macrocarpa* var. *Crippsii* Masters, in Journ. Hort. Soc., XXXI, 344 (1896).

Cupressus goveniana Gord.   **Gowen Cypress.**

Syn.—*Cupressus Goveniana* Gordon, in Journ. Hort. Soc. London, IV, 295 (1849).
Cupressus Californica Carrière, Trait. Conif., 1 éd., 127 (1855).
Juniperus aromatica Hort. ex Carr., l. c. (1855).
Cupressus Bourgeauii Hort. ex Gordon, Pinetum, 2d ed., 79 (1875).
Cupressus macrocarpa ? var. *farallonensis*[1] Masters, in Journ. Linn. Soc., XXXI, 344 (1896).

VARIETIES DISTINGUISHED IN CULTIVATION.

Cupressus goveniana parva nom. nov.

Syn.—*Cupressus Goveniana* var. *pigm* a Lemmon, Handb. West. Am. Cone-b., 77 (1895), not *C. Laws. pygmæa* Gord. (1875).
Cupressus Goveniana compacta Andre, in Rev. Hort. 1896, 8, f. 1 (1896), not *C. Nut. compacta* Gord. (1875).

---

[1] A form collected on the Farallones Islands, off the California coast.

Cupressus goveniana huberiana Carr.

SYN.—*Cupressus Goweniana Huberiana* Carrière, Trait. Conif.,
nouv. éd., 170 (1867).
Cupressus excelsa Hort. Huber ex Carr., l. c. (1867).

Cupressus goveniana glaucifolia nom. nov.

SYN.—*Cupressus Goweniana glauca* Carrière, Trait. Conif., nouv.
éd., 171 (1867), not *C. glauca* Lam. (1786).

Cupressus goveniana gracilis (Nels.) Carr.

SYN.—*Cupressus Californica gracilis* Nelson, Pinac., 70 (1866), in
part.
Cupressus Goweniana gracilis Carrière, Trait. Conif., nouv.
éd., 171 (1867).

Cupressus goveniana cornuta Carr.

SYN.—*Cupressus cornuta* Carrière, in Rev. Hort., 1866, 250 (1866).
Cupressus Goweniana cornuta Carr., Trait. Conif., nouv. éd.,
171 (1867).
Cupressus Goweniana monstrosa Carr., Mss. ex Carr., l. c.
(1867).

Cupressus goveniana viridis Carr.

SYN.—*Cupressus Goweniana viridis* Carrière, Trait. Conif., nouv.
éd., 171 (1867).

Cupressus goveniana attenuata (Gord.) Carr.

SYN.—*Cupressus attenuata* Gordon, Pinetum, ed. 1, 57 (1858).
Cupressus nivea Hort. ex Gord., l. c. (1858).
Cupressus Goweniana attenuata Carrière, Trait. Conif., nouv.
éd., 172 (1867).
Cupressus Kæmpferi Hort. ex Carr., l. c. (1867

Cupressus macnabiana Murr. **Macnab Cypress.**

SYN.—*Cupressus Macnabiana* Murray, in Edinburgh New Phil.
Journ., new ser., I, 293 (1855).
Cupressus glandulosa Hooker ex Gordon, Pinetum, ed. 1, 64
(1858).
Juniperus Mac-Nabiana Lawson ex Gordon, l. c. (1858).
Cupressus Californica gracilis Nelson, Pinac., 70 (1866), in
part.
Cupressus nivalis Lindley, MSS. ex Masters, in Journ. Linn.
Soc., XXXI, 348 (1896).
Cupressus glandulosa Hooker, MSS. ex Masters, l. c. (1896).
Cupressus Coulteri Hort. Glasnevin ex Masters, l. c. (1896),
not Forb. (1839).

Cupressus macnabiana Murr.—Continued.

COMMON NAMES.

SYN.—Cypress (Oreg., Wash.).
White Cedar (Oreg., Wash.).
Shasta Cypress (Cal.).
MacNab's Cypress (cult. Eng., Eu.).
California Mountain Cypress (Cal. lit.).

Cupressus guadalupensis Watson.　　　　　**Arizona Cypress.**

SYN.—*Cupressus macrocarpa?* Watson, in Proc. Am. Acad. Sci., XI,
119 (1876), not Hartweg (1847).
Cupressus Guadalupensis Wats., l. c., XIV, 300 (1879).
Cupressus Arizonica Greene, in Bull. Torr. Bot. Club, IX,
64 (1882).
Cupressus Benthami var. *arizonica* Masters, in Journ. Linn.
Soc., XXXI, 340, f. 14, 17 (1896).
Cupressus macrocarpa var. *guadalupensis* Masters, l. c., 343,
f. 18, 20 (1896).

COMMON NAMES.

Yew (Ariz.).
Arizona Cypress (Ariz.).
Red-bark Cypress (Ariz.).
Arizona Red-bark Cypress (*C. Arizonica*) (Cal. lit.).
Guadalupe Cypress (Cal. lit.).

CHAMÆCYPARIS Spach, Hist. Vég., XI, 329 (1842).

Chamæcyparis thyoides (L.) B. S. P.　　　　　**White Cedar.**

SYN.—*Cupressus thyoides* Linnæus, Spec. Pl., ed. 1, II, 1003 (1753).
Cupressus palustris Salisbury, Prodr., 398 (1796).
Thuya sphæroidea Sprengel, Syst. Veg., III, 889 (1826).
Thuya sphæroidalis Richard, Conif., 45, t. 8, f. 2 (1826).
CHAMÆCYPARIS SPHÆROIDEA Spach, Hist. Vég., XI, 331
(1842).
Chamæcyparis squarrosa Hort. ex Carrière, Trait. Conif.,
1 éd., 65 (1855), not Sieb. & Zucc., in Endl. (1847).
Chamæcyparis variegata Hort. ex. Carr., l. c., 133 (1855).
Chamæcyparis pseudo-squarrosa Parlatore, in de Candolle,
Prodr., XVI, sect. 2, 467 (1868).
Chamæcyparis thyoides (L.) B. S. P., Prelim. Cat. Anth.
Pter., 71 (1888).

COMMON NAMES.

White Cedar (Mass., R. I., N. Y., N. J., Pa., Del., N. C.,
S. C., Ala., Fla., Miss.).

Chamæcyparis thyoides (L.) B. S. P.—Continued.

SYN.—Swamp Cedar (Del.).
Post Cedar (Del.).
Juniper (Ala., N. C., Va.).

VARIETIES DISTINGUISHED IN CULTIVATION.

Chamæcyparis thyoides glauca (Endl.) nom. nov.

SYN.—*Chamæcyparis sphæroidea glauca* Endlicher, Syn. Conif.;
62 (1847).ʼ
Chamæcyparis sphæroidea Kewensis Carr., Man. des Pl., IV,
328 (1854).
Chamæcyparis Kewensis Hort. ex Carrière, Trait. Conif.,
1 éd., 133 (1855).
Cupressus sphæroidea Kewensis Knight ex Carr., l. c., (1855).
Cupressus sphæroidea pendula Hort. ex Gordon, Pinetum,
ed. 1, 50 (1858).
Cupressus thyoides Kewensis[1] Hort. ex Gord., l. c., (1858).
Thuia sphæroidea glauca Hort. ex Carrière, Trait. Conif.,
nouv. éd., 123 (1867).
Chamæcyparis Kewensis glauca Hort. ex Carr., l. c., (1867).
Cupressus thyoides glauca Hort. ex Beissner, Handb.
Nadelh., 68 (1891).

Chamæcyparis thyoides crocea nom. nov.

SYN.—*Chamæcyparis sphæroidea aurea* Hort. ex Gordon, Pinetum,
2d ed., 423 (1875), not *Ch. Bours. aurea* Carr. (1867).
Cupressus Thyoides aurea Hort. ex Beissner, Handb. Nadelh.,
69 (1891).

Chamæcyparis thyoides variegata (Loud.) nom. nov.

SYN.—*Cupressus thyoides variegata* Loudon, Enc. Trees, 1075 (1842).
Chamæcyparis sphæroidea variegata Endlicher, Syn. Conif.,
62 (1847).
Thuia sphæroidea variegata Hort. ex Carrière, Trait. Conif.,
1 éd., 133 (1855).

Chamæcyparis thyoides atrovirens (Knight) nom. nov.

SYN.—*Chamæcyparis sphæroidea atrovirens* Knight, Syn. Conif., 20
(1850).

---

[1] In the Hand-list of Coniferæ in the Royal Kew Garden (1896) this form is listed
as distinct from var. *glauca*, but Gordon (l. c.), who first stated the characters of the
form var. *kewensis*, held the two to be the same.

Chamæcyparis thyoides atrovirens (Knight) nom. nov.—Cont'd.

SYN.—*Chamæcyparis atrovirens* Hort. ex Gordon, Pinetum, ed. 1, 50 (1858).
Cupressus thyoides atrovirens Lawson ex Gord., l. c. (1858).

Chamæcyparis thyoides pyramidata (Beissn.) nom. nov.

SYN.—*Chamæcyparis sphæroidea pyramidata* Hort ex Beissner, Handb. Nadelh., 69 (1891).

Chamæcyparis thyoides fastigiata cinereo-folia nom. nov.

SYN.—*Chamæcyparis sphæroidea fastigiata glauca* Hort. ex Beissner, Handb. Nadelh., 69 (1891), not *Ch. sphær. glauca* Endl. (1847).

Chamæcyparis thyoides penduliformis nom. nov.

SYN.—*Chamæcyparis sphæroidea pendula* Hort. ex Beissner, Handb. Nadelh., 69 (1891), not *Ch. pendula* Maxim. (1866), nor Beissn. (1887).

Cupressus thyoides pendula Hort. ex Beissn., l. c. (1891), not *C. pendula* Thunb. (1784).

Chamæcyparis thyoides hoveyi (Veitch) nom. nov.

SYN.—*Cupressus thyoides Hoveyi* Veitch, Man. Conif., 238 (1881).
Chamæcyparis sphæroidea Hoveyi Hort. ex Beissner, Handb. Nadelh., 69 (1891).

Chamæcyparis thyoides nana (Loud.) nom. nov.

SYN.—*Cupressus thyoides nana* Loudon, Enc. Trees, 1075 (1842).
Chamæcyparis sphæroidea nana Endlicher, Syn. Conif., 62 (1847).
Thuja sphæroidea nana Hort. ex Gordon, Pinetum, ed. 1, 50 (1858), not *Th. nana* Carr (1855).
Chamæcyparis nana Hort. ex Parlatore, in de Candolle, Prodr., XVI, sect. 2, 464 (1868).

Chamæcyparis thyoides pumila (Carr.) nom. nov.

SYN.—*Chamæcyparis pumila* Hort. ex Carrière, Trait. Conif., nouv. éd., 124 (1867).
Chamæcyparis sphæroidea pygmæa Hort. ex Beissner, Handb. Nadelh., 69 (1891).

Chamæcyparis thyoides leptoclada (Gord.) nom. nov.

SYN.—*Retinospora squarrosa leptoclada* Gordon, Pinetum, ed. 1, Suppl., 91 (1862).
Chamæcyparis sphæroidea Andelyensis Carrière, Trait. Conif., nouv. éd., 123 (1867).
Retinospora andelyensis Carr., in Rev. Hort. 1880, 178 (1880).

Chamæcyparis thyoides ericoides (Knight) nom. nov.

SYN.— *Widdringtonia ericoides* Knight, Syn. Conif., 13 (1850).
Chamæcyparis ericoides Carrière, Trait. Conif., 1 éd., 140 (1855).
Retinospora ericoides ex Carr., l. c. (1855), not Hort.
Wriddingtonia ericoides Knight ex Carr., l. c., 141 (1855).
Cupressus ericoides Hort. ex Beissn., l. c. (1891).
Chamæcyparis sphæroidea ericoides Beissner & Hochstetter, in Beissn., Handb. Nadelh., 67, f. 14 (1891).
Juniperus ericoides Hort. ex Masters, in Journ. Linn. Soc., XVIII, 495 (1881), not Nois. (1839).
Frenela ericoides Hort. ex Beissn., l. c. (1891), not Endl. (1847).

Chamæcyparis nootkatensis (Lamb.) Spach.   **Yellow Cedar.**

SYN.—*Cupressus Nootkatensis* Lambert, Gen. Pinus, ed. 1, II, 18 (1824).
Thuya excelsa Bongard, in Mém. l'Acad. Pétersburg, sér. 6, II, 164 (1831).
Cupressus Nutkatensis Hooker, Fl. Bor.-Am., II, 165 (1840).
CHAMÆCYPARIS NUTKAËNSIS Spach, Hist. Vég., XI, 333 (1842).
Cupressus Americana Trautvetter, Pl. Imag. Fl. Russ., 12, t. 7 (1844).
Thuiopsis borealis Hort. ex Carrière, Trait. Conif., 1 éd., 113 (1855).
Thuiopsis Tchugatskoy Hort. ex Carr., l. c. (1855).
Thuiopsis Tchugatskoyæ Hort. ex Carr., l. c. (1855).
Cupressus Nutkaensis var. *glauca* Walpers, Ann., V., 769 (1857).
Thuiopsis cupressoides Carrière, in Jacques & Her., Man. Pl., IV, 324 (1862).
Chamæcyparis Nootkaënsis Carr., Trait. Conif., nouv. éd., 127 (1867).
Thuiopsis troubetskoyana Hort., ex Hand-list Conif. Roy. Gard. Kew, 43 (1896).

COMMON NAMES.

Yellow Cedar (Oreg.).
Sitka Cypress (Oreg., Cal.).
Yellow Cypress (Oreg., Wash.).
Nootka Cypress (Cal. lit.).
Nootka Sound Cypress (cult. Eng.).
Alaska Ground Cypress (Cal. lit.).
Alaska Cypress (Cal. lit.).

VARIETIES DISTINGUISHED IN CULTIVATION.

Chamæcyparis nootkatensis viridifolia nom. nov.

SYN.—*Chamæcyparis nutkaensis viridis* Hort. ex Beissn., Handb.
Nadelh., 82 (1891), not *Ch. Laws. erect. viridis* Beissn., l. c.
(1891), ante!
Thuyopsis borealis viridis Hort. ex Beissn., l. c. (1891).

Chamæcyparis nootkatensis cinerascens nom. nov.

SYN.—*Cupressus nutkaënsis glauca* Veitch, Man. Conif., 235 (1881).
Chamæcyparis nutkaensis glauca Hort. ex Beissner, Handb.
Nadelh., 82 (1891), not *Ch. sphær. glauca* Endl. (1847).
Thuyopsis borealis glauca Hort. ex Beissn., l. c. (1891).

Chamæcyparis nootkatensis cinerascens genuina nom. nov.

SYN.—*Chamæcyparis nutkaensis glauca vera* Hort. ex Beissner,
Handb. Nadelh., 82 (1891), not *Ch. Laws. pend. vera*
Beissn., l. c. (1891), ante!

Chamæcyparis nootkatensis cinerascens aureo-discolor nom. nov.

SYN.—*Chamæcyparis nutkaenis glauca aureo-variegata* Hort. ex
Beissner, Handb. Nadelh., 82 (1891), not *Ch. Laws. aureo-variegata* Beissn., l. c. (1891), ante!

Chamæcyparis nootkatensis argenteo-varians nom. nov.

SYN.—*Cupressus nutkaënsis argenteo-variegata* Veitch, Man. Conif.,
235 (1881), not *C. Laws. argenteo-variegata* Veitch, l. c.
(1881), ante!
Chamæcyparis nutkaensis argenteo-variegata Hort. ex Beissner, Handb. Nadelh., 82 (1891), not *Ch. Laws. argenteo-variegata* Beissn., l. c. (1891), ante!
Thuyopsis borealis argenteo-variegata Hort. ex Beissn., l. c.
(1891).

Chamæcyparis nootkatensis aureo-versicolor nom nov.

SYN.—*Cupressus nutkaënsis aureo-variegata* Veitch, Man. Conif.,
235 (1881).
Chamæcyparis nutkaensis aureo-variegata Hort. ex Beissner,
Handb. Nadelh., 82 (1891), not *Ch. Laws. aureo-variegata*
Beissn., l. c. (1891), ante!
Thuyopsis borealis aureo-variegata Hort. ex Beissn., l. c.
(1891).

Chamæcyparis nootkatensis zanthophylla nom. nov.

SYN.—*Chamæcyparis nutkaensis aurea* Hort. ex Beissner, Handb.
Nadelh., 82 (1891), not *Ch. Bours. aurea* Carr. (1867).
Cupressus nootkatensis var. *lutea*,[1] ex Hand-list Conif. Roy.
Gard. Kew, 45 (1896).

---

[1] A catalogue name published without characters.

Chamæcyparis nootkatensis pendens nom. nov.

SYN.—*Chamæcyparis nutkaensis pendula* Hort. ex Beissner, Handb.
Nadelh., 83 (1891), not *Ch. Laws. pendula* Beissn. (1887)
and l. c. (1891), ante!
Thuyopsis borealis pendula Hort. ex Beissn., l. c. (1891).
Cupressus nootkatensis var. *pendula*,[1] ex Hand-list Conif. Roy.
Gard. Kew, 45 (1896), not *C. pendula* Thunb. (1784).

Chamæcyparis nootkatensis compacta (Veitch) Beissn.

SYN.—*Cupressus nutkaënsis compacta* Veitch, Man. Conif., 235
(1881).
Chamæcyparis nutkaensis compacta Beissner, Handb. Conif.,
34 (1887); Handb. Nadelh., 83 (1891).
Chamæcyparis nutkaensis compacta glauca Hort. ex Beissn.,
l. c. (1887).
Thuyopsis borealis compacta Hort. ex Beissner, Handb.
Conif., 34 (1887); Handb. Nadelh., 83 (1891).
Cupressus nootkatensis var. *compacta*,[1] ex Hand-list Conif.
Roy. Gard. Kew, 43 (1896).

Chamæcyparis nootkatensis compressa Beissn.

SYN.—*Chamæcyparis nutkaensis gracilis* Hort. ex Beissner, Handb.
Nadelh., 83 (1891), not *Ch. Laws. gracilis* Beissn. (1887);
and l. c. (1891), ante!
Chamæcyparis Nutkaensis compressa Hort. ex Beissn. l. c.
(1887).
Thuyopsis borealis gracilis Hort. ex Beissn., l. c. (1887).
Thuyopsis borealis compressa Hort. ex Beissn., l. c. (1887).
Cupressus nootkatensis var. *gracilis*,[1] ex Hand-list Conif.
Roy. Gard. Kew, 43 (1896).

Chamæcyparis nootkatensis nidiformis Beissn.

SYN.—*Chamæcyparis nutkaensis nidiformis* Hort. ex Beissner, Handb.
Nadelh., 83 (1891).
Cupressus nootkatensis var. *nidifica*,[1] ex Hand-list Conif.
Roy. Gard. Kew, 45 (1896).

Chamæcyparis nootkatensis albo-picta[2] nom. nov.

SYN.—*Cupressus nootkatensis* var. *albo-variegata*,[1] ex Hand-list
Conif. Roy. Gard. Kew, 43 (1896), not *C. Lawsoniana
albo-variegata*[3] Veitch (1881).

---

[1] A catalogue name published without characters.
[2] A variety distinguished by its variously white-spotted or variegated branchlets.
[3] This varietal term is further barred from use in the genus *Chamæcyparis* by the existence of *Ch. lawsoniana albo-variegata* Beissn. (1887).

Chamæcyparis nootkatensis aureo-viridis[1] (Gard. Kew) nom. nov.

SYN.—*Cupressus nootkatensis* var. *aureo-viridis*,[2] ex Hand-list
Conif. Roy. Gard. Kew, 43 (1896).

Chamæcyparis nootkatensis picta nom. nov.

SYN.—*Cupressus Nutkaensis variegata* Hort. ex Gordon, Pinetum,
2d ed., 95 (1875), not *C. thyoides variegata* Loud. (1842).
Thuiopsis Borealis variegata Hort. ex Gord., l. c. (1875).
Cupressus Nutkaensis argentea Hort. ex Gord., l. c. (1875), not
C. Laws. argentea Gord. (1862).
Cupressus nootkatensis var. *variegata*,[2] ex Hand-list Conif.
Roy. Gard. Kew, 45 (1896).

Chamæcyparis lawsoniana (Murr.) Parl.   **Port Orford Cedar.**

SYN.—*Cupressus Lawsoniana* Murray, in Edinburgh New Phil.
Journ., new ser., I, 292, t. 9 (1855).
Cupressus fragrans Kellogg, in Proc. Cal. Acad., I, 103
(1857); reprint, ed. 2, 115 (1873).
Chamæcyparis Lawsoniana Parlatore, in Ann. Mus. Stor.
Nat. Fir., I, 181 (1864).
Chamæcyparis Boursierii Carrière, Trait. Conif., nouv. éd.,
125 (1867).
Chamæcyparis Nutkanus Torrey, in Bot. Wilkes's Exped., II,
t. 16 (1874).

COMMON NAMES.

Port Orford Cedar (Oreg., Cal.).
Oregon Cedar (Oreg., Cal.).
White Cedar (Oreg., Cal.).
Ginger Pine (Cal.).
Lawson's Cypress (Cal., Oreg.).

VARIETIES DISTINGUISHED IN CULTIVATION.

Chamæcyparis lawsoniana erecta (Gord.) nom. nov.

SYN.—*Cupressus Lawsoniana erecta* Hort. ex. Gordon, 2d ed., 87
(1875).
Cupressus Lawsoniana stricta Hort. ex Gord., l. c. (1875).
Cupressus Lawsoniana pyramidalis Hort. ex Gord., l. c.
(1875), not *C. pyramidalis* Targ.-Tozz. (1810).
Chamæcyparis Lawsoniana pyramidalis Beissner, Handb.
Conif., 33 (1887); Handb. Nadelh., 73 (1891).

---

[1] A variety distinguished by the pale or deep golden-green color of its foliage.
Little known and apparently not published before.
[2] A catalogue name published without characters.

Chamæcyparis lawsoniana erecta viridis (Veitch) Beissn.

SYN.—*Cupressus Lawsoniana viridis* Hort. ex Gordon, Pinetum, 2d ed., 87 (1875). *Cupressus Lawsoniana erecta viridis* Veitch, Man. Conif., 232 (1881). *Chamæcyparis Lawsoniana erecta viridis* Hort. ex Beissner, Handb. Conif., 33 (1887); Handb. Nadelh., 72 (1891).

Chamæcyparis lawsoniana erecta glaucifolia nom. nov.

SYN.—*Chamæcyparis Lawsoniana erecta glauca* Beissner, Handb. Conif., 33 (1887); Handb. Nadelh., 73 (1891), not *Ch. sphæroidea glauca* Endl. (1847). *Cupressus lawsoniana* var. *erecta viridis argentea*,[1] ex Handlist Conif. Roy. Gard. Kew, 41 (1896), not *C. Laws. argentea* Gord. (1862), nor *C. Nutk. argentea* Gord. (1875).

Chamæcyparis lawsoniana erecta glaucescens nom. nov.

SYN.—*Chamæcyparis Lawsoniana erecta alba* Kees ex Beissner, Handb. Nadelh., 73 (1891), not Beissn. (1887).

Chamæcyparis lawsoniana pyramidalis leucophylla nom. nov.

SYN.—*Chamæcyparis Lawsoniana pyramidalis alba* Beissner, Handb. Conif., 33 (1887); Handb. Nadelh., 73 (1891), not *Ch. Laws. parva alba* (Veitch, 1881) Sudworth.

Chamæcyparis lawsoniana pyramidalis flaveola nom. nov.

SYN.—*Chamæcyparis Lawsoniana pyramidalis lutea*[2] Beissner, Handb. Conif., 33 (1887); Handb. Nadelh., 73 (1891).

Chamæcyparis lawsoniana pyramidalis luteo-tenuis nom. nov.

SYN.—*Chamæcyparis Lawsoniana pyramidalis lutea gracilis* Hort. ex Beissner, Handb. Nadelh., 73 (1891), not *Ch. Laws. gracilis* Beissn. (1887).

Chamæcyparis lawsoniana rosenthali Beissn.

SYN.—*Chamæcyparis Lawsoniana Rosenthali* Beissner, Handb. Conif., 33 (1887).

Chamæcyparis lawsoniana worlei Beissn.

SYN.—*Chamæcyparis Lawsoniana Worlei* P. Smith u. Cie ex Beissner, Handb. Nadelh., 73 (1891).

Chamæcyparis lawsoniana alumi Beissn.

SYN.—*Chamæcyparis Lawsoniana Alumi* Beissner, Handb. Conif., 33 (1887).

---

[1] A catalogue name published without characters.
[2] This term is preoccupied for *Chamæcyparis* in *Ch. laws. lutea* (Gord.) Beissn.

Chamæcyparis lawsoniana alumi Beissn.—Continued.

SYN.—*Cupressus lawsoniana Alumi*, ex Hand-list Conif. Roy. Gard. Kew, 41 (1896).

Chamæcyparis lawsoniana monumentalis nova Beissn.

SYN.—*Chamæcyparis Lawsoniana monumentalis nova* Hort. ex Beissner, Handb. Nadelh., 73 (1891).

Chamæcyparis lawsoniana monumentalis albescens nom. nov.

SYN.—*Chamæcyparis Lawsoniana monumentalis glauca* Hort. ex Beissner, Handb. Nadelh., 74 (1891), not *Ch. sphær. glauca* Endl. (1847).

Chamæcyparis lawsoniana fraseri Beissn.

SYN.—*Chamæcyparis Lawsoniana glauca* Beissner, Handb. Conif., 33 (1887), not *Ch. sphær. glauca* Endl. (1847).
Chamæcyparis Lawsoniana Fraseri Beissn., l. c. (1887).
Chamæcyparis Lawsoniana Fraseri glauca Hort. ex Beissn., Handb. Nadelh., 74 (1891).
Cupressus lawsoniana var. *Fraseri.* ex Hand-list Conif. Roy. Gard. Kew, 41 (1896).

Chamæcyparis lawsoniana robusta Beissn.

SYN.—*Chamæcyparis Lawsoniana robusta* Beissner, Handb. Conif., 33 (1887); Handb. Nadelh., 74 (1891).

Chamæcyparis lawsoniana robusta aurifolia nom. nov.

SYN.—*Chamæcyparis Lawsoniana robusta aurea* Hort. ex Beissner, Handb. Nadelh., 74 (1891), not *Ch. Boursierii aurea* Carr. (1867).

Chamæcyparis lawsoniana robusta cinerea nom. nov.

SYN.—*Chamæcyparis Lawsoniana robusta glauca* Hort. ex Beissner, Handb. Nadelh., 74 (1891), not *Ch. sphær. glauca* Endl. (1847).

Chamæcyparis lawsoniana robusta argentifolia nom. nov.

SYN.—*Chamæcyparis Lawsoniana robusta argentea* Hort. ex Beissner, Handb. Nadelh., 74 (1891), not *Ch. Boursierii argentea* Carr. (1867.)

Chamæcyparis lawsoniana atroviridis nom. nov.

SYN.—*Chamæcyparis Lawsoniana atrovirens* Hort. ex Beissner, Handb. Nadelh., 74 (1891), not *Ch. sphær. atrovirens* Knight (1850).

Chamæcyparis lawsoniana cyanea nom. nov.

SYN.—*Chamæcyparis Lawsoniana glauca* Hort. ex Beissner, Handb. Nadelh., 74 (1891), not *Ch. sphær glauca* Endl. (1847).
Cupressus Lawsoniana var. *glauca*, ex Hand-list Conif. Roy. Gard. Kew, 41 (1896).

Chamæcyparis lawsoniana cyanea pendens[1] nom. nov.

SYN.—*Cupressus lawsoniana* var. *glauca pendula*,[2] ex Hand-list Conif. Roy. Gard. Kew, 41 (1896), not *C. pendula* Thunb. (1787).

Chamæcyparis lawsoniana beissneriana Smith & Cie.

SYN.—*Chamæcyparis Lawsoniana Beissneriana* Smith u. Cie ex Beissner, Handb. Nadelh., 74 (1891).

Chamæcyparis lawsoniana nivea Beissn.

SYN.—*Chamæcyparis Lawsoniana nivea* Beissner, Handb. Conif., 33 (1887); Handb. Nadelh., 75 (1891).

Chamæcyparis lawsoniana lutea (Gord.) Beissn.

SYN.—*Cupressus Lawsoniana lutea* Rollisson ex Gordon, Pinetum, 2d ed., 88 (1875).
Chamæcyparis Lawsoniana lutea Hort. ex Beissner, Handb. Nadelh., 75 (1891).

Chamæcyparis lawsoniana lutea flavescens (Gord.) nom. nov.

SYN.—*Cupressus lawsoniana lutea flavescens* Cripps ex Gordon, Pinetum, 2d ed., 88 (1875).
Chamæcyparis Lawsoniana lutescens Hort. ex Beissner, Handb. Nadelh., 75 (1891).
Cupressus lawsoniana var. *ochroleuca*, ex Hand-list Conif. Roy. Gard. Kew, 41 (1896).

Chamæcyparis lawsoniana aurea (Gord.) Beissn.

SYN.—*Cupressus Lawsoniana aurea* Gordon, Pinetum, ed. 1, Suppl., 24 (1862).
Chamæcyparis Boursierii aurea Carrière, Trait. Conif., nouv. éd., 125 (1867).
Chamæcyparis Lawsoniana aurea Beissner, Handb. Conif., 33 (1887); Handb. Nadelh., 75 (1891).

---

[1] A variety distinguished mainly by its blue-glaucous foliage and pendulous branchlets.

[2] The subvarietal term *pendula* is also unavailable in combination with *Chamæcyparis* on account of the *Ch. pendula* Maxim. (1866), a form of an Asiatic species.

Chamæcyparis lawsoniana aurea magnifica (Beissn.) nom. nov.

SYN.—*Chamæcyparis Lawsoniana magnifica aurea* Hort. ex Beissner, Handb. Nadelh., 75 (1891), not *Ch. Bours. aurea* Carr. (1867).

Chamæcyparis lawsoniana westermanni Beissn.

SYN.—*Chamæcyparis Lawsoniana Westermanni* Hort. ex Beissner, Handb. Nadelh., 75 (1891).

Chamæcyparis lawsoniana versicolor Beissn.

SYN.—*Chamæcyparis Lawsoniana versicolor* Conink ex Beissner, Handb. Nadelh., 75 (1891).

Chamæcyparis lawsoniana argenteo-variegata (Veitch) Beissn.

SYN.—*Cupressus Lawsoniana argenteo-variegata* Veitch, Man. Conif., 232 (1881).
Chamæcyparis Lawsoniana argenteo-variegata Beissner, Handb. Conif., 33 (1887); Handb. Nadelh., 75 (1891).

Chamæcyparis lawsoniana argenteo-variegata novicia nom. nov.

SYN.—*Chamæcyparis Lawsoniana argenteo-variegata nova* Beissner, Handb. Nadelh., 75 (1891), not *Ch. Laws. monumentalis nova* Beissn., l. c. (1891), ante!

Chamæcyparis lawsoniana aureo-variegata (Veitch) Beissn.

SYN.—*Cupressus Lawsoniana aureo-variegata* Veitch, Man. Conif., 232 (1881).
Chamæcyparis Lawsoniana aureo-variegata Beissner, Handb. Conif., 33 (1887); Handb. Nadelh., 76 (1891).

Chamæcyparis lawsoniana aureo-spica Beissn.

SYN.—*Chamæcyparis Lawsoniana aureo-spica* Juriss. ex Beissner, Handb. Nadelh., 76 (1891).

Chamæcyparis lawsoniana albo-spica (Gord.) Beissn.

SYN.—*Cupressus Lawsoniana alba spica* Hort. ex Gordon, Pinetum, 2d ed., 87 (1875).
Chamæcyparis Lawsoniana albo-spica Beissner, Handb. Conif., 33 (1887); Handb. Nadelh., 76 (1891).
Cupressus lawsoniana var. *albo-spica*, ex Hand-list Conif. Roy. Gard. Kew, 41 (1896).
Cupressus lawsoniana var. *albo-maculata*,[1] ex Hand-list, l. c. (1896).
Cupressus lawsoniana var. *albo-picta*,[1] ex Hand-list, l. c. (1896).

---

[1] A variety often cultivated as distinct from *Ch. laws. albo-spica*, but the distinction is difficult to maintain.

Chamæcyparis lawsoniana overeynderi Beissn.

SYN.—*Chamæcyparis Lawsoniana Overeynderi* Hort. ex Beissner, Handb. Nadelh., 76 (1891).

Chamæcyparis lawsoniana nutans nom. nov.

SYN.—*Chamæcyparis Lawsoniana pendula* Beissner, Handb. Conif., 33 (1887); Handb. Nadelh., 76 (1891), not *Ch.* pendula Maxim. (1866).

Chamæcyparis lawsoniana nutans vera (Beissn.) nom. nov.

SYN.—*Chamæcyparis Lawsoniana pendula vera* Hesse. (Auth.?), Gartenflora, 449 (1890); ex Beissner, Handb. Nadelh., 76 (1891).
Cupressus Lawsoniana var. *pendula vera*, ex Hand-list Conif. Roy. Gard. Kew, 43 (1896).

Chamæcyparis lawsoniana nutans alba (Gord.) nom. nov.

SYN.—*Cupressus Lawsoniana pendula alba* Paul ex Gordon, Pinetum, 2d ed., 89 (1875).
Chamæcyparis Lawsoniana alba pendula Hort. ex Beissner, Handb. Nadelh., 76 (1891), not *Ch. pendula* Maxim. (1866).
Chamæcyparis Lawsoniana alba elegans pendula Hort. ex Beissn., l. c. (1891).

Chamæcyparis lawsoniana filiformis (Veitch) Beissn.

SYN.—*Cupressus Lawsoniana filiformis* Veitch, Man. Conif., 232 (1881).
Chamæcyparis Lawsoniana filiformis Beissner, Handb. Conif., 34 (1887); Handb. Nadelh., 76 (1891).
Chamæcyparis Lawsoniana filiformis elegans Hort. ex Beissn., Handb. Nadelh., 76 (1891).
Chamæcyparis Lawsoniana filifera Hort. ex Beissn., l. c. (1891).
Chamæcyparis Lawsoniana filifera gracilis Hort. ex Beissn., l. c. (1891), not *Ch. Laws, gracilis* Beissn. (1887).

Chamæcyparis lawsoniana filiformis globosa Beissn.

SYN.—*Chamæcyparis Lawsoniana filiformis compacta*[1] Hort. ex Beissner, Handb. Nadelh., 77 (1891).
Chamæcyparis Lawsoniana filiformis globosa Hort. ex Beissn., l. c. (1891).

---

[1] The subvarietal term *compacta* is unavailable for this form on account of *Ch. nootkatensis compacta* (Veitch) Beissn., founded on *Cupressus nutkaensis compacta* Veitch (1881).

Chamæcyparis lawsoniana intertexta (Veitch) Beissn.

SYN.—*Cupressus Lawsoniana intertexta* Veitch. Man. Conif., 232 (1881).
Chamæcyparis Lawsoniana intertexta Beissner, Handb. Conif., 34 (1887); Handb. Nadelh., 77 (1891).

Chamæcyparis lawsoniana gracilis (Gord.) Beissn.

SYN.—*Cupressus Lawsoniana gracilis* Hort. ex Gordon, Pinetum, 2d ed.. 88 (1875).
Cupressus Lawsoniana gracilis pendula Veitch, Man. Conif., 232 (1881). not *Cup. pendula* Thunb. (1784).
Chamæcyparis Lawsoniana gracilis pendula Hort. ex Beissner, Handb. Nadelh., 77 (1891). not *Ch. pendula* Maxim. (1866).
Chamæcyparis Lawsoniana gracilis Beissn., Handb. Conif., 34 (1887); Handb. Nadelh., 77 (1891).
Chamæcyparis Lawsoniana gracillima ex Beissn., l. c. (1891).
Cupressus Lawsoniana gracilis gracillima, ex Hand-list Conif. Roy. Gard. Kew, 43 (1896).

Chamæcyparis lawsoniana gracilis pusilla nom. nov.

SYN.—*Chamæcyparis Lawsoniana gracilis nana* Hort. ex Beissner, Handb. Nadelh.. 77 (1891), not *Ch. sphær. nana* Endl. (1847).

Chamæcyparis lawsoniana laxa Beissn.

SYN.—*Chamæcyparis Lawsoniana laxa* Hort. ex Beissner. Handb. Nadelh.. 77 (1891).

Chamæcyparis lawsoniana crispa Beissn.

SYN.—*Chamæcyparis Lawsoniana crispa* Conink ex Beissner, Handb. Nadelh., 77 (1891).

Chamæcyparis lawsoniana casuarinifolia Beissn.

SYN.—*Chamæcyparis Lawsoniana casuarinifolia* Hort. ex Beissner, Handb. Nadelh.. 77 (1891).

Chamæcyparis lawsoniana tortuosa Beissn.

SYN.—*Chamæcyparis Lawsoniana tortuosa* Hort. ex Beissner, Handb. Nadelh., 77 (1891).

Chamæcyparis lawsoniana compacta[1] recens nom. nov.

SYN.—*Chamæcyparis Lawsoniana compacta nova* Hort. ex Beissner, Handb. Nadelh., 78 (1891), not *Ch. Laws. monumentalis nova* Beissn., l. c. (1891), ante!

---

[1] This presupposes an existing trinomial, which has not been found, but if found, and of an earlier date than 1881, it would conflict with *Ch. nootkatensis compacta*, a form of a different species, causing one or the other name to fall into synonymy, and necessitating further change.

Chamæcyparis lawsoniana fragrans[1] (Gord.) Beissn.

SYN.—*Cupressus Lawsoniana* var. *fragrans*[1] Standish ex Gordon,
Pinetum, 2d ed., 88 (1875), not *C. fragrans*, Kell. (1857).
Cupressus Lawsoniana aromatica Hort. ex Gord., l. c. (1875).
Chamæcyparis Lawsoniana fragrans in Hort. Amer. ex Beiss-
ner, Handb. Nadelh., 78 (1891).

Chamæcyparis lawsoniana fragrans argyropsis nom. nov.

SYN.—*Chamæcyparis Lawsoniana fragrans argentea* (in) Hort. Kew
ex Beissner, Handb. Nadelh.. 78 (1891), not *Ch. Bours.
argentea* Carr. 1867).
Cupressus lawsoniana var. *fragrans argentea*, ex Hand-list
Conif. Roy. Gard. Kew, 43 (1896).

Chamæcyparis lawsoniana fragrans conica Beissn.

SYN.—*Chamæcyparis Lawsoniana fragrans conica* ex Beissner,
Handb. Nadelh., 78 (1891).

Chamæcyparis lawsoniana parva nom. nov.

SYN.—*Chamæcyparis Lawsoniana Boursierii nana* Carrière, Trait.
Conif., nouv. éd., 126 (1867), not *Ch. sphær. nana* Endl.
(1847).
Cupressus Lawsoniana nana Hort. ex Carr., l. c. (1867).
Cupressus Lawsoniana glauca nana Hort. ex Gordon, Pine-
tum, 2d ed., 88 (1875).
Cupressus Lawsoniana pumila[2] Hort. ex Gord., l. c. (1875).
Chamæcyparis Lawsoniana nana Beissner, Handb. Conif.,
34 (1887); Handb. Nadelh., 78 (1891).

Chamæcyparis lawsoniana parva candida nom. nov.

SYN.—*Cupressus Lawsoniana nana alba* Veitch, Man. Conif., 233
(1881), not *C. Laws. pend. alba* Paul ex Gord. (1875).

Chamæcyparis lawsoniana parva albo-variegata (Gord.) nom. nov.

SYN.—*Cupressus Lawsoniana alba variegata* Lawson ex Gordon,
Pinetum, 2d ed., 87 (1875), not *Ch. sphær. variegata* Endl.
(1847).
Cupressus Lawsoniana albo-variegata Veitch, Man. Conif.,
232 (1881).
Chamæcyparis Lawsoniana nana albo-variegata Beissner,
Handb. Conif., 34 (1887); Handb. Nadelh., 78 (1891).

---

[1]If at any time the species and varieties of *Chamæcyparis* should be referred to the
genus *Cupressus*, of which *Chamæcyparis* is by some botanists made a subgenus, the
present garden variety would have to receive a new name, on account of the preex-
isting *Cupressus fragrans* Kell. (1857), a synonym of the type species.
[2]The varietal term *pumila* is preoccupied for *Chamæcyparis* in *Ch. pumila* Carr.
(1867), a form of *Ch. thyoides*.

Chamæcyparis lawsoniana parva albo-spiciformis nom. nov.

SYN.—*Chamæcyparis Lawsoniana nana albo-spicata* Beissner, Handb. Conif., 34 (1887); Handb. Nadelh., 78 (1891), not *Ch. Laws. albo-spica* Beissn., l. c. (1887) and (1891), ante!

Chamæcyparis lawsoniana parva densa nom. nov.

SYN.—*Chamæcyparis Lawsoniana nana compacta* Hort. ex Beissner, Handb. Nadelh., 78 (1891), not *Ch. Laws. compacta*[1] *nova* Beissn., l. c. (1891), ante!

Chamæcyparis lawsoniana shawi Beissn.

SYN.—*Chamæcyparis Lawsoniana Shawi* Hort. ex Beissner, Handb. Nadelh., 78 (1891).
Cupressus lawsoniana var. *Shawii*,[2] ex Hand-list Conif. Roy. Gard. Kew, 43 (1896).

Chamæcyparis lawsoniana minima (Gord.) nom. nov.

SYN.—*Cupressus Lawsoniana minima* Hort. ex Gordon, Pinetum, 2d ed., 89 (1875).
Cupressus Lawsoniana pygmæa Hort. ex Gord., l. c. (1875).
Chamæcyparis Lawsoniana nana argentea Beissner, Handb. Conif., 34 (1887); Handb. Nadelh., 78 (1891), not *Ch. Boursierii argentea* Carr. (1867).
Cupressus Lawsoniana nana glauca Veitch, Man. Conif., 233 (1881).
Chamæcyparis Lawsoniana nana glauca Beissner, Handb. Conif., 34 (1887); Handb. Nadelh., 78 (1891), not *Ch. sphær. glauca* Endl. (1847).

Chamæcyparis lawsoniana argentea (Gord.) Beissn.

SYN.—*Cupressus Lawsoniana argentea* Gordon, Pinetum, ed. 1, Suppl., 24 (1862).
Chamæcyparis Boursierii argentea Carrière, Trait. Conif., nouv. éd., 126 (1867).
Chamæcyparis Lawsoniana argentea Beissner, Handb. Conif., 33 (1887); Handb. Nadelh., 74 (1891).

Chamæcyparis lawsoniana argentea depauperata nom. nov.

SYN.—*Chamæcyparis Lawsoniana minima glauca* Beissner, Handb. Conif., 34 (1887); Handb. Nadelh., 79 (1891), not *Ch. sphær. glauca* Endl. (1847).

---

[1] See footnote, p. 87.
[2] Catalogue name published without characters.

Chamæcyparis lawsoniana argentea minuta nom. nov.

SYN.—*Chamæcyparis Lawsoniana pygmæa argentea* Hort. ex Beissner, Handb. Conif., 34 (1887); Handb. Nadelh., 79 (1891),
not *Ch. Boursierii argentea* Carr. (1867).

Chamæcyparis lawsoniana argentea prostrata (Beissn.) nom. nov.

SYN.—*Chamæcyparis Lawsoniana prostrata glauca* Hort. ex Beiss-
ner, Handb. Nadelh., 79 (1891), not *Ch. sphær. glauca*
Endl. (1847).
Chamæcyparis prostrata glauca Hort. ex Beissn., l. c. (1891).

Chamæcyparis lawsoniana forstekiana Beissn.

SYN.—*Chamæcyparis Lawsoniana Forstekiana* Beissner, Handb.
Conif., 34 (1887); Handb. Nadelh., 79 (1891).

Chamæcyparis lawsoniana weisseana Mœll.

SYN.—*Chamæcyparis Lawsoniana Weisseana* Hort. (auth.?) ex
Mœller's Deutsch. Gærtnerz., 245 (1890).

Chamæcyparis lawsoniana silver-queen Beissn.

SYN.—*Chamæcyparis Lawsoniana Silver Queen* Hort. ex Beissner,
Handb. Nadelh., 75 (1891).
Cupressus lawsoniana var. *Silver Queen*,[1] ex Hand-list Conif.
Roy. Gard. Kew, 41 (1896).

Chamæcyparis lawsoniana amabilis[1] (Gard. Kew) nom. nov.

SYN.—*Cupressus lawsoniana* var. *amabilis*,[2] ex Hand-list Conif.
Roy. Gard. Kew, 43 (1896).

Chamæcyparis lawsoniana bowleri[1] (Gard. Kew) nom. nov.

SYN.—*Cupressus lawsoniana* var. *Bowleri*,[2] ex Hand-list Conif. Roy.
Gard. Kew, 43 (1896).

Chamæcyparis lawsoniana californica[1] (Gard. Kew) nom. nov.

SYN.—*Cupressus lawsoniana* var. *californica*,[2] ex Hand-list Conif.
Roy. Gard. Kew, 43 (1896).

Chamæcyparis lawsoniana darleyensis[1] (Gard. Kew) nom. nov.

SYN.—*Cupressus lawsoniana* var. *darleyensis*,[2] ex Hand-list Conif.
Roy. Gard. Kew, 43 (1896).

---

[1] Nothing is known of the distinctive characters of this variety as, so far as known,
it has never been described. The form is said to be cultivated in the Royal Kew
Gardens, England, and the name is given here to complete a full enumeration of
these garden forms.
[2] Catalogue name published without characters.

JUNIPERUS Linn., Spec. Pl., 1038 (1753).

Juniperus virginiana Linn.   **Red Juniper.**

SYN.—*Juniperus Virginiana* Linnæus, Spec. Pl., ed. 1, II, 1039 (1753).
Juniperus Caroliana Miller, Gard. Dict., ed. 8, No. 4 (1768).
Juniperus arborescens Mœnch. Meth., 699 (1794).
Juniperus Barbadensis Michaux, Fl. Bor.-Am., II, 246 (1803), not Linn. (1753).
Juniperus Virginiana var. *Hermanni* Persoon, Syn. Pl., II, 632 (1807).
Juniperus Hermanni Persoon, l. c. (1807).
Juniperus fœtida var. *Virginiana* Spach, in Ann. Sc. Nat., 2 sér., XVI, 298 (1841).
Juniperus australis Endlicher, Syn. Conif., 26 (1847).
Juniperus Virginiana vulgaris Endl., l. c., 28 (1847).
Juniperus andina Nuttall, Sylva, III, 95, t. 110 (1849).
Juniperus Sabina var. *Virginiana* Antoine, Kupress., t. 83, 84 (1860).
Juniperus dioica Hort. ex Carrière, Trait. Conif., nouv. éd., 45 (1867).

COMMON NAMES.

Red Cedar (N. H., Vt., Mass., R. I., N. Y., N. J., Pa., Del., Va., W. Va., N. C., S. C., Ga., Fla., Ala., Miss., La., Tex., Ariz., Ky., Mo., Ill., Ind., Wis., Iowa, Mich., Minn., Ohio, Ont., Nev., Idaho, Utah, Colo., S. Dak.).
Cedar (Conn., Pa., N. J., S. C., Ky., Ill., Iowa, Ohio, Mont.).
Savin (Mass., R. I., N. Y., Pa., N. Mex., Colo., Idaho, Minn.).
Juniper (N. Y., Pa., Mont.).
Juniper Bush (Minn.).
Cedre (La.).

VARIETIES DISTINGUISHED IN CULTIVATION.

Juniperus virginiana caroliniana (Marsh.) Willd.

SYN.—*Juniperus caroliniana* Marshall, Arb. Am., 71 (1785).
Juniperus virginiana var. *Caroliniana* Willdenow, Berl. Baumz., ed. 1, 196 (1796).
Juniperus Gossainthanea Lodd. Cat. ex Loudon, Enc. Trees, 1090 (1842).
Juniperus Bedfordiana Hort. ex Loudon, l. c. (1842).
Juniperus Virginiana β australis Endlicher, Syn. Conif., 28 (1847).
Juniperus gracilis Endl., l. c., 31 (1847).

Juniperus virginiana caroliniana (Marsh.) Willd.—Continued.

SYN.—*Juniperus Virginiana Bedfordiana* Knight, Syn. Conif., 12 (1850).
Juniperus Virginiana Barbadensis Loudon ex Gordon, Pinetum, ed. 1, 114 (1858), not *J. Barbadensis* Linn. (1753).
Juniperus Virginiana Gossainthanea Carrière, Trait. Conif., nouv. éd., 45 (1867).

Juniperus virginiana pyramidiformis nom. nov.

SYN.—*Juniperus Virginiana pyramidalis* Hort. ex Carrière, Trait. Conif., nouv. éd., 47 (1867), not *J. pyramidalis* Carr., l. c., 22 (1855).

Juniperus virginiana pyramidiformis glaucifolia nom. nov.

SYN.—*Juniperus virginiana pyramidalis glaucá* Beissner, Handb. Conif., 39 (1887); Handb. Nadelh., 125 (1891), not *J. glauca* Salisb. (1796).

Juniperus virginiana pyramidiformis viridifolia nom. nov.

SYN.—*Juniperus virginiana pyramidalis viridis* Beissner, Handb. Conif., 39 (1887); Handb. Nadelh., 125 (1891), not *J. virg. viridis* Gord. (1875).

Juniperus virginiana cannarti (Koch) Beissn.

SYN.—*Juniperus Cannarti* Hort. ex Koch, Dendrol., zw. Th. zw. Ab., 140 (1873).
Juniperus Virginiana Cannarti Hort. ex Beissner, Handb. Nadelh., 125 (1891).

Juniperus virginiana polymorpha Beissn.

SYN.—*Juniperus Virginiana polymorpha* Hort. ex Beissner, Handb. Nadelh., 125 (1891).
Juniperus polymorpha Hort. ex Beissn., l. c. (1891).

Juniperus virginiana pendula Carr.

SYN.—*Juniperus Virginiana pendula* Carrière, Trait. Conif., 1 éd., 45 (1855).
Juniperus Virginiana viridis pendula Hort. ex Beissner, Handb. Nadelh., 125 (1891).

Juniperus virginiana smithi penduliformis nom. nov.

SYN.—*Juniperus Virginiana Smithi pendula* Hort. ex Beissner, Handb. Nadelh., 125 (1891), not *J. Virg. pendula* Carr. (1855).
Juniperus Smithi pendula Hort. ex Beissn., l. c. (1891).

Juniperus virginiana chamberlayni Carr.

SYN.—*Juniperus Chamberlayni* Carrière, Man. des Pl., IV, 313 (1854).
Juniperus Virginiana Chamberlayni Carr., Trait. Conif., 1 éd., 46 (1855).

Juniperus virginiana nutans Beissn.

SYN.—*Juniperus Virginiana nutans* Hort. ex Beissner, Handb. Nadelh., 125 (1891).
Juniperus nutans Hort. ex Beissn., l. c. (1891).

Juniperus virginiana interrupta (Wend.) Beissn.

SYN.—*Juniperus interrupta* Wendland (Mss.) ex Carrière, Trait. Conif., 1 éd., 23 (1855).
Juniperus Virginiana interrupta Hort. ex Beissner, Handb. Nadelh., 125 (1891).

Juniperus virginiana dumosa Carr.

SYN.—*Juniperus Virginiana dumosa* Carrière, Trait. Conif., 1 éd., 45 (1855).

Juniperus virginiana pumila Gord.

SYN.—*Juniperus Virginiana humilis* Hort. ex Gordon, Pinetum, 2d ed., 156 (1875), not *J. humilis* Salisb. (1796).
Juniperus Virginiana pumila Hort. ex Gord., l. c. (1875).
Juniperus Virginiana globosa Hort. ex Beissner, Handb. Nadelh., 126 (1891).
Juniperus Virginiana nana compacta Hort. ex Beissner, l. c. (1891).
Juniperus virginiana var. *compacta,* ex Hand-list Conif. Roy. Gard. Kew, 31 (1896).

Juniperus virginiana pumila[1] nivea (Beissn.) nom. nov.

SYN.—*Juniperus Virginiana nana[1] nivea* Hort. ex Beissner, Handb. Nadelh., 126 (1891).

Juniperus virginiana schotti Gord.

SYN.—*Juniperus Schotti* Hort. ex Gordon, Pinetum, ed. 1, 122 (1858).
Juniperus Virginiana Schotti Hort. ex Gord., l. c., 2d ed., 157 (1875).

---

[1] The term *pumila* is here substituted for *nana*. The existence of *J. nana* Willd. (Berl. Baumz., 159, 1796), a different plant, should have precluded the use of the term *nana* in Beissner's name for this form, and necessitates a change in the structure of the above name. The presupposed combination *Juniperus virginiana nana* has not been found, and probably was never published, the varietal and subvarietal members *nana nivea* most likely having been added to *Juniperus virginiana* at once.

Juniperus virginiana schotti Gord.—Continued.

SYN.—*Juniperus Virginiana viridis* Hort. ex Gord., l. c. (1875).
Juniperus Scholli Hort. ex Beissner, Handb. Nadelh., 126 (1891).

Juniperus virginiana tripartita Gord.

SYN.—*Juniperus Virginiana tripartita* Smith ex Gordon, Pinetum, ed. 1, 122 (1858).
Juniperus tripartita Hort. ex Gord., l. c. (1858).

Juniperus virginiana tripartita aureo-versicolor nom. nov.

SYN.—*Juniperus Virginiana tripartita aureo-variegata* Hort. ex Beissner, Handb. Nadelh., 126 (1891), not *J. Virg. aureo-variegata* Veitch (1881).
Juniperus tripartita aureo-variegata Hort. ex Beissn., l. c. (1891).

Juniperus virginiana kosteriana Beissn.

SYN.—*Juniperus Virginiana Kosteriana* Hort. ex Beissn., Handb. Nadelh., 126 (1891).

Juniperus virginiana glaucescens nom. nov.

SYN.—*Juniperus glauca* Willdenow, Enum. Plant., Suppl., 67 (1813), not *J. glauca* Salisb. (1796).
Juniperus Virginiana glauca Carrière, Trait. Conif., 1 éd., 45 (1855).

Juniperus virginiana cinerascens Carr.

SYN.—*Juniperus Virginiana cinerascens* Hort. ex Carrière, Trait. Conif., 1 éd., 45 (1855).
Juniperus Virginiana argentea Hort. ex Carr., l. c. (1855).
Juniperus argentea Hort. ex Gordon, Pinetum, ed. 1, 113 (1858).
Juniperus cinerascens Hort. ex Koch, Dendrol., zw. Th. zw. Ab., 141 (1873).

Juniperus virginiana plumosa alba (Carr.) Beissn.

SYN.—*Juniperus alba* Knight, Syn. Conif., 13 (1850); ex Carrière, Trait. Conif., 1 éd., 58 (1855).
Juniperus Virginiana plumosa alba Hort. ex Beissner, Handb. Nadelh., 127 (1891).
Juniperus Virginiana plumosa argentea Hort. ex Beissn., l. c. (1891), not *J. virg. argentea* Carr. (1855).

Juniperus virginiana plumosa candida nom. nov.

SYN.—*Juiperus Virginiana plumosa nivea* Schwerdt ex Beissner, Handb. Nadelh., 127 (1891), not *J. Virg. nana nivea* Beissn., l. c. (1891), ante!

Juniperus virginiana albo-spica Beissn.

SYN.—*Juniperus Virginiana albo-spica* Hort. ex Beissner, Handb. Nadelh., 127 (1891).
Juniperus Virainiana albo-spicata Hort. ex Beissn., l. c. (1891).

Juniperus virginiana albo-variegata Beissn.

SYN.—*Juniperus Virginiana albo-variegata* Hort. ex Beissner, Handb. Nadelh., 127 (1891).

Juniperus virginiana aureo-spica Beissn.

SYN.—*Juniperus Virginiana aureo-spica* Hesse. ex Beissner, Handb. Nadelh., 127 (1891).

Juniperus virginiana aureo-variegata Veitch.

SYN.—*Juniperus Virginiana aureo-rariegata* Veitch, Man. Conif., 284 (1881).

Juniperus virginiana aurea superba nom. nov.

SYN.—*Juniperus Virginiana aurea elegans* Hort. ex Beissner, Handb. Nadelh., 127 (1891), not *J. elegans* Hort. ex Gord. (1875).

Juniperus virginiana elegantissima Beissn.

SYN.—*Juniperus rirginiana elegans* Nicholson, Dic. Gard., IV, 214 (1889), not *J. elegans* Hort. ex Gord. (1875).
Juniperus Virginiana elegantissima Hort. ex Beissner, Handb. Nadelh., 128 (1891).

Juniperus virginiana[1] **horizontaliformis**[2] nom. nov.

SYN.—*Juniperus rirginiana horizontalis*, ex Hand-list Conif. Roy. Gard. Kew, 31 (1896), not *J. horizontalis* Moench (1794).

Juniperus virginiana triomphe d'angers Beissn.

SYN.—*Juniperus Virginiana Triomphe d'Angers* Hort. ex Beissner, Handb. Nadelh., 127 (1891).
Juniperus Triomphe d'Angers Hort. ex Beissn., l. c. (1891).

---

[1] The following varieties are named as forms cultivated in the Royal Kew Gardens. They appear to be unknown to American nurserymen, and nothing is known of the distinctive characters, as the names are not accompanied by any descriptions.
Juniperus occidentalis Burkei Nicholson, Dic. Gard., IV, 213, (1889).
Juniperis occidentalis var. *fragrans*, ex Hand-list, l. c. (1896), not *J. fragrans* Knight (1850).

[2] The Kew Garden name of this variety was published without characters, and so far as known this form has never received a name before. The name was doubtless applied to a form of our Red Juniper peculiar for its straight, horizontally disposed branches, not uncommon in the wild state, especially when the species occurs on limestone.

Juniperus occidentalis Hook.   **Western Juniper.**

SYN.—*Juniperus excelsa* Pursh, Fl. Am. Sept., II, 647 (1814), not Bieb. (1800).
Juniperus occidentalis Hooker, Fl. Bor.-Am., 11, 166 (1840).
Juniperus dealbata Hort. ex Knight, Syn. Conif., 12 (1850), not Loud. (1842).
Chamæcyparis Boursierii Decaisne, in Bull. Soc. France, 1, 70 (1854).
Juniperus occidentalis var. *α pleiosperma* Engelmann, in Trans. Acad. Sci. St. Louis, III, 590 (1877).
Juniperus pyriformis Lindley, in Gard. Chron. 1855, 420 (1855).
Cupressus bacciformis Knight ex Gordon, Pinetum, ed. 1, Suppl., 38 (1862), not Willd. (1816)—*nomen nudum.*

COMMON NAMES.

Juniper (Oreg., Cal., Colo., Utah, Nev., Mont., Idaho, N. Mex.).
Cedar (Idaho, Mont.).
Yellow Cedar (Colo., Mont.).
Western Cedar (Idaho).
Western Red Cedar.
Western Juniper (Cal. lit.).

Juniperus occidentalis monosperma Engelm.   **One-seed Juniper.**

SYN.—*Juniperus occidentalis* var. *β monosperma* Engelmann, in Trans. Acad. Sci. St. Louis, III, 590 (1877).
Juniperus occidentalis var. (c) *gymnocarpa* n. var. Lemmon, Handb. West. Am. Cone-b., 80 (1895).

COMMON NAMES.

One-seeded Juniper.
Naked-seeded Juniper (Cal. lit.—var. *gymnocarpa*).

Juniperus occidentalis conjugens Engelm.   **Mountain Juniper.**

SYN.—*Juniperus occidentalis* var. ? *γ. conjugens* Engelmann, in Trans. Acad. Sci. St. Louis, III, 590 (1877).

COMMON NAMES.

Juniper Cedar (Tex.).
Mountain Cedar (Tex.).
Juniper.
Mountain Juniper.

Juniperus californica Carr.   **California Juniper.**

SYN.—*Juniperus tetragona* Torrey, in Sitgreaves's Rep., 173 (1853), not Schlecht. (1838).

Juniperus californica Carr.—Continued.

SYN.—*Juniperus Californica* Carrière. in Rev. Hort., sér. IV, III,
352 (1854).
Juniperus tetragona var. *osteosperma* Torrey. in Pacif. R. R.
Rep., IV, 141 (1857).
Juniperus Cerrosianus Kellogg, in Proc. Calif. Acad. Sci.,
II, 37 (1863).
Sabina Californica Antoine. Kupress.. 52, t. 72 (1860).

COMMON NAMES.

White Cedar.
Juniper (Cal.).
Sweet-fruited Juniper (Cal.).
California Juniper (Cal. lit.).
Sweet-berried Cedar.

Juniperus californica utahensis Engelm.   **Utah Juniper.**

SYN.—*Juniperus occidentalis* Watson, in King's Rep., V, 336 (1871),
in part: not Hook. (1840).
Juniperus Californica var. *Utahensis* Engelmann, in Trans.
Acad. Sci. St. Louis, III, 588 (1877).
Juniperus occidentalis var. *Utahensis* Veitch, Man. Conif.,
289 (1881).
Juniperus Utahensis Lemmon, in Third Bienn. Rep. Cal. St.
Bd. For., 183 (1890).
Juniperus occidentalis Utahensis Beissner, Handb. Nadelh.,
129 (1891).

COMMON NAMES.

Juniper (Utah).
Western Red Cedar.
Desert Juniper (Cal. lit.).
Utah Juniper.

Juniperus pachyphlœa Torr.   **Alligator Juniper.**

SYN.—*Juniperus No. 1* Torrey, in Seagreaves's Rep.. 173 (1853).
Juniperus pachyphlœa Torrey, in Pacif. R. R. Rep., IV, 142
(1857).
Juniperus Sabina pachyphlœa Antoine, Kupress., 39 (1857–
1860).

COMMON NAMES.

Juniper (Ariz., N. Mex.).
Oak-barked Cedar (Ariz.).
Alligator Juniper (Ariz.).
Oakbark Juniper (Ariz.).

Juniperus pachyphlœa Torr.—Continued.

SYN.—Mountain Cedar (Tex.).
Thick-barked Juniper (Cal. lit.).

Juniperus flaccida Schlect.   **Drooping Juniper.**

SYN.—*Juniperus flaccida* Schlectendahl, in Linnæa, XII, 495 (1838).

Juniperus communis[1] Linn.   **Common Juniper.**

SYN.—*Juniperus communis* Linnæus, Spec. Pl., ed. 1, II, 1040 (1753).
Juniperus difformis Gilbert, Exerc. Phytol., II, 416 (1792).
Juniperus borealis Salisbury, Prodr., 397 (1796).
Juniperus dealbata Loudon, Enc. Trees, 1090 (1842).
Juniperus fœtidissima Hort. ex Endlicher, Syn. Conif., 30
    (1847), not Willd. (1805).
Juniperus davurica Hort. ex Lindley & Gordon, in Journ.
    Hort. Soc., V., 200 (1850), not Pall. (1784).
Juniperus Taurica Hort. ex Lindl. & Gord., l. c. (1850).
Juniperus intermedia Schur, in Verh. Sieben. Ver. Naturw.,
    II, 169 (1850).
Juniperus occidentalis Hort. ex Carrière, Trait. Conif., 1 éd.,
    54 (1855), not Hook. (1840).
Juniperus Withmanniana Hort. ex Carrière, Trait. Conif.,
    nouv. éd., 18 (1867).
Juniperus Wittmanniana Stev. ex Parlatore, in de Candolle,
    Prodr., XVI, sect. 2, 479 (1868), not Fisch. ex Lindl. &
    Gord. (1850).
Juniperus argœa Bal. ex Parlatore, in de Candolle, l. c., 480
    (1868).
Juniperus echinoformis Rinz ex Bolse in Koch, in Wochen-
    schr., XI, 284 (1868), not Hort. ex Carr. (1867).
Juniperus elliptica Hort. ex Koch, Dendrol. zw. Th. zw. Ab.,
    118 (1873), not Hort. ex Carr. (1855).

VARIETIES DISTINGUISHED IN CULTIVATION.

Juniperus communis cracovia (Koch) Beissn.

YN.—*Juniperus Cracovia* Koch, Dendrol., zw. Th. zw. Ab., 115
    (1873).
Juniperus communis cracovica Beissner, Handb. Conif., 41
    (1887); Handb. Nadelh., 136 (1891).

---

[1] This species, heretofore not known to be strictly arborescent, is here introduced on the authority of Prof. Robert Ridgeway, who reports it from the lower
Wabash Valley, 25 feet in height and 18 inches in diameter.—(Additional Notes on
Trees of Lower Wabash Valley, in Proc. U. S. Nat. Mus., XVII [No. 1010], 415 (1894).
I have not seen specimens of Professor Ridgeway's tree, but suspicion that it may
prove to be only a form of *Juniperus virginiana* with primary or plumose foliage, not an
uncommon state, and one in which the two species may be said to resemble each
other.

Juniperus communis suecica (Mill.) Loud.

Syn.—*Juniperus suecica* Miller, Gard. Dict., ed. 8. No. 2 (1768).
Juniperus communis Suecica Loudon, Enc. Trees. 1081 (1842).

Juniperus communis hibernica (Lodd.) Gord.

Syn.—*Juniperus hibernica* Loddiges ex Loudon. Arb. Frut., IV, 2490 (1838).
Juniperus Hispanica Booth ex Endlicher, Syn. Conif., 15 (1847), not Mill. (1768).
Juniperus fastigiata Hort. ex Knight, Syn. Conif.. 11 (1850).
Juniperus pyramidalis Hort. ex. Carrière. Trait. Conif., 1 éd., 22 (1855).
Juniperus stricta Hort. ex Carr., l. c. (1855).
Juniperus communis hibernica Gordon, Pinetum. ed. 1, 94 (1858).
Juniperus communis stricta Carrière, Trait. Conif.. nouv. éd., 18 (1867).
Juniperus communis fastigiata Nicholson. Dic. Gard., IV, 212 (1889).

Juniperus communis hibernica compressa Carr.

Syn.—*Juniperus communis hibernica compressa* Carrière, Man. Pl., IV, 309 (1854).
Juniperus compressa Hort. ex Carr., Trait. Conif.. 1 éd., 22 (1855).
Juniperus communis Hispanica Lawson ex Gordon, Pinetum, ed. 1, 94 (1858).
Juniperus communis compressa Carrière, Trait. Conif., nouv. éd., 18 (1867).

Juniperus communis oblonga (Bieb.) Loud.

Syn.—*Juniperus oblonga* Bieberstein, Fl. Taur. Cauc., II, 426 (1808).
Juniperus communis oblonga Loudon. Arb. Frut., IV, 2489 (1838).
Thuyæcarpus juniperinus Trautvetter, Plant. Imag., 11, t. 6 (1844).
Juniperus communis caucasica Endlicher, Syn. Conif., 16 (1847).
Juniperus Caucasica Fischer ex Gordon. Pinetum, ed. 1, 98 (1858).

Juniperus communis oblongo-pendula (Loud.) Carr.

Syn.—*Juniperus oblongo-pendula* Loudon. Enc. Trees, 1082, f. 201 (1842).
Juniperus communis oblongo-pendula Carrière, Man. des Pl., IV, 310 (1854).

Juniperus communis oblongo-pendula (Loud.) Carr.—Continued.

SYN.—*Juniperus reflexa* Hort. ex Gordon, Pinetum, ed. 1, 98 (1858).
Juniperus communis reflexa Parlatore, in de Candolle, Prodr.,
XVI, sect. 12, 479 (1868).

Juniperus communis pendens nom. nov.

SYN.—*Juniperus communis pendula* Hort. ex Gordon, Pinetum, ed.
1, Suppl., 31 (1862), not *J. virg. pendula* Carr. (1855).

Juniperus communis hemisphærica (Presl.) Parl.

SYN.—*Juniperus hemisphærica* Presl., Delic. Prag., 142 (1822).
Juniperus nana hemisphærica Carrière, Trait. Conif., nouv.
éd., 16 (1867).
Juniperus communis hemisphærica Parlatore, in de Candolle,
Prodr., XVI, sect. 2, 479 (1868).

Juniperus communis echinoformis (Knight) Beissn.

SYN.—*Juniperus Oxycedrus echinoformis* Knight, Syn. Conif., 11
(1850).
Juniperus echinoformis Hort. ex Carrière, Trait. Conif., nouv.
éd., 13 (1867).
Oxycedrus echinoformis Hort. ex Carr., l. c. (1867).
Juniperus communis echinoformis Beissner, Handb. Conif.,
41 (1887); Handb. Nadelh., 137 (1891).

Juniperus communis variegata aurea Carr.

SYN.—*Juniperus communis variegata aurea* Carrière, Trait. Conif.,
nouv. éd., 19 (1867).
Juniperus communis aureo-variegata Hort. ex Beissner,
Handb. Conif., 41 (1887); Handb. Nadelh., 138 (1891).

Juniperus communis sibirica (Burgsd.) Rydberg.

SYN.—*Juniperus sibirica* Burgsdorff, Anleit. Holz., No. 272 (1787).
Juniperus communis montana Aiton, Hort. Kew, ed. 1, III,
414 (1789).
Juniperus nana Willdenow, Sp. Pl., IV, Par. II, 854 (1805).
Juniperus Alpina Gray, Nat. Arr. Brit. Pl., II, 226 (1821).
Juniperus communis alpina Gaudin, Fl. Helv., VI, 301 (1830).
Juniperus Canadensis Loddiges, Cat., 47 (1836).
Juniperus montana Hort. ex Lindley & Gordon, in Journ.
 · Hort. Soc., V, 200 (1850).
Juniperus saxatilis Hort. ex Lindl. & Gord., l. c. (1850).
Juniperus communis canadensis Nicholson, Dic. Gard., IV,
212 (1889).
Juniperus communis sibirica (Burgsd.) Rydberg, in Contr.
U. S. Nat. Herb., III, 533 (1896).

Juniperus communis argyrophylla [1] nom. nov.

SYN.—*Juniperus communis* var. *glauca*, ex Hand-list Conif. Roy.
Gard. Kew, 33 (1896)—*nomen nudum ;* not *J. glauca* Salisb.
(1796).

Juniperus communis pygmæa (Koch) nom. nov.

SYN.—*Juniperis pygmæa* Koch, in Linnæa, XXIII, 302 (1849).

TUMION [2] Raf., Amen. Nat., 63 (1840).

Tumion taxifolium (Arn.) Greene.   **Florida Torreya.**

SYN.—TORREYA TAXIFOLIA Arnott, in Ann. Nat. Hist., 1, 130
(1838).
Caryotaxus taxifolia Henkel & Hochstetter, Nadelhölz., 365
(1865).
Fætataxus montana Nelson, Pinac., 167 (1866).
Tumion taxifolium Greene, in Pittonia, II, pt. 10, 194 (1891).

COMMON NAMES.

Stinking Cedar (Fla.).
Savin (Fla.).
Torrey Tree (Fla.).
Stinking Savin (Fla.).
Fetid Yew (Eng. lit.).

Tumion californicum (Torr.) Greene.   **California Torreya.**

SYN.—TORREYA CALIFORNICA Torrey, in New York Journ. Pharm.,
III, 49 (1854).
Torreya Myristica Hooker, in Bot. Mag., t. 4780 (1854).
Caryotaxus Myristica Henkel & Hochstetter, Nadehölz., 368
(1865).
Fætataxus Myristica Nelson, Pinac., 168 (1866).
Tumion Californicum Greene, Pittonia, II, pt. 10, 195 (1891).
Tumion Californicum var. *littoralis*, Lemmon, Handb. West.
Am. Cone-b., 84 (1895).

COMMON NAMES.

California Nutmeg (Cal., Idaho).
Stinking Cedar (Idaho).
Yew (Idaho).
California False Nutmeg (Cal. lit.).
Coast Nutmeg (Cal. lit.).

---

[1] A variety distinguished by its very glaucous or silvery leaves.
[2] *Tumion*, Rafinesque, Amen. Nat., 63 (1840) = *Torreya*, Arnott, in Ann. Nat. Hist.,
I, 130 (1838), not Raf. (1817 and 1818), nor Sprengel (1821), nor Eaton (1833).

TAXUS Linn., Spec. Pl., 1040 (1753).

Taxus brevifolia Nutt.   **Pacific Yew.**

SYN.—*Taxus baccata* Hooker, Fl. Bor.-Am., II, 167 (1840), in part; not Linn. (1753).
Taxus brevifolia Nuttall, Sylva, III, 86, t. 108 (1849).
Taxus Boursierii Carrière, in Rev. Hort. 1854, 228 (1854).
Taxus Lindleyana Murray, in Edinburgh New Phil. Journ., new ser., I, 294 (1855).
Taxus Canadensis Bigelow, in Pacific R. R. Rep., IV, 25 (1857), not Willd. (1805).
Taxus baccata var. *Canadensis* Bentham, Pl. Hartweg., 338 (1857).

COMMON NAMES.

Yew (Cal., Idaho, Oreg.).
Mountain Mahogany (Idaho).
Western Yew (Cal.).
Pacific Yew (Cal. lit.).

Taxus floridana Nutt.   **Florida Yew.**

SYN.—*Taxus montana* Nuttall, Sylva, III, 92 (1849), not Willd. (1805).
Taxus Floridana Nuttall, l. c. (1849).

COMMON NAMES.

Yew (Fla.).
Savin (Fla.).

MONOCOTYLEDONES.

Family PALMACEÆ.

THRINAX Linn. f. ex Swartz, Prodr. Veg. Ind., 57 (1788).

Thrinax parviflora Swartz.   **Silktop Palmetto.**

SYN.—*Thrinax parviflora* Swartz, Prodr., 57 (1788).
Thrinax Garberi Chapman, in Bot. Gaz., III, 12 (1878).

COMMON NAMES.

Silktop Palmetto.
Silver Thatch (Fla.).

Thrinax argentea (Jacq.) Desf.   **Silvertop Palmetto.**

SYN.—*Palma argentea* Jacquin, Fragm. 38, No. 125, t. 43, f. 1 (1809).
Thrinax argentea Loddiges ex Desfontaines, Cat., 3d. ed., 31 (1829).

COMMON NAMES.

Silvertop Palmetto (Fla.).
Prickly Thatch (Fla.).
Brittle Thatch (Fla.).

Thrinax microcarpa[1] Sargent.   **Littlefruit Palmetto.**

Syn.—*Thrinax (Porothrinax) microcarpa* Sargent. n. sp., in Gard. and For.. IX, 162 (1896).

SABAL Adanson. Fam. Pl., II, 495 (1863).

Sabal palmetto (Walt.) Rœm. & Sch.   **Cabbage Palmetto.**

Syn.—*Corypha Palmetto* Walter. Fl. Caroliniana, 119 (1788).
Chamærops Palmetto Michaux, Fl. Bor.-Am., I, 206 (1803).
Sabal Palmetto Loddiges ex Rœmer & Schultes, Syst., VII, 1487 (1830).

COMMON NAMES.

Cabbage Palmetto (N. C.. S. C.).
Bank's Palmetto (N. C.).
Palmetto (N. C., S. C.).
Cabbage Tree (Miss.. Fla.).
Latanier (La.).
Tree Palmetto (La.).

Sabal mexicana Mart.   **Mexican Palmetto.**

Syn.—*Sabal giganteum* Fulchir. ex Steudel. Nom. Bot., ed. sec., II, 489 (1841)— ?
Sabal Mexicana Martius. Hist. Nat. Palm.. III, 246, t. 8 (1850).
Sabal Palmetto Hort. ex Mart.. l. c.. 246 (1850). not Lodd. ex Rœm. & Sch. (1830).
Sabal umbraculiferum Hort. ex Mart.. l. c. (1850). not l. c., 245 (1850).

PSEUDOPHŒNIX Wendland. in Gard. and For.. I. 352 (1888).

Pseudophœnix sargentii Wend.   **Sargent Palm.**

Syn.—*Pseudophœnix Sargentii* Wendland, in Gard. and For.. I. 352 (1888).

COMMON NAMES.

Florida Palm (Fla.).
Sargent's Palm.

---

[1] A tree 30 feet or less in height, first detected by Mr. A. H. Curtiss in 1879 on No Name Key and Boca Chica Key, off the southern coast of Florida. It occurs also on Bahia Honda Key and along the shores of Sugar Loaf Sound. Formerly referred to *Thrinax argentea*.

OREODOXA Willdenow, in Mém. Acad. Berl. 1804, 34 (1807).

Oreodoxa regia H., B. K.   **Royal Palm.**

SYN.—*Oreodoxa regia* Humboldt, Bonpland & Kunth., Nov. Genera
Spec., I, 305 (1815).
Œnocarpus regia Sprengel, Syst. Veg., II, 140 (1825).
Oreodoxa oleracea? Cooper, in Smithsonian Rep. 1860, 440
(1861).

COMMON NAME.

Royal Palm (Fla.).

NEOWASHINGTONIA[1] nom. nov.

Neowashingtonia filamentosa (Wend.) nom. nom.

Fanleaf Palm.

SYN.—*Brahea dulcis?* Cooper, in Smithsonian Rep. 1860, 442 (1861),
not Martius (1850).
Prichardia filamentosa Wendland, in Bot. Zeit., XXXIV,
807 (1876).
Brahea filamentosa ex Watson, in Proc. Am. Acad. Sci., XI,
147 (1876).
Washingtonia filifera Wendland, in Bot. Zeit., XXXVII, 68
(1879).
Washingtonia filamentosa Kuntze, Rev. Gen. Pl., Par. II, 737
(1891).

COMMON NAMES.

Washington Palm (Cal.).
California Fan Palm (Cal.).
Arizona Palm (Cal.).
Wild Date (Cal.).
Fanleaf Palm (Cal.).

Family LILIACEÆ.

YUCCA Linn., Spec. Pl., 319 (1753).

Yucca treculeana Carr.   **Spanish Bayonet.**

SYN.—*Yucca Treculeana* Carrière, in Rev. Hort., VII, 280 (1858).

---

[1] *Neowashingtonia* nom. nov. Sudworth = *Washingtonia* Wendland, Bot. Zeit.,
XXXVII, 68 (1879), not Winslow (1854), nor Carr. (1867).

Yucca treculeana—Carr.—Continued.

SYN.—*Yucca canaliculata* Hooker, in Bot. Mag., XVI. 3d ser., t. 5201 (1860).

COMMON NAME.

Spanish Bayonet (Tex.).

Yucca arborescens (Torr.) Trelease.   Joshua Yucca.

SYN.—*Yucca Draconis* (?) var. *arborescens* Torrey, in Pacific R. R. Rep.. IV, 147 (1857).

YUCCA BREVIFOLIA Engelmann, in King's Rep.. V. 496 (1871), not Schott (1858).

Yucca arborescens (Torr.) Trelease, in Third Rep. Mo. Bot. Gard., 163 (1892).

COMMON NAMES.

Tree Yucca (Cal.).
Yucca Cactus (Cal.).
The Joshua (Utah).
Joshua Tree (Utah, Ariz., N. Mex.).

Yucca brevifolia Schott.   Schott Yucca.

SYN.—*Yucca puberula* Torrey, in Bot. Mex. Bound. Surv., 221 (1858),[1] not Haw. (1828).

Yucca brevifolia Schott, MSS. ex Torrey, l. c. (1858).

Yucca Schottii n. sp. Engelmann, in Trans. Acad. Sci. St. Louis, III, 46 (1873).

Yucca macrocarpa Engelm., in Bot. Gaz., VI. 224 (1881), not *Y. baccata* var. *macrocarpa* Torr. (1858).

Yucca radiosa (Engelm.) Trelease.   Spanish Bayonet.

SYN.—*Yucca angustifolia* var. *radiosa* Engelmann. in King's Rep., V. 496 (1871).

Yucca angustifolia var. *elata* Engelm.. in Trans. Acad. Sci. St. Louis, III, 50 (1873).

YUCCA ELATA Engelm., in Bot. Gaz.. VII, 17 (1882).

Yucca radiosa (Engelm.) Trelease, in Third Rep. Mo. Bot. Gard.. 163 (1892).

COMMON NAME.

Spanish Bayonet (Ariz.).

Yucca macrocarpa (Torr.) Sargent.   Broadfruit Yucca.

SYN.—*Yucca baccata* var. *macrocarpa* Torrey, in Bot. Mex. Bound. Surv., 222 (1858).[1]

---

[1] The first title-page of this report (Volume II) bears the date 1859; likewise the second title-page, or that of "Part I, Botany of the Boundary," has the date 1859; but the third title-page, which is identical with the first, with the omission of "34th Cong., 1st session, House of Representatives, Ex. Doc. No. 135," bears the date 1858.

Yucca macrocarpa (Torr.) Sargent—Continued.

SYN.—*Yucca baccata* β *australis* Engelmann, in Trans. Acad. Sci.
St. Louis, III, 44 (1873), in part.
Yucca baccata Sargent, in Tenth Cent. U. S., IX (Cat. For.
Trees U. S.), 219 (1884), in part.
Yucca filifera Trelease, in Third Rep. Mo. Bot. Gard., 162
(1892), in part; not Chabaud (1876).
Yucca australis (Engelm.) Trelease, l. c. (1892); l. c., IV, 190,
(1893), in part.
Yucca macrocarpa Sargent, in Gard. and For., IX, 104 (1896).

Yucca mohavensis Sargent.   **Mohave Yucca.**

SYN.—*Yucca filamentosa?* Wood, in Proc. Acad. Sci. Phila., 1868,
167 (1868), not Linn. (1753).
Yucca baccata Engelmann, in Trans. Acad. Sci. St. Louis,
III, 44 (1873), in part; not Torr. (1858).
Yucca macrocarpa Merriam, in N. A. Fauna, No. 7, 358, t. 14
(1893), not *Y. baccata* var. *macrocarpa* Torr. (1858), nor
Engelm. (1881).
Yucca Mohavensis Sargent, in Gard. and For., IX, 104
(1896).

Yucca australis (Engelm.) Trelease.   **Southern Yucca.**

SYN.—*Yucca baccata* var. β *australis* Engelmann, in Trans. Acad.
Sci. St. Louis, III, 44 (1873).
Yucca australis (Engelm.) Trelease, in Third Rep. Mo. Bot.
Gard., 162 (1892.)

DICOTYLEDONÆ.

Family JUGLANDACEÆ.

JUGLANS Linn., Spec. Pl., 997 (1753).

Juglans cinerea Linn.   **Butternut.**

SYN.—*Juglans cinerea* Linnæus, Syst. Nat., ed. 10, 1272 (1759).
Juglans oblonga Miller, Gard. Dict., ed. 8, No. 3 (1768).
Juglans oblonga alba Marshall, Arb. Am., 67 (1785).
Juglans nigra β Schœpf, Mat. Med. Am., 139 (1787).
Juglans cathartica Michaux f., Hist. Arb. Am., 1, 165, t. 2
(1812).
Carya cathartica Barton, Compend. Fl. Phila., II, 178 (1818).
Wallia cinerea Alefeld, in Bonplandia, 336 (1861).

Juglans cinerea Linn.—Continued.

COMMON NAMES.

Butternut (Me., N. H., Vt., Mass., R. I., Conn., N. Y., N. J., Pa., Del., W. Va., N. C., S. C., Ala., Ark., Ky., Mo., Ill., Iowa, Ind., Mich., Minn., Wis., Kans., Nebr., Ont., Ohio).
White Walnut (Del., Pa., Va., W. Va., N. C., S. C., Ala., Ark., Ky., Mo., Ill., Ind., Wis., Iowa, Nebr., Minn., S. Dak.).
Walnut (Minn.).
Oil Nut (Me., N. H., S. C.).
Buttnut (N. J.).

Juglans nigra Linn. **Black Walnut.**

SYN.—*Juglans nigra* Linnæus, Spec. Pl., ed. 1, II, 997 (1753).
Juglans nigra oblonga Marshall, Arb. Am., 67 (1785), not *J. oblonga* Mill. (1768).
Juglans Pitteursii Morren, in Ann. Soc. Roy. Ag. Bot. Gand., IV, 179, t. 197 (1848).
Wallia nigra Alefeld, in Bonplandia. IX. 336 (1861).
Wallia fraxinifolia Alefeld, l. c. (1861).
Wallia nigra microcarpa Alefeld. l. c. (1861).
Wallia nigra macrocarpa Alefeld, l. c. (1861).

COMMON NAMES.

Black Walnut (N. H., Vt., Mass., R. I., Conn., N. Y., N. J., Del., Pa., Va., W. Va., N. C., Ga., Fla., Ala., Miss., Tex., La., Ark., Ky., Mo., Ind., Ill., Kans., Nebr., Iowa, Mich., Ohio, Ont., S. Dak., Minn.).
Walnut (N. Y., Del., W. Va., Fla., Ky., Mo., Ohio, Ind., Iowa).
Walnut Tree (Pa., S. C.).
Dent-soo-kwa-no-ne (Round Nut of New York Indians).

Juglans rupestris Engelm. **Western Walnut.**

SYN.—*Juglans rupestris* Engelmann, in Sitgreaves's Rep., 171, t. 15 (1853).
Juglans rupestris var. *major* Torrey, in Sitgreaves l. c., t. 16 (1853).
Juglans Californica Rothrock, in Wheeler's Rep., VI. 249 (1878), not Wats. (1875).

COMMON NAMES.

Western Walnut (Tex.).
Dwarf Walnut (Tex.).
Little Walnut (Tex.).
California Walnut (Ariz.).
Walnut (N. Mex., Ariz.).

109

Juglans californica Watson.   **California Walnut.**

SYN.—*Juglans Californica* Watson, in Proc. Am. Acad. Sci., X, 349 (1875).

COMMON NAMES.

Walnut (Cal.).
California Walnut (Cal.).

HICORIA[1][2] Raf., Med. Rep., V, 352 (1808).

Hicoria pecan (Marsh.) Britton.   **Pecan (Hickory).**

SYN.—*Juglans Pecan* Marshall, Arb. Am., 69 (1785).
Juglans Illinoinensis Wangenheim, Beitr. Holz., 54, t. 18, f. 43 (1787).
Juglans angustifolia Aiton, Hort. Kew, ed. 1, III, 361 (1789).
Juglans alba ε pacana Castiglioni, Viag. negli. Stati Uniti, II, 262 (1790).
Juglans cylindrica Poiret, in Lamarck, Enc. Méth. Bot., IV, 505 (1797).
Juglans olivæformis Michaux, Fl. Bor.-Am., II, 192 (1803).
Hicorius oliveformis Rafinesque, Fl. Ludovic., 109 (1817)— *nomen nudum.*
CARYA OLIVÆFORMIS Nuttall, Genera, II, 221 (1818).
Carya angustifolia Sweet, Hort. Brit., ed. 1, 97 (1827).
Carya tetraptera Liebmann, in Vidensk. Medd. For. Kjöbenh. 1850, 80 (1850).
Hicoria Texana Le Conte, in Proc. Phila. Acad. Sci., VI, 402 (1853).
Carya texana de Candolle, in Ann. Sc. Nat., sér. 4, XVIII, 33 (1862).—?
Carya illinoënsis Koch, Dendrol. erst. Th., 593 (1869).
Hicoria Pecan (Marsh.) Britton, in Bull. Torr. Bot. Club, XV, 282 (1888).
Hicorius Pecan Sargent, in Gard. and For., II, 460 (1889).

COMMON NAMES.

Pecan (Va., N. C., S. C., Ga. (cult.), Ala., Miss., Tex., La., Ark., Mo., Ill., Ind., Iowa, Kans.).

---

[1] Rafinesque's name (l. c.) is spelled "*Scoria*," but it is assumed that this was a misprint for *Hicoria*, since later (Fl. Ludovic., 109, 1817) he wrote *Hicorius*, which antedates Nuttall's *Carya* (Gen., II, 221, 1818).

[2] The only other species not native in the United States is the following Mexican Hickory, found on the high mountains of Alvarez, at an elevation of 8,000 feet, 20 miles southeast of San Luis Potosi (Trelease):

Hicoria mexicana (Engelm.) Britton.   **Mexican Hickory.**
SYN.—*Carya Mexicana* Engelmann, in Hemsley Bot. Biol. Am. Cent., III, 162 (1882).
Hicoria Mexicana (Engelm.) Britton, in Bull. Torr. Bot. Cl., XV, 283 (1888).

Hicoria pecan (Marsh.) Britton—Continued.

Syn.—Pecan Nut (La.).

Pecanier (La.).

Pecan Tree (La.).

Hicoria pecan × minima[1] Trelease.

Syn.—*Hicoria Pecan × minima* (Galloway) Trelease, in Seventh Rep. Mo. Bot. Gard., 46, pl. 16, f. 12, 13, 14; pl. 20 (1896).
Hicoria Pecan × minima (Reppert) Trelease, l. c., pl. 16, f. 15, 16 (1896).

Hicoria pecan × alba[2] Trelease.

Syn.—*Hicoria Pecan × alba* (Schneck) Trelease, in Seventh Rep. Mo. Bot. Gard., 46, pl. 21 (1896).
Hicoria Pecan × alba (Reppert) Trelease, l. c., pl. 23, f. 2–5 (1896).

Hicoria pecan × laciniosa[3] Trelease.   **Nussbaum Hybrid (Hickory).**

Syn.—*Hicoria Pecan × laciniosa* (Schneck) Trelease, in Seventh Rep. Mo. Bot. Gard., 46, pl. 22, 23, f. 6 (1896).
Hicoria Pecan × laciniosa (Nussbaumer) Trelease, l. c., f. 7–9 (1896).
Hicoria pecan Corsa. Nut Cult. U. S., pl. 9, f. 6 (1896), not Britton (1888).
Hicoria laciniosa[4] Sargent, Silva, VII, t. CCCXLIX, f. 4 (1896).

Hicoria minima (Marsh.) Britton.   **Bitternut (Hickory).**

Syn.—*Juglans alba minima* Marshall, Arb. Am., 68 (1785).
Juglans cordiformis Wangenheim, Beitr. Holz., 25, t. 10, f. 25 (1787).
Juglans sulcata Willdenow. Berl. Baumz., ed. 1, 154, t. 7 (1796).

---

[1] Supposed to be a hybrid between *Hicoria pecan* and *H. minima*. Mr. S. J. Galloway reported the discovery of a single tree (Gardening, Apr. 1, 1894) near Eaton, Ohio. Mr. F. Reppert has also detected several trees supposed to be of similar parentage near Muscatine, Iowa.

[2] Supposed to be a hybrid between *Hicoria pecan* and *H. alba*. In 1894 Dr. J. Schneck reported the existence of a tree in Wabash County, Ill. In the same year Mr. F. Reppert reported a tree of supposed similar parentage from Muscatine, Iowa.

[3] Supposed to be a hybrid between *Hicoria pecan* and *H. laciniosa*. The first tree was found by Mr. J. J. Nussbaum between Mascoutah and Fayetteville, Ill.; the first account of it appearing in 1884 (A. S. Fuller, in American Agriculturist, XLIII, 546, f. 1), under the name "Nussbaum's Hybrid." Mr. R. M. Floyd, of Cedar Rapids, Iowa, has raised a tree with similar fruit (see A. S. Fuller, in New York Weekly Tribune, July 9, 1892). Dr. J. Schneck (1895) reported a similar tree from Posey County, Ind.

[4] Professor Sargent (l. c.) indicates the supposed parentage of this form, but includes it in his plate under *H. laciniosa*.

Hicoria minima (Marsh.) Britton—Continued.

SYN.—*Juglans angustifolia* Poiret, in Lamarck, Enc. Méth. Bot., IV,
504 (1797), not Ait. (1789).
Juglans amara Michaux f., Hist. Arb. Am., I, 177, t. 4 (1812).
Hickorius amara Rafinesque, Fl. Ludovic., 109 (1817).
Juglans minima Borkhausen, Handb. Forstb., 1760 (1800).
CARYA AMARA Nuttall, Genera, II, 222 (1818).
Carya minima (Marsh.) B. S. P., Prelim. Cat. Anth. Pter., 49
(1888).
Hicoria minima (Marsh.) Britton, in Bull. Torr. Bot. Club,
XV, 284 (1888).
Hicorius minimus Sargent, in Gard. and For., II, 460 (1889).
Scoria minima (Marsh.) MacMillan, Metasperm. Minn. Val.,
178 (1892).

COMMON NAMES.

Bitternut (N. H., Mass., R. I., N. Y., N. J., Pa., Del., Va.,
W. Va., N. C., S. C., Ala., Fla., Miss., La., Tex., Ark.,
Mo., Ill., Kans., Nebr., Mich., Minn., Ohio, Ont.).
Swamp Hickory (Del., Pa., N. C., S. C., Miss., Tex., Ark.,
Iowa, Minn.).
Pig Hickory (Ill.).
Pig Nut (N. Y., W. Va., Mo., Ill., Iowa, Kans.).
Bitter Pecan Tree (La.).
Pecanier Amer (La.).
Pecanier Sauvage (La.).
Bitter Pig Nut (N. Y., N. J.).
Hickory (Nebr.).
Bitter Hickory (N. H.).
Pig Walnut (N. H.).
Bitter Walnut (Vt.).
Noyer Dur (Quebec).
White Hickory (Tex.).

Hicoria myristicæformis (Michx. f.) Britton.   **Nutmeg (Hickory).**

SYN.—*Juglans myristicæformis* Michaux f., Hist. Arb. Am., I, 211,
t. 10 (1812).
CARYA MYRISTICÆFORMIS Nuttall, Genera, II, 222 (1818).
Hicoria myristica Rafinesque, Alsograph. Am., 66 (1838).
Carya amara var. (?) *myristicæformis* Cooper, in Smithsonian
Rep. 1858, 255 (1859).
Hicoria myristicæformis (Michx. f.) Britton, in Bull. Torr.
Bot. Club, XV, 284 (1888).
Hicorius myristicæformis Sargent, in Gard. and For., II,
460 (1889).

Hicoria myristicæformis (Michx. f.) Britton—Continued.

Syn.—*Hicoria Fernowiana*[1] Sudworth, in Arboresc. Flor. Washington. D. C., 6 (1891).

COMMON NAMES.

Nutmeg Hickory (N. C., S. C.. Ala.).
Bitter Waternut (La.).

Hicoria aquatica (Michx. f.) Britton. **Water (Hickory).**

Syn.—*Juglans aquatica* Michaux f., Hist. Arb. Am.. I. 182. t. 5 (1812).
Hicorius integrifolia Rafinesque, Fl. Ludovic., 109 (1817).
CARYA AQUATICA Nuttall, Genera, II, 222 (1818).
Carya integrifolia Sprengel, Syst., Veg. III, 849 (1826).
Hicoria aquatica (Michx. f.) Britton, in Bull. Torr. Bot. Club, XV, 284 (1888).
Hicorius aquaticus Sargent, in Gard. and For., II, 460 (1889).

COMMON NAMES.

Water Hickory (N. C., Ala., Fla., Miss., La., Tex., Mo.).
Swamp Hickory (S. C., Fla., Miss.. La.).
Bitter Pecan (Miss., La., Tex.).
Water Bitternut (S. C., Tenn.).

Hicoria ovata (Mill.) Britton. **Shagbark (Hickory).**

Syn.—*Juglans ovata* Miller, Gard. Dict., ed. 8, No. 6 (1768).
Juglans alba ovata Marshall, Arb. Am., 69 (1785).
Juglans ovalis Wangenheim, Nordam Holz., 24, t. 10, f. 23 (1787).
Juglans compressa Gærtner. Fruct., II, 51, t. 89, f. 1 (1791).
Juglans procera Salisbury, Prodr., 392 (1796).
Juglans obcordata Poiret, in Lamarck, Euc. Méth. Bot., IV, 504 (1797).
Juglans alba Michaux. Fl. Bor.-Am., II, 193 (1803), not Linn. (1753).
Juglans squamosa Michaux f., Hist. Arb. Am., I. 190, t. 7 (1812), not Poir. (1797).
CARYA ALBA Nuttall. Genera, II, 221 (1818).
Carya compressa Don, in Loudon, in Hort. Brit., 384 (1830).
Hicoria alba Rafinesque, Alsograph. Am., 66 (1838), in part.
Juglans glabra Gmelin, Syst., 755 (1867).
Carya ovata Koch, Dendrol., erst. Th., 598 (1869).
Hicoria ovata (Mill.) Britton, in Bull. Torr. Bot. Club, XV, 283 (1888).

---

[1] The form upon which this species was founded is very different as seen in cultivation from the true *H. myristicæformis* as found growing wild. The foliage of the cultivated tree is much larger, with fruit one-third larger than in the wild form.

Hicoria ovata (Mill.)Britton—Continued.

SYN.—*Hicorius ovatus* Sargent, in Gard. and For., II, 460 (1889).
Scoria ovata (Mill.) MacMillan, Metasperm. Minn. Val., 178 (1892).

COMMON NAMES.

Shellbark Hickory (Vt., N. H., Mass.,R. I., N. Y., Pa., Del., Va., W. Va., N. C., S. C., Ala., Ga., Miss., La., Tex., Ark., Ky., Mo., Ind., Ill., Wis., Iowa, Kans., Nebr., Ohio, Ont., Mich.).
Shagbark Hickory (Vt., N. H., Mass., R. I., Conn., N. Y., N. J., Pa., Del., S. C., Ala., Miss., Tex., Ark., Mo., Ill., Wis., Mich., Minn., Kans., Nebr., Iowa).
Shellbark (R. I., N. Y., Pa., N. C.).
Upland Hickory (Ill.).
Hickory (Vt., Ohio).
Scalybark Hickory (W. Va., S. C.).
Shagbark (R. I., Ohio).
Shellbark Tree (Del.).
White Walnut (N. J.).
Walnut (Vt., N. Y.).
White Hickory (Iowa, Ark.).
Shagbark Walnut (Vt.).
Sweet Walnut (Vt.).
Redheart Hickory (Miss.).

Hicoria laciniosa[1] (Michx. f.) Sargent.   **Shellbark (Hickory).**

SYN.—*Juglans laciniosa* Michaux f., Hist. Arb. Am., I, 199, t. 8 (1812).
Juglans ambigua Michaux, l. c., 203 (1812).
Juglans amara Muehlenberg, Cat., 88 (1813) *nomen nudum;* not Michx. (1812).
Juglans pubescens Willdenow, Enum. Hort. Berol. Suppl., 64 (1813).
Juglans sulcata Pursh, Fl. Am. Sep., II, 637 (1814), not Willd. (1796).
CARYA SULCATA Nuttall, Genera, II, 221 (1818).
Carya pubescens Sweet, Hort. Brit., ed. 1, 97 (1827).
Carya laciniosa Loudon, Hort. Brit., ed. 1, 384 (1830)—*nomen nudum.*
Carya cordiformis Koch, Dendrol., erst., Th., 597 (1869).
Hicoria sulcata (Willd.) Britton, in Bull. Torr. Bot. Club, XV, 283 (1888).
Hicorius sulcatus Sargent, in Gard. and For., II, 460 (1889).

[1] Mr. B. Shimek reports (Bull. Lab. Nat. Hist. Univ. Iowa, III, 210, 1896) a possible hybrid between this species and the Pecan, found in Muscatine County, Iowa.

Hicoria laciniosa (Michx. f.) Sargent—Continued.

SYN.—*Hicoria acuminata* Dippel, Handb. Laubh., zw. T., 336 (1892).
Hicoria laciniosa (Michx. f.) Sargent ex Coulter, in List
Pter. Sperm. N. E. N. Am. (Mem. Torr. Bot. Club. V),
Append., 354 (1892); Silva, VII, 157 t. CCCXLIX (excl.
f. 4) (1895).

COMMON NAMES.

Big Shellbark (R. I., Pa., W. Va., Ky., Mo., Ill., Kans.).
Bottom Shell Bark (Ill.).
Western Shell Bark.
Shellbark (R. I., Ky.).
Thick Shellbark (S. C., Tenn., Ind.).
Thick Shellbark Hickory (N. C., Ark.).
King Nut (Tenn.).

Hicoria alba (Linn.) Britton.　　　　　**Mockernut (Hickory).**

SYN.—*Juglans alba* Linnæus, Spec. Pl., ed. 1, II, 997 (1753).
Juglans rubra Gærtner, Fruct., II, 51, t. 89, f. 1 (1791).
Juglans tomentosa Poiret, in Lamarck, Enc. Méth. Bot., IV,
504 (1797).
CARYA TOMENTOSA Nuttall, Genera Pl., II, 221 (1818).
Carya tomentosa β maxima Nutt., l. c. (1818).
Hicoria maxima Rafinesque, Alsograph. Am., 67 (1838).
Carya tomentosa var. *integrifolia* Torrey, Bot. N. York, II,
182, t. (1843), not *C. integrifolia* Spreng. (1826).
Hicoria alba (L.) Britton, in Bull. Torr. Bot. Club, XV, 283
(1888).
Hicoria alba var. *maxima* (Nutt.) Britton, l. c. (1888).
Hicorius albus Sargent, in Gard. and For., II, 460 (1889).
Carya alba Koch, Dendrol., erst. Th., 596 (1896), not Nutt.
(1818).

COMMON NAMES.

Mocker Nut (Mass., R. I., N. Y., N. J., Del., Ala., Miss., La.,
Tex., Ark., Ill., Iowa, Kans.).
Whiteheart Hickory (R. I., N. Y., Pa., Del., N. C., Tex., Ill.,
Ont., Iowa, Kans., Minn., Nebr.).
Bullnut (N. Y., Fla., Miss., Tex., Mo., Ohio, Ill., Minn.).
Black Hickory (Tex., Miss., La., Mo.).
Big Bud (Fla.).
Red Hickory (Fla.).
Hickory (Ala., Tex.).
Hardbark Hickory (Ill.).
Hickory (Pa., S. C., Nebr.).
Common Hickory (N. C.).
White Hickory (Pa., S. C.).

Hicoria alba (Linn.) Britton—Continued.

SYN.—Hickory Nut (Ky.).
Big Hickory Nut (W. Va.).
Hog Nut (Del.).

Hicoria glabra (Mill.) Britton.　　　　　**Pignut (Hickory).**

SYN.—*Juglans glabra* Miller, Gard. Dict., ed. 8, No. 5 (1768).
Juglans alba acuminata Marshall, Arb. Am., 68 (1785).
Juglans squamosa Poiret, in Lamarck, Enc. Méth. Bot., IV, 504 (1797).
Juglans obcordata Muehlenberg & Willdenow, in Neue Schr. Gesell. Berl., III, 392 (1801), not Poir. (1797).
Juglans porcina Michaux f., Hist. Arb. Am., I, 206, t. 9 (1812).
Juglans porcina α *obcordata* Pursh, Fl. Am. Sept., II, 638 (1814).
Juglans porcina var. *ficiformis* Pursh, l. c. (1814).
CARYA PORCINA Nuttall, Genera, II, 222 (1818).
Hicoria porcina Rafinesque, Alsograph. Am., 66 (1838), in part.
Carya glabra Sweet, Hort. Brit., ed. 1, 97 (1827).
Carya obcordata Sweet, l. c. (1827).
Carya amara var. *porcina* Darby, Bot. S. States, 513 (1855).
Hicoria glabra (Mill.) Britton, in Bull. Torr. Bot. Club, XV, 284 (1888).
Hicorius glaber Sargent, in Gard. and For., II, 460 (1889).
Scoria glabra Kuntze, Rev. Gen. Pl., Par. II, 638 (1891).

COMMON NAMES.

Pignut (N. H., Vt., Mass., Conn., R. I., N. Y., N. J., Pa., Del., W. Va., N. C., S. C., Fla., Ala., Miss., La., Tex., Ark., Ky., Mo., Ill., Ind., Wis., Iowa, Kans., Nebr., Minn., Ohio, Ont.).
Bitternut (Ark., Ill., Iowa, Wis.).
Black Hickory (Miss., La., Ark., Mo., Ind., Iowa).
Broom Hickory (Mo.).
Brown Hickory (Del., Miss., Tenn., Minn., Tex.).
Hard Shell (W. Va.).
Red Hickory (Del.).
Switch-bud Hickory (Ala.).
White Hickory (N. H., Iowa).

Hicoria glabra villosa[1] Sargent.　　　**Woolly Pignut (Hickory).**

SYN.—*Hicoria glabra* var. *villosa* Sargent, Silva, VII, 167 (1895).

---

[1] A variety recently detected in Missouri and characterized chiefly by the soft pubescence on the branchlets, petioles, and under surface of the leaflets.

Hicoria odorata (Marsh.) Sargent. **Small Pignut (Hickory).**

Syn.—*Juglans alba odorata* Marshall, Arb. Am., 68 (1785).
 Carya microcarpa Nuttall, Genera, Pl., II, 221 (1818)?
 Juglans squamosa β microcarpa Barton. Compend. Fl.
   Phila., II, 179 (1818).
 Hicoria microcarpa (Nutt.) Britton, in Bull. Torr. Bot. Club,
   XV, 283 (1888).
 Hicorius odoratus Sargent, in Gard. and For., II. 460 (1889).
 Hicoria odorata Dippel. Handb., Laubh., zw. T., 332 (1892).
 Hicoria glabra var. *odorata* Sargent. Silva, VII, 167,
   t. CCCLIV, (1895).
 Hicoria glabra var. *microcarpa*[1] (Nutt.) Sargent ex Trelease,
   in Seventh Ann. Rep. Mo. Bot. Gard. 37: Pl. 10; 14, f. 2;
   17, f. 7, 8 (1896).

COMMON NAMES.

Small Pignut (Md.).
Little Pignut (Md.).
Little Shagbark (Md.).

Family MYRICACEÆ.

MYRICA[2] Linn., Spec. Pl., 1024 (1753).

Myrica cerifera Linn. **Wax Myrtle.**

Syn.—*Myrica cerifera* Linnæus, Spec. Pl., ed. 1, 1024 (1753).
 Myrica cerifera β Lamarck, Enc. Méth. Bot., II, 592 (1786).
 Myrica cerifera α angustifolia Aiton. Hort. Kew., ed. 1, II,
   396 (1789).
 Myrica cerifera α arborescens Castiglioni. Viag. Stati Uniti,
   II, 302 (1790).

---

[1] Our understanding of this plant, which I take to be a form of *H. odorata*, appears to be as yet very imperfect. The forms that may be grouped here vary greatly in the character of the bark and fruit. Individuals from Maryland present the same fruit characters as Professor Trelease's *H. glabra microcarpa* (l. c., pl. 17, f. 7–8), but with the bark of his *H. glabra odorata* (l. c., pl. 8), or still others have the bark of the latter and a form of fruit not figured and probably not yet seen by Professor Trelease. For the present, therefore, it is thought best to retain the forms under the oldest name which is taken to indicate a more or less distinct type.

[2] The following species is a low shrub:

Myrica cerifera pumila Michx. **Dwarf Wax Myrtle.**

Syn.—*Myrica cerifera γ pumila* Michaux, Fl. Bor.-Am., II, 228 (1803).
 Myrica cerifera β Willdenow, Sp. Pl., IV, Par. II, 746 (1805).
 Myrica sessilifolia Rafinesque, Alsograph. Am., 10 (1838).
 Myrica pusilla Raf., l. c. (1838).

Myrica cerifera Linn.—Continued.

SYN.—*Lacistema Berterianum* Schultes, in Roemer & Schultes,
Syst. Mant., I, 66 (1822).
Lacistema alternum Sprengel, Syst. Veg., I, 124 (1825).
Myrica heterophylla Rafinesque, Alsograph. Am., 9 (1838).
Cerophora lanceolata Raf., l. c., 11 (1838).
Myrica microcarpa Grisebach, Fl. Brit. W. Ind., 177 (1864),
in part; not Benth. (1846).
Myrica Carolinensis Richard, Fl. Cub., III, 231 (1853), not
Mill. (1768).
Myrica altera C. de Candolle, Prodr., XVI, sect. 2, 595 (1868).
Myrica cerifera sempervirens Hort. ex-Sargent, in Tenth Cen.
U. S. IX (Cat. for Trees U. S.), 136 (1884).

COMMON NAMES.

Wax Myrtle (R. I., N. J., Del., N. C., S. C., Ala., Fla.).
Bayberry (Mass., R. I., N. J., N. Y., Pa., Del., N. C., S. C.,
Ala., Fla.).
Waxberry (R. I., Pa., S. C.).
Cirier (La.).
Candleberry (Fla.).
Myrtle (Fla.).
Myrtletree (Fla.).
Puckerbush (Fla.).

Myrica inodora Bartr.   **Odorless Myrtle.**

SYN.—*Myrica inodora* Bartram, Travels, 405 (1791).
Cerophora inodora Rafinesque, Alsograph. Am., 11 (1838).
Myrica obovata C. de Candolle, Prodr., XVI, sect. 2, 150
(1864).
Myrica Laureola C. de C., l. c., 154 (1864).

Myrica californica Cham.   **California Wax Myrtle.**

SYN.—*Myrica Californica* Chamisso, in Linnæa, VI, 535 (1831).
Gale Californica Greene, Man. Bot. Bay Reg., 298 (1894).

COMMON NAMES.

California Bayberry (Nev.).
Myrtle (Nev.).
Bayberry (Cal.).
California Myrtle (Cal.).
Wax Myrtle (Cal.).

Family LEITNERIACEÆ.

LEITNERIA Chapm.. Fl. S. St.. 428 (1860).

Leitneria floridana[1] Chapm.   **Corkwood.**

Syn.—*Leitneria Floridana* Chapman. Fl. S. States. ed. 1. 428 (1860).

Family SALICACEÆ.

SALIX Linn., Spec. Pl., 1015 (1753).

Salix nigra Marsh.   **Black Willow.**

Syn.—*Salix nigra* Marshall. Arb. Am.. 139 (1785).
Salix pentandra Walter. Fl. Caroliniana. 243 (1788). not Linn. (1753).—?
Salix Caroliniana Michaux. Fl. Bor.-Am., II, 226 (1803).
Salix ligustrina Michaux f.. Hist. Arb. Am.. III, 326. t. 5, f. 2 (1813).
Salix Houstoniana Pursh. Fl. Am. Sept.. II, 614 (1814).
Salix ambigua Pursh. l. c., 617 (1814).—?
Salix flavo-virens Hornemann. Cat. Hort. Hafn. Suppl.. II, 11 (1819).
Salix virgata Forbes, Sal. Wob., 23 (1829).
Salix nigra var. *falcata* Torrey, Fl. N. York, II, 209 (1843).
Salix nigra a *angustifolia* α *falcata* Andersson, in Sven. Vet. Akad. Handl., ser. 4. VI (Monog. Salic.). 20 (1867).
Salix nigra a *angustifolia* β *longifolia* Anderss., l. c. (1867).
Salix nigra b *latifolia* Anderss.. l. c., 21 (1867).
Salix nigra b *latifolia* α *brevijulis* Anderss., l. c. (1867).
Salix nigra b *latifolia* β *longijulis* Anderss.. l. c. (1867).
Salix nigra b *latifolia* γ *brevifolia* Anderss., l. c. (1867).
Salix nigra b *latifolia* γ *brevifolia testacea* Anderss.. l. c. (1867).
Salix nigra subsp. *marginata* Anderss., l. c. (1867).

COMMON NAMES.

Black Willow (N. H.. Vt., R. I.. N. Y.. Pa., Del.. S. C.. Ala.. Fla., Miss.. La., Tex.. Ariz.. Oreg.. Cal.. N. Mex.. Utah, Ill., Wis.. Mich.. Minn.. Nebr.. Kans.. Ohio, Ont., N. Dak.).

---

[1] Until recent observation (Rep. Mo. Bot. Gard., V. 150: VI) has shown it to be a tree, this plant has not been included among our arborescent species. As now known, it is a small tree 15 to 20 feet high and 3 to 5 inches in diameter, usually forming thickets; very local in distribution, being known only in the swampy region near Apalachicola, Fla., and in Butler and Duncan counties, Mo.

Salix nigra Marsh—Continued.

SYN.—Swamp Willow (N. C., S. C.).
Willow (N. Y., Pa., N. C., S. C., Miss., Tex., Cal., Ky., Mo., Nebr.).

Salix nigra falcata (Pursh) Torr.　　**Crescentleaf Willow.**

SYN.—*Salix falcata* Pursh, Fl. Am. Sept., II, 614 (1814).
Salix Purshiana Sprengel, Syst. Veg., V, 608 (1828).
Salix nigra var. *falcata* Torrey, Fl. N. Y., II, 209 (1843).

Salix nigra × amygdaloides[1] Glatfelter.

SYN.—*Salix nigra × amygdaloides* Glatfelter, in Trans. Acad. Sci. St. Louis, VI, 427 (1894).

Salix nigra × alba Bebb.

SYN.—*Salix nigra × alba*[2] Bebb, in Gard. and For.. VIII, 423, f. 58 (1895).

Salix wardi Bebb.　　**Ward Willow.**

SYN.—*Salix cordata β angustata* 1° *discolor* Andersson, in de Candolle, Prodr., XVI, sect. II, 252 (1868), not *S. discolor* Muehl (1803).
Salix nigra Marsh var. *Wardi* Bebb, (MSS.) in Bull. No. 22, U. S. Nat. Mus. (Flor. Wash., D. C., and Vicinity), 114 (1881).
Salix Wardi Bebb, in Gard. and For., VIII, 363 (1895).

Salix occidentalis Koch.　　**Western Willow.**

SYN.—*Salix occidentalis* Koch, Sal. Europ. Com., 16 (1828).

Salix occidentalis longipes (Anderss.) Bebb.　　**Longstalk Willow.**

SYN.—*Salix longipes* (Shuttleworth in herb. Hooker) Andersson, in Öfv. Vet. Akad. Förh., XV, 114 (1858).
Salix Wrightii Anderss., l. c., 115 (1858).
Salix nigra subsp. *longipes gongylocarpa* (Shuttl.) Anderss., in Sven. Vet. Akad. Handl. (Monog. Salic.), VI, 22 (1867).
Salix nigra subsp. *longipes venulosa* Anderss., l. c. (1867).
Salix nigra subsp. *Wrightii* Anderss., l. c. (1867).
Salix nigra var. *venulosa* Bebb, in Bot. Gaz., XVI, 102 (1891).
Salix occidentalis Koch var. *longipes* (Anderss.) Bebb, in Gard. and For., VIII, 363 (1895).
Salix occidentalis var. *longipes forma venulosa* Bebb, l. c. (1895).

[1] A hybrid.
[2] A hybrid discovered by Mr. E. L. Hankenson, near Newark, Wayne County, N. Y., where several individuals are in existence. It reaches a height of 30 to 40 feet and 1 to 2 feet in diameter. The tree has the foliage of *S. nigra* and the flowers of *S. alba*.

Salix amygdaloides Anderss. **Almondleaf Willow.**

SYN.—*Salix amygdaloides* Anderss.. Öfv. Vet. Akad. Förh.. XV, 114 (1858).
Salix nigra subsp. *amygdaloides* Andersson, in Sven. Vet. Akad. Handl.. ser. 4. VI (Monog. Salic.), 21 (1867).

COMMON NAMES.

Willow (Cal., Nev., Oreg., Colo.. Mont.. Utah).
Black Willow (Mo., Idaho).
Common Willow (Mont.).

Salix lævigata Bebb. **Smoothleaf Willow.**

SYN.—*Salix lævigata* Bebb, in Am. Nat., VIII, 202 (1874).

COMMON NAMES.

Willow (Cal., Nev.. Utah).
Black Willow.

Salix lævigata angustifolia Bebb. **Narrowleaf Willow.**

SYN.—*Salix lævigata* var. *angustifolia* Bebb, in Watson. Bot. Cal.. II, 84 (1880).

Salix lævigata congesta Bebb.

SYN.—*Salix lævigata* var. *congesta* Bebb, in Watson, Bot. Cal., II, 84 (1880).

Salix lasiandra Benth. **Western Black Willow.**

SYN.—*Salix Hoffmanniana* Hooker & Arnott, in Bot. Beechey's Voyage, 159 (1833), not Smith (1828).
Salix lasiandra Bentham, Pl. Hartweg., 335 (1857).
Salix arguta Andersson, in Sven. Vet. Akad. Handl., ser. 4, VI (Monog. Salic.), 33 (1867).

COMMON NAMES.

Willow (Cal., Nev., Utah, Oreg., Colo.).
Black Willow.

Salix lasiandra lyalli Sargent. **Lyall Willow.**

SYN.—*Salix speciosa* Nuttall. Sylva, I, 58, t. 17 (1842), not Hook. & Arn. (1841).
Salix lucida angustifolia [forma] *lasiandra* Andersson, in Öfv. Vet. Akad. Förh., XV, 115 (1858).
Salix lucida macrophylla Anderss.. in Sven. Vet. Akad., Handl., ser. 4, VI (Monog. Salic.). 32 (1867). not Kern. (1860).
Salix lancifolia Anderss., l. c.,34, f. 23 (1867), not Doell (1859).

Salix lasiandra lyalli Sargent—Continued.

Syn.—*Salix lasiandra* var. *lancifolia* Bebb, in Watson, Bot. Cal.,
II, 84 (1880).
Salix lasiandra var. *Lyalli* Sargent, in Gard. and For., VIII,
463 (1895).

Salix lasiandra caudata (Nutt.) Sudworth.

Syn.—*Salix pentandra β caudata* Nuttall, Sylva, I, 61, t. 18 (1842).
Salix Fendleriana Andersson, in Öfv. Vet. Akad. Föhr.,
XV, 115 (1858).
Salix arguta Andersson, in Kongl. Sven. Vet. Akad. Handl.,
VI (Monog. Salic.), 32 (1867).
Salix lasiandra var. *Fendleriana* Bebb, in Watson, Bot. Cal.,
II, 84 (1880).
Salix lasiandra var. *caudata* (Nutt.) Sudworth, in Bull. Torr.
Bot. Club, XX, 43 (1893).

Salix bonplandiana[1] H., B. K.   **Bonpland Willow.**

Syn.—*Salix Bonplandiana* Humboldt, Bonpland & Kunth, Nov.
Gen. Sp., II, 24, t. 101, 102 (1817).
Salix pallida H., B., & K., l. c., 25 (1817).
Salix Bonplandiana subsp. *pallida* Andersson, Sven. Vet.
Akad. Handl., ser. 4, VI (Monog. Salic.), 18 (1867).
Salix Bonplandiana β pallida Anderss., in de Candolle,
Prodr., XVI, sect. II, 200 (1868).

Salix lucida[2] Muehl.   **Shining Willow.**

Syn.—*Salix lucida* Muehlenberg, in Neue Schr. Gesell. Nat. Fr.
Berl., IV, 239, t. 6, f. 7 (1803).
Salix lucida latifolia Andersson, Öfv. Vet. Akad. Förh., XV,
115 (1858).
Salix lucida oratifolia Anderss., l. c. (1858).
Salix lucida pilosa Anderss., l. c. (1858).
Salix lucida var. *angustifolia forma pilosa* Anderss., in Proc.
Am. Acad., IV, 54 (1858).
Salix lucida rigida Anderss., in Sven. Vet. Akad. Handl.,
ser. 4, VI (Monog. Salic.), 32 (1867).
Salix lucida tenuis Anderss., l. c. (1867).

---

[1] A species formerly not known to occur north of Mexico, and therefore not hitherto included in our flora. Discovered growing in a canyon at the base of Santa Catalina Mountains, by Prof. C. S. Sargent and J. W. Toumey, in 1894. Professor Toumey states that "it is a frequent tree in the canyons and along the washes in the foothills throughout southern Arizona."

[2] Formerly not included in the arborescent flora, as it was not known to occur as a tree.

Salix fluviatilis Nutt.   **Longleaf Willow.**

SYN.—SALIX LONGIFOLIA Muchlenberg, Neue Schr. Gesell. Nat.
Fr. Berl., IV, 238. t. 6. f. 6 (1803), not Lam. (1778).
Salix rubra Richardson, in Franklin Journ. Append., 7, 765
(1823), not Hudson (1762).
Salix fluviatalis Nuttall, Sylva, I, 73 (1842).
Salix longifolia pedicellata Andersson, in Sven. Vet. Akad.
Handl., ser. 4, VI (Monog. Salic.). 55, f. 35 (1867).
Salix Nevadensis Watson, in Am. Nat., VII, 302 (1873).
Salix longifolia forma integerrima Kuntze. Rev. Gen. Pl.,
par. II, 643 (1891).
Salix longifolia forma paucidenticulata Kuntze, l. c. (1891).
Salix longifolia forma multidenticulata Kuntze, l. c. (1891).

COMMON NAMES.

Sandbar Willow (R. I.. Miss.. Cal., Wis., Kans.. Nebr., Minn.,
S. Dak.. Ont.).
Longleaf Willow (Ala., Mich., Kans.).
Long-leaved Willow (Colo.. Cal., Tenn., Idaho., Nebr., Minn.,
Wash.).
Narrow-leaved Willow (Nebr.).
Shrub Willow (Nebr.).
White Willow (Mo.).
Red Willow (Mont.).
Osier Willow (Mont.).
Willow (Vt.. N. Y., Tex., Miss.. Ky., Cal., Ind.. Nev.. Utah.,
Mont.).

Salix fluviatilis exigua (Nutt.) Sargent.

SYN.—*Salix exigua* Nuttall, Sylva, I, 75 (1842).
Salix longifolia angustissima Andersson, in Öfv. Vet. Akad.
Förh., XVI, 116 (1858).
Salix longifolia var. *exigua* Bebb, in Watson. Bot. Cal.. II,
85 (1880).

Salix fluviatilis argyrophylla (Nutt.) Sargent.

SYN.—*Salix argyrophylla* Nuttall. Sylva, I, 71. t. 20 (1842).
Salix brachycarpa Nutt., l. c., 69 (1842).
Salix longifolia opaca Andersson, in Sven. Vet. Akad. Handl..
ser. 4, VI (Monog. Salic.). 55 (1867).
Salix fluviatilis var. *argyrophylla* Sargent, Silva, IX, 124
(1896).

Salix sessilifolia Nutt.   **Silver Willow.**

SYN.—*Salix sessilifolia* Nuttall. Sylva, I. 68 (1842).
Salix Hindsiana Bentham, Pl. Hartweg, 335 (1857).

Salix sessilifolia Nutt.—Continued.

SYN.—*Salix sessilifolia Hindsiana* Andersson, in Öfv. Vet. Akad.
Förh., XV, 117 (1858).
Salix Hindsiana tenuifolia Anderss., in Sven. Vet. Akad.
Handl., ser. 4, VI (Monog. Salic.), 56 (1867).
Salix sessilifolia β villosa Anderss., in de Candolle, Prodr.,
XVI, sect., II, 215 (1868).

COMMON NAMES.

Willow (Cal., Oreg.).
Silver Willow (Cal.).

Salix taxifolia[1] H., B. K.   **Yewleaf Willow.**

SYN.—*Salix taxifolia* Humboldt, Bonpland, & Kunth, Nov. Gen.
Sp., II, 22 (1817).
Salix microphylla Chamisso & Schlectendal, in Linnæa, VI,
354 (1831).
Salix taxifolia var. *a. seriocarpa* Andersson, in Sven. Vet.
Akad. Handl., ser. 4, VI (Monog. Salic.), 57 (1867).
Salix taxifolia var. *b. lejocarpa* Anderss., l. c. (1867).

Salix bebbiana[2] Sargent.   **Bebb Willow.**

SYN.—SALIX ROSTRATA Richardson, Arct. Exp., 753 (1823), not
Thuill. (1799).
Salix vagans b. occidentalis Andersson, in Öfv. Vet. Akad.
Förh., XV, 122 (1858), not *S. occidentalis* Koch (1828).
Salix vagans subsp. *rostrata* Anderss., Sven. Vet. Akad.
Handl., ser. 4, VI (Monog. Salic.), 87 (1867), not of prev.
auths.
Salix vagans subsp. *rostrata latifolia* Anderss., l. c. 88 (1867).
Salix vagans subsp. *rostrata lanata* Anderss., l. c. (1867), not
of prev. auths.
Salix vagans subsp. *rostrata obovata* Anderss., l. c. (1867),
not of prev. auths.
Salix vagans subsp. *rostrata lanceolata* Anderss., l. c. (1867),
not of prev. auths.
Salix vagans β rostrata Anderss., in de Candolle, Prodr.,
XVI, sect., II, 227 (1868).
Salix Bebbiana Sargent, in Gard. and For., VIII, 463 (1895).

---

[1] A species formerly not included in our arborescent flora. Mr. Charles Wright first
collected specimens near El Paso, Tex. Mr. C. G. Pringle collected it on the Rillita
River, southern Arizona, in 1883, and Prof. J. W. Toumey has recently found it in
the San Rita Mountains, Arizona; it is also in Mexico, Guatemala, and Lower California. It is said to be a moderate-sized tree, 30 to 50 feet in height, with delicate
sprays, the lower branches drooping.

[2] Formerly a doubtfully arborescent species not hitherto included in our forest
flora; at present known to reach 20 to 25 feet in height, with the habit of a tree.

Salix discolor Muehl. **Glaucous Willow.**

Syn.—*Salix discolor* Muehlenberg, in Neue Schr. Gesell. Nat. Fr.
Berl.. IV, 234, t. 6, f. 1 (1803).
Salix sensitiva Barratt, Sal. Am.. No. 8 (1840)

COMMON NAMES.

Glaucous Willow (R. I., N. Y., Pa., Miss., Mich., Minn.. Ont.).
Pussy Willow (N. J., Minn.).
Silver Willow (Kans.).
Swamp Willow (N. J.).
Willow (Vt., N. Y.. S. C.. Mo.).

Salix discolor eriocephala (Michx.) Anderss.

Syn.—*Salix eriocephala* Michaux, Fl. Bor.-Am., II, 225 (1803).
Salix crassa Barratt, Sal. Am., No. 7 (1840).
Salix discolor subsp. *eriocephala* Andersson, in Sven. Vet.
Akad. Handl., ser. 4, VI (Monog. Salic.), 85 (1867).
Salix discolor subsp. *eriocephala* var. *parvifolia* Anderss.,1. c.
(1867).
Salix discolor subsp. *eriocephala* var. *rufescens* Anderss.. l. c.
(1867).

Salix discolor prinoides (Pursh) Anderss.

Syn.—*Salix prinoides* Pursh, Fl. Am. Sept., II, 613 (1814).
Salix discolor subsp. *prinoides* Andersson, Sven. Vet. Akad.
Handl.. ser. 4. VI (Monog. Salic.), 86 (1867).

Salix cordata mackenzieana[1] Hook. **Mackenzie Willow.**

Syn.—*Salix cordata* γ *Mackenzieana* Hooker, Fl. Bor.-Am., II. 149
(1839).
Salix cordata × *ragans* Andersson. in Öfv. Vet. Akad. Förh.,
XV, 125 (1858).
Salix cordata × *rostrata* Anderss., in Proc. Am. Acad.. IV,
65 (1858).
Salix cordata subsp. *Mackenziana*[2] Barratt ex Andersson,
Sven. Vet. Akad. Handl.. ser. 4, VI (Monog. Salic.), 160
(1867).

Salix cordata lutea[3] (Nutt.) Bebb. **Yellow Willow.**

Syn.—*Salix lutea* Nuttall. Sylva, I, 63, t. 19 (1842).

---

[1] A plant hitherto not included in our arborescent flora. Originally found only on
Great Slave Lake and Mackenzie River; later collected in the plains of Idaho by Dr.
Edward Palmer, southward in the mountains to Lake County, Cal., by Dr. C. L.
Anderson. A small tree with twigs the color of pipe-clay and highly polished.
[2] Andersson's exact spelling of this subspecific term is retained.
[3] A little known variety not formerly placed among our arborescent flora. It
occurs in the Rocky Mountains of Montana and northward, and is described by
Nuttall as "A small arborescent willow—remarkable for its smooth, bright-yellow
branches."

Salix cordata lutea (Nutt.) Bebb—Continued.

SYN.—*Salix cordata* subsp. *angustata vitillina* Andersson, in Sven.
Vet. Akad. Handl., ser. 4, VI (Monog. Salic.), 159 (1867).
Salix cordata var. *lutea* Bebb, in Gard. and For., VIII, 473
(1895).

Salix missouriensis[1] Bebb. **Missouri Willow.**

SYN.—*Salix cordata* subsp. *rigidad. vestita* Andersson, in Sven.Vet.
Akad. Handl., ser. 4, VI (Monog. Salic.), 159 (1867), not
S. vestita Pursh (1814).
SALIX CORDATA var. VESTITA Sargent, in Tenth Cens. U.
S., IX (Cat. For. Trees U. S.), 170 (1884).
Salix Missouriensis n. sp. Bebb, in Gard. and For., VIII,
373 (1895).

Salix lasiolepis Benth. **Bigelow Willow.**

SYN.—*Salix Bigelovii*[2] Torrey, in Pacif. R. R. Rep., IV, Pt. V, 139
(1856; Jan., 1857).
Salix (Discolora) *lasiolepis* Bentham, Pl. Hartweg., 335
(Feb., 1857.)
Salix Bigelovii **a** *latifolia* Andersson, Sven. Vet. Akad,
Handl., ser. 4, VI (Monog. Salic.), 163 (1867).
Salix Bigelovii **b** *angustifolia* Anderss., l. c. (1867).
Salix Bigelovii subsp. *fuscior* Anderss., l. c., f. 94 (1867).
Salix ——? Watson, in King's Rep., V, 325 (1871).
Salix lasiolepis var. *Bigelovii* Bebb, in Watson, Bot. Cal.,
II, 86 (1880).
Salix lasiolepis var. (?) *fallax* Bebb, l. c. (1880).

COMMON NAME.

Willow (Cal., Idaho).

Salix nuttallii Sargent. **Nuttall Willow.**

SYN.—SALIX FLAVESCENS Nuttall, Sylva, I, 65 (1842), not Host
(1828).
Salix Nuttallii Sargent, in Gard. and For., VIII, 463 (1895).

---

[1] A specimen of this species was first collected at Fort Osage, on the Missouri
River, and described by Andersson (l. c.). Recently it has been rediscovered by Prof.
C. S. Sargent and B. F. Bush near Courtney, Jackson County, Mo., within 20
miles of the original locality. It has also been collected near St. Louis (1895), and
in Nebraska (ex-Governor Furnas). It attains a height of 30 to 50 feet, and a diameter of 10 to 12, or rarely 18 inches.

[2] *Salix bigelovii* would appear to be an older name than *S. lasiolepis* Bentham.
The title-page of the report (cited) in which Torrey's paper was published, bears
the date 1856, and his introduction the date January, 1857. But Torrey cites species
of Salix (in close connection with *S. bigelovii*) described in Bentham's Plantæ Hart-
wegianæ with page reference in February, 1857, clearly indicating that Bentham's
work must have appeared before Torrey's paper.

Salix nuttallii Sargent—Continued.

Syn.—Mountain Willow (Mont.).
Willow (Cal., Oreg., Utah).
Black Willow.

Salix nuttallii brachystachys (Benth.) Sargent.

Syn.—*Salix Scouleriana* (Barratt mst.) Hooker, Fl. Bor.-Am., II,
145 (18 9), in part.—?
Salix brachystachys Bentham, Pl. Hartweg., 336 (1857).
Salix capreoides Andersson, in Öfv. Vet. Akad. Förh., XV,
120 (1858).
Salix brachystachys subsp. *Scouleriana* Anderss., Sven. Vet.
Akad. Handl., ser. 4, VI (Monog. Salic.), 83 (1867).
Salix brachystachys subsp. *Scouleriana tenuijulis* Anderss.,
l. c. (1867).
Salix brachystachys β *Scouleriana crassijulis* Anderss., in de
Candolle, Prodr., XVI, sect. II, 225 (1868).
Salix flavescens Bebb, in Watson, Bot. Cal., II, 86 (1880),
in part.
Salix flavescens var. Scouleriana Bebb, in Bot.
Gaz., VII, 129 (1882).
Salix Nuttallii var. *capreoides* Sargent, in Gard. and For.,
VIII, 463 (1895).
Salix flavescens var. *capreoides* Bebb, l. c., 373 (1895).

Salix piperi[1] Bebb.   Piper Willow.

Syn.—*Salix Piperi* n. sp. Bebb, in Gard. and For., VIII, 482 (1895).

Salix hookeriana Barratt.   Hooker Willow.

Syn.—*Salix Hookeriana* (Barratt mst.) in Hooker, Fl. Bor.-Am.,
II. 145, t. 180 (1839).

Salix alba[2] Linn.   White Willow.

Syn.—*Salix alba* Linnæus, Spec. Pl., ed. 1. II, 1021 (1753).
Salix flexibilis Gilbert, Exerc., II, 406 (1792).
Salix pallida Salisbury, Prodr.. 394 (1796.)

---

[1] A new species recently discovered near Seattle. Wash., by Prof. C. V. Piper.
Arborescence questionable; only a few individuals known at present, said to be 18 to
20 feet in height, several stems rising from the same root, and little branched except
at the top. Growing "in swamp near Lake Union, and in sphagnum bog on high
ground."
[2] This European species has been so long and widely introduced by cultivation,
that in some localities it has become thoroughly naturalized, maintaining itself in an
almost wild state.

Salix alba Linn.—Continued.

SYN.—*Salix heterophylla* Bray, in Denkschr. Bot. Gesell. Reg., 1, 5 (1815).
Salix splendens Opiz, Böhm. Gewäch., 110 (1823).

Salix alba × lucida[1] Bebb.

SYN.—*Salix alba* subsp. *Pameachiana* Andersson, Sven. Vet. Akad. Handl., ser. 4, VI (Monog. Salic.), 48 (1867), not *S. Pameachiana* Barratt (1840).
Salix alba × lucida Bebb, in Gard. and For., VIII, 423, f. 57 (1895).

Salix fragilis[2] Linn.   **Crack Willow.**

SYN.—*Salix fragilis* Linnæus, Spec. Pl., ed. 1, II, 1017 (1753).
Salix decipiens Hoffmann, Hist. Sal., II, Fasc. 1, 9, t. 31 (1791).
Salix persicifolia Schleicher, Cat. Pl., ed. 2, 30 (1807).
Salix Wargiana Lejeune, Fl. Spa., II, 332 (1813).
Salix fragilior Host, Hist. Salix, I, 6, t. 20, 21 (1828).
Salix fragilissima Host, l. c. (1828).
Salix monspeliensis Forbes, Salic. Wob., 59 (1829).
Salix excelsa Koch, Syn. Fl. Germ., I. 643 (1837).

Salix babylonica[3] Linn.   **Weeping Willow.**

SYN.—*Salix Babylonica* Linnæus, Spec. Pl., ed. 1, II, 1017 (1753).
Salix pendula Moench, Meth., 336 (1794).
Salix propendens Seringe, Ess. Monog. Saul., 73 (1815).
Salix Japonica Blume, Bijdrag. Tot. Flor., 516 (1825), not Thunb. (1784).
Salix babylonica Napoliona Hort. ex Loudon, Arb. Frut., III, 1514 (1838).

Salix sitchensis Sanson.   **Silky Willow.**

SYN.—*Salix Sitchensis* Sanson Mss. in Bongard, in Mém. Acad. Pétersb., II, 162 (1831).
Salix Scouleriana (Barratt mst.) in Hooker, Fl. Bor.-Am., II, 145 (1839), in part.
Salix cuneata Nuttall, Sylva, 1, 66 (1842).

[1] Professor Bebb (Gard. and For., VIII, 423, 1895) describes this form as "vacillating between that of the two parents; sometimes a branching shrub, as in *S. lucida*, and others a small tree 25 to 30 feet in height, with a distinct trunk, as in *S. alba*." Found near Amherst, Mass., Westville, Conn., Providence, R. I., Ithaca, N. Y., Newark, N. Y.
[2] Naturalized in Eastern North America; native in Europe.
[3] Naturalized in a few localities in the Atlantic States; native in Europe.

Salix sitchensis Sanson—Continued.

SYN.—*Salix Coulteri* Andersson, in Proc. Am. Acad. Sci. (Salic.
Bor.-Am.), IV, 58 (1858).
Salix sitchensis congesta Anderss., Sven. Vet. Akad. Handll.,
ser. 4, VI (Monog. Salic.), 107 (1867).
Salix sitchensis denudata Anderss., l. c. (1867).

COMMON NAMES.

Silky Willow (Oreg.).
Sitka Willow (Germ. lit.).

POPULUS Linn., Spec. Pl., 1034 (1753).

Populus tremuloides Michx.   **Aspen.**

SYN.—*Populus tremula* Marshall, Arb. Am., 107 (1785), not Linn.
(1753).—?
Populus tremula var. Burgsdorf, Anleit. Anpfl., Pt. II, 174
(1787).
Populus tremuloides Michaux, Fl. Bor.-Am., II, 243 (1803).
Populus trepida Willdenow, Sp. Pl., IV, Par. II, 803 (1805).
Populus tremuliformis Emerson, Trees and Shr. Mass., 243
(1846).
Populus Atheniensis Koch, Dendrol., zw. Th. erst. Ab., 486
(1872).
Populus Græca Lauche, Deutsche Dend., zw. Ausg., 316
(1883), not Ait. (1789).

COMMON NAMES.

Aspen (N. H., Mass., R. I., Conn., N. Y., N. J., Pa., Del.,
Ill., Ind.. Wis., Mich., Minn., N. Dak., Nebr., Ohio, Ont.,
Oreg., Utah, Idaho, Nev., Mont., Colo., Cal.).
Quaking Asp (N. Y.. Pa., Del., Cal., N. Mex., Idaho, Colo.,
Ariz.. Ill., Iowa, Minn., Mont., Nebr., Utah, Oreg., Nev.).
Mountain Asp (Mont.).
American Aspen (Vt.).
Aspen Leaf (Pa.).
White Poplar (Mass.).
Trembling Poplar (Minn., Colo.).
American Poplar (Minn.. Colo.).
Poplar (Vt., N. Y., Ill., Ind.. Minn., Mont.).
Popple (Wis.. Iowa, Mont.).
Tremble (Quebec).
Trembling Aspen (Iowa).
Aspen Poplar (Cal., Mont.).

VARIETY DISTINGUISHED IN CULTIVATION.

Populus tremuloides pendens nom. nov.   **Weeping Aspen.**

SYN.—*Populus tremuloides* a *pendula* Dippel, Handb. Laubh., zw.
T., 198 (1892), not *P. pendula* Burgsd. (1787).

Populus grandidentata Michx.   **Largetooth Aspen.**

SYN.—*Populus grandidentata* Michaux, Fl. Bor.-Am., II, 243 (1803).

COMMON NAMES.

Large-toothed Aspen (N. J., Pa., Del., S. C., Mich., Minn.).
Poplar (Me., N. H., Vt., Mass., R. I., Conn., N. Y., N. J., Pa.,
W. Va., N. C., S. C., Ga., Ill., Wis., Ohio, Ont.).
Large-toothed Poplar (N. C., Minn.).
Large Poplar (Tenn.).
White Popple (Mass.).
Popple (Me.).
White Poplar (Mass.).
Large American Aspen (Ala.).

VARIETY DISTINGUISHED IN CULTIVATION.

Populus grandidentata penduliformis nom. nov.

Weeping Largetooth Aspen.

SYN.—*Populus grandidentata* β *pendula* Nuttall, Genera, II, 239
(1818), not *P. pendula* Burgsdorf (1787).
Populus Græca pendula Nurs. ex Bailey, in Bull. 68, Corn.
Un. Ag. Ex. Stn., 232 (1894).
Populus nigra pendula Nurs. ex Bailey, l. c. (1894).
Parosol de St. Julien Nurs. ex Bailey, l. c. (1894).

Populus heterophylla Linn.   **Swamp Cottonwood.**

SYN.—*Populus heterophylla* Linnæus, Spec. Pl., ed. 1, II, 1034
(1753).
Populus balsamifera Miller, Gard. Dict., ed. 8, No. 5 (1768),
in part; not Linn. (1753).
Populus Caroliniensis Fougeroux, in Mém. Agr. Par., 90
(1786).
Populus Athemsay Hort. ex Fouger., l. c., 96 (1786).
Populus Athemse Hort. ex Fouger., l. c. (1786).
Populus heterophilla Fouger., l. c. (1786).
Populus cordifolia Burgsdorf, Anleit. Anpfl., ed. 1, 177
(1787).

Populus heterophylla Linn.—Continued.

SYN.—*Populus argentea* Michaux f., Hist. Arb. Am., III, 290, t. 9
(1813).
Populus heterophylla β *argentea* Wesmæl, in de Candolle,
Prodr., XVI, sect. II, 326 (1868).

COMMON NAMES.

River Cottonwood (R. I., Miss., La., Ohio).
Swamp Cottonwood (S. C., Miss., Del.).
Black Cottonwood (Ala.).
Cottonwood (N. Y., Va., W. Va., N. C., S. C., Miss.).
Downy Poplar (Tenn., Ala., Ark.).
Swamp Poplar (N. J.).
Cotton Tree (N. C.).
Balm of Gilead (N. Y.).
Liar (La.).
Langues de femmes (La.).

Populus balsamifera Linn.   **Balsam.**

SYN.—*Populus balsamifera* Linnæus, Spec. Pl., ed. 1, II, 1034 (1753).
Populus Tacamahaca Miller, Gard. Dict., ed. 8, No. 6 (1768).
Populus balsamifera lanceolata Marshall, Arb. Am., 108
(1785).
Populus Lindleyana Booth ex Loudon, Arb. Frut., IV, 2651
(1838).
Populus pseudo-balsamifera Turcz., in Bull. Soc. Nat. Mosc.,
I, 101 (1838).
Populus Simonii Carrière, in Rev. Hort., 360 (1867).
Populus balsamifera α *genuina* Wesmæl, in de Candolle,
Prodr., XVI, sect. II, 329 (1868).
Populus balsamifera a *suaveolens* α *pendula* Dippel, Handb.
Laubh., zw. T., 207 (1892), not *P. pendula* Burgsd. (1787).

COMMON NAMES.

Balsam (N. H., N. Y., Wis., Mich., Minn., Nebr., Mont., Ohio,
Ont.).
Balm of Gilead (Me., N. H., Vt., Mass., R. I., Conn., N. Y.,
Mich., Nebr., Minn., N. Dak., Ont.).
Cottonwood (Idaho).
Poplar (Wis., Minn.).
Balsam Poplar (N. H., Vt., Nebr., Minn.).
Tacamahac (Minn.).
Baumier (Quebec).
Rough-barked Poplar (Hudson Bay region).

Populus balsamifera candicans (Ait.) Gray.   **Balm of Gilead.**

SYN.—*Populus candicans* Aiton, Hort. Kew, ed. 1, III, 406 (1789).
Populus Caroliniensis Borkhausen, Vers. Forstb. Holz. (1790),
not Moench (1785).
Populus latifolia Moench, Meth., 338 (1794).
Populus Ontariensis Desfontaines, Cat. Hort. Reg. Par.
(1829), and in Hort.
Populus macrophylla Lindley, in Loudon, Enc. Pl., 840 (1829),
and in Hort.
Populus balsamifera var. *candicans* Gray, Man. Bot. N. U.
S., ed. 2, 419 (1856), not *P. candicans* Loud. (1838).
Populus Acladesca ex Koch, in Wochenschr., VIII, 238
(1865), not Loud. (1829).
Populus heterophylla Hort. ex. Koch, Dendrol., zw. Th. erst.
Ab., 496 (1872), not Linn. (1753).
Populus tristis ex Koch, l. c. (1872).
Populus candicans a *elongata* Dippel, Handb. Laubh., zw.
T., 204 (1892).
Populus balsamifera var. *Canadensis* (Moench) Sudworth,
in Bull. Torr. Bot. Club, XX, 44 (1893).
Populus balsamifera Jackson, Index Kew, III, 605 (1894),
not Linn. (1753).

COMMON NAMES.

Balm of Gilead (Me., N. H., Vt., Mass., N. Dak., Minn., Ont.).
Balsam (Mich.; Md. and Va.—cult.).

VARIETIES DISTINGUISHED IN CULTIVATION.

Populus balsamifera intermedia[1] Loud.

SYN.—*Populus balsamifera intermedia* London, Arb. Frut., III, 1674
(1838), not Loud., l. c., 1640 (1838); Pallas, Fl. Ross., I,
Par. II, t. XLI, A (as *P. balsamifera*) (1784).
Populus laurifolia Ledebour, Fl. Alt., IV, 297 (1833).
Populus Sibirica pyramidalis Green, in Bull. 9, Un. Minn.
Ag. Ex. Stn., 42 (1889); Hort. ex Bailey, in Bull. 68, Corn.
Un. Ag. Exp. Stn., 218 (1894); not *P. pyramidalis* Salisb.
(1796).
Populus balsamifera c *oblongata* Dippel, Handb. Laubh., zw.
T., 208 (1892).

Populus balsamifera viminalis[1] Loud.

SYN.—*Populus viminalis* Loddiges Cat. ed. (1836) ex Loudon, Arb.
Frut., III, 1673 (1838).

---

[1]The status of these forms is far from satisfactory, as their origin is exceedingly
obscure, being deeply involved by long cultivation and possible hybridization. The
exact relationship indicated to *Populus balsamea* may even be questioned, although
so far as can be determined at present, this relationship seems probable. They are,
therefore, introduced here mainly for the purpose of approximating as nearly as possible a full history of these cultivated forms, with the understanding that future
study may necessitate some changes at least in the present synonymy.

Populus balsamifera viminalis Loud.—Continued.

SYN.—*Populus balsamifera* var. *viminalis* Loud., l. c. (1838); Enc. Trees, 830, t. 1510 (1842).

Populus Longifolia Fischer ex Loud., l. c. (1838); Pallas, Fl. Ross, I, Par. II. t. XLI. B (as *P. balsamifera*) (1784).

Populus salicifolia Hort. ex Loud., l. c. (1838).

Populus viminea Bon Jard. 1845, 565 ex Wesmæl, in de Candolle, Prodr., XVI, sect. 2, 330 (1868), not Du Mont. Cours. (1811).

Populus Lindleyana Booth. in Rev. Hort. 1867, 380 (1867).

Populus crispa Hort. ex Dippel. Handb. Laubh., zw. T., 209 (1892), not *P. canad. crispa* Dipp. l. c. 200 (1892).

Populus Dudleyi Green, in Bull. 9, Un. Minn. Ag. Ex. Stn., 42 (1889); Hort. ex Bailey, in Bull. 68, Corn. Un. Ag. Exp. Stn., 218 (1894).

Populus pyramidalis suaveolens Hort. ex Bailey, l. c. (1894), not *P. bal. suaveolens* Loud. (1838).

Populus laurifolia Hort. ex Dippel. Handb. Laubh., zw. T., 209 (1892), not Ledebour (1833), nor Turzan ex Dippel, l. c., 208 (1892).

Populus balsamifera latifolia[1] (Moench) Loud.

SYN.—*Populus latifolia* Moench, Meth., 328 (1794).

Populus balsamifera var. *latifolia* Loudon, Enc. Trees, 830 (1842).

Populus balsamifera b *Wobstii* Regel, Russ. Dend. 2 Aufl., 151 (1876?), not Linn. (1753).

Populus Wobsky Green, in Bull. 9, Un. Minn. Ag. Ex. Stn., 41, 42 (1889), Hort. ex Bailey, in Bull. 68, Corn. Un. Ag. Exp. Stn., 218 (1894).

Populus Wobstii Schröder ex Dippel, Handb. Laubh., zw. T., 207 (1892).

Populus acuminata[2] Rydberg. **Lanceleaf Cottonwood.**

SYN.—*Populus acuminata* n. sp. Rydberg, in Bull. Torr. Bot. Club, XX. 50. pl. 141 (1893).

Populus angustifolia James. **Narrowleaf Cottonwood.**

SYN.—*Populus angustifolia* James. in Bot. Long's Exped., I, 497 (1823).

Populus salicifolia Rafinesque, Alsograph. Am., 43 (1838).

Populus Canadensis γ *angustifolia* Wesmæl, in de Candolle, Prodr., XVI, sect. II, 329 (1868).

Populus balsamifera var. *angustifolia* Watson, in King's Rep., V, 327 (1871).

---

[1] See page 131.

[2] A new and little known species detected in the Black Hills, South Dakota, western Nebraska to the Rocky Mountains of Colorado.

133

Populus angustifolia James—Continued.

SYN.—*Populus balsamifera* γ *laurifolia* Kuntze, Rev. Gen. Pl.,
Par. II, 642 (1891), not Westm. (1868).
Populus atheniensis × *balsamifera* Kuntze, l. c., 643 (1891).

COMMON NAMES.

Black Cottonwood (N. Mex., Utah, Colo.).
Narrow-leaved Cottonwood (Colo., Utah).
Narrow-leaved Poplar (Mont., Utah).
Balsam (Mont.).
Cottonwood (Idaho, Colo.).
Willow Cottonwood (Idaho).
Bitter Cottonwood (Idaho, Iowa).
Willow-leaved Cottonwood (Mont.).

Populus trichocarpa Torr. & Gr.   **Black Cottonwood.**

SYN.—*Populus balsamifera* γ Hooker, Fl. Bor.-Am., II, 154 (1839).
Populus trichocarpa Torrey & Gray, in Hooker, Icon., X,
878 (1852).
Populus angustifolia Newberry, in Pacific R. R. Rep., VI,
Pt. III, 89 (1857), not James (1823).
Populus balsamifera Lyall, in Journ. Linn. Soc., VII, 134
(1864), not Linn. (1753).
Populus balsamifera var.? *Californica* Watson, in Am. Journ.
Sci., ser. 3, XV, 135 (1878).
Populus trichocarpa var. *cupulata* Watson, l. c., 136 (1878).

COMMON NAMES.

Black Cottonwood (Idaho).
Cottonwood (Idaho, Oreg., Cal.).
Balsam Cottonwood (Idaho).
Balm (Oreg.).
Balm Cottonwood (Cal.).

Populus deltoides[1] Marsh.   **Cottonwood.**

SYN.—*Populus heterophylla* Du Roi, Harbk. Baumz., II, 150 (1772),
not Linn. (1753).
Populus deltoide Marshall, Arb. Am., 106 (1785).
Populus nigra Marsh., l. c., 107 (1785), not Linn. (1753).
Populus Carolinensis Moench, Verz. Weiss., 81 (1785).
Populus Canadensis Moench, l. c. (1785).
Populus Virginiana Fougeroux, in Mém. Agric. Par., 87
(1786).

---

[1]Marshall's original spelling "*deltoide*" was corrected to *deltoides* (Sudworth, in
Bull. Torr. Bot. Club, XX, 43, 1893). The correction to "*deltoidea*" (suggested by Professor Sargent, l. c.) seems unlikely, being an uncommon adjective form.

Populus deltoides Marsh.—Continued.

Syn.—POPULUS MONILIFERA Aiton, Hort. Kew ed. 1, III, 406 (1789).

Populus lævigata Aiton, l. c. (1789).

Populus angulata Aiton, l. c., 407 (1789).

Populus nigra β Virginiana Castiglioni, Viag. Stati Uniti, II, 334 (1790).

Populus glandulosa Moench, l. c., 339 (1794).

Populus dilatata β Caroliniensis Willdenow, Berl. Baumz., ed. 1, 230 (1796).

Populus angulosa Michaux, Fl. Bor.-Am., II, 243 (1803).

Populus nigra B *Helvetica* Poiret, in Lamarck, Enc. Méth. Bot., V, 234 (1804).

Populus Marylandica Poir., l. c., Suppl., IV, 378 (1816).

Populus Acladesca Lindley, in Loudon Enc. Pl., 840 (1829).

Populus macrophylla Loddiges Cat. ed. (1836) ex Loudon, Arb. Frut., III, 1671 (1838), not Lindl. in Loud. (1829).

Populus a 3 Medusæ Loud., l. c. (1838).

Populus a 2 nova Audibert ex Loud., l. c. (1838).

Populus Medusæ Booth ex Loud., l. c., IV, 2651 (1838).

Populus serotina Hartig, Forst. Cult. Deutschl., 437 (1851).

Populus nova Audibert ex Baxt. in Loudon, Hort. Brit. Suppl., II, 660 (1839).

Populus neglecta Hort. ex Wesmæl, in de Candolle, Prodr., XVI, sect. 2, 329 (1868).

Populus Canadensis β discolor Wesm., l. c. (1868).

Populus silicica Kotschy ex Wesm., l. c., XVII, sect. 2, 325 (1873).

Populus Petrowskiana Schrœder, Regel. Russ. Dend., 149 (1876?).—?

Populus canadensis a Petrowskiana[1] Dippel, Handb. Laubh., zw. T., 200 (1892).—?

Populus angulata a serotina Dippel, l. c. (1892).

Populus Carolina Hort. ex Bailey, in Bull. 68, Corn. Un. Ag. Ex. Stn., 225 (1894).

Populus deltoidea[2] Marshall ex Sargent, Silva, IX, 179, t. CCCCXCIV, CCCCXCV (1896).

COMMON NAMES.

Cottonwood (N. H., Vt., Mass., R. I., N. Y., N. J., W. Va., N. C., Ala., Fla., Miss., La., Tex., Cal., Ky., Mo., Ill., Wis., Kans., Nebr., Iowa, Minn., Mich., Ohio, Ont., Colo., Mont., N. Dak., S. Dak.).

Big Cottonwood (Miss., Nebr.).

---

[1] Cultivated in some European gardens as a distinct variety.

[2] See footnote p. 133.

Populus deltoides Marsh.—Continued.

SYN.—Yellow Cottonwood (Ark., Iowa, Nebr.).
Cotton Tree (N. Y.).
Carolina Poplar (Pa., Miss., La., N. Mex., Ind., Ohio).
Necklace Poplar (Tex., Colo.).
Vermont Poplar (Vt.).
White Wood (Iowa).
Broad-leaved Cottonwood (Colo.).

VARIETIES DISTINGUISHED IN CULTIVATION.

Populus deltoides aurea (Nichol.) nom. nov.
Goldenleaf Cottonwood.

SYN.—*Populus monilifera aurea* Nicholson, Dic. Gard., III, 200
(1887).
Populus canadensis aurea Nichol., l. c. (1887).
Populus Van Geertii ex Bailey, in Bull. 68, Corn. Un. Ag.
Exp. Stn., 226 (1894).

Populus deltoides erecta (Selys) nom. nov.
SYN.—*Populus monilifera erecta* Selys, in Bull. Soc. Bot. Belg., III,
11 (1864).
Populus canadensis b *erecta* Dippel, Handb. Laubh., zw. T.,
200 (1892).

Populus deltoides crispa (Dippel) nom. nov.
SYN.—*Populus Lindleyana* Hort. ex Steudel, Nom. Bot. ed. sec., II,
381 (1841)?—not Booth ex Loud. (1838).
Populus canadensis c *crispa* Dippel, Handb. Laubh., zw. T.,
200 (1892).
Populus monilifera Lindleyana ex Dippel, l. c. (1892).

Populus fremontii Wats.   **Fremont Cottonwood.**
SYN.—*Populus monilifera* Torrey, in Sitgreaves's Rep., 172 (1853),
not Ait. (1789).
Populus Canadensis Wesmæl, in de Candolle, Prodr., XVI,
sect. 2, 329 (1868), in part; not Mœnch (1785).
Populus Fremontii Watson, in Proc. Am. Acad. Sci., X, 350
(1875).
Populus Fremontii var. ? *Wislizeni* Watson, in Am. Journ.
Sci., ser. 3, XV, 137 (1878).
Populus Canadensis var. *Fremontii* Kuntze (Wats.), Rev.
Gen. Pl. Par., II, 643 (1891).

COMMON NAMES.

Cottonwood (Cal., Utah).
White Cottonwood (N. Mex.).

Populus alba[1] Linn.   **White Poplar.**

Syn.—*Populus alba* Linnæus, Spec. Pl., ed. 1, II, 1034 (1753).
Populus nigra Miller, Gard. Dict., ed. 8, No. 3 (1768), not
   Linn. (1753).
Populus major Miller, l. c., No. 4 (1768).
Populus excelsa Salisbury, Prodr., 395 (1796).
Populus viminalis Hort. Par. ex Poiret, in Lamarck, Enc.
   Méth. Bot., V, 238 (1805).
Populus ⹂Egyptiaca Hort. ex Baxt., in Loudon, Hort. Brit.,
   Suppl., II, 660 (1832).
Populus Bachofenii Wierzb. ex Reichenbach, Ic. Fl. Germ.,
   XXI, 29, t. 616 (1834).
Populus hybrida Reichenb., l. c., 29 (1834).
Populus intermedia Mertens ex Loudon, Arb. Frut., III, 1640
   (1838).
Populus grisea Loddiges Cat. ed. (1836) ex Loud., l. c. (1838).
Populus alba belgica Loud., l. c. (1838).
Populus belgica Loddiges ex Loud., l. c. (1838).
Populus alba ægyptiaca Hort. ex Loud., l. c. (1838).
Populus alba pallida Hort. ex Loud., l. c. (1838).

Populus alba nivea[2] (Willd.) Loud.   **Snowy Poplar.**

Syn.—*Populus nivea* Willdenow, Berl. Baumz., ed. 1, 227 (1796).
Populus alba acerifolia Loudon, Arb. Frut., III, 1640 (1838).
Populus palmata Hort. ex Loud., l. c. (1838).
Populus alba arambergica Loud., l. c. (1838).
Populus arambergica Loddiges ex Loud., l. c. (1838).
Populus alba nivea Loud., l. c. (1838).
Populus acerifolia Loddiges ex Loud., l. c. (1838).
Populus quercifolia Hort. ex Loud., l. c. (1838).
Populus nivea Salomonii Carrière, in Rev. Hort. 1867, 340
   (1867).
Populus argentea Koch, Dendrol., zw. Th. erst. Ab., 484
   (1872), not Fouger. (1786).
Populus Arembergiana Koch, l. c. (1872).
Populus argentea vera Hort. ex. Dippel, Handb. Laubh., zw.
   T., 191 (1892).
Populus alba macrophylla Hort. ex Dippel, l. c. (1892), not
   P. macrophylla Lodd. ex Loud. (1838).
Populus alba nivea aureo-intertexta Dippel, l. c. (1892).

---

[1]The following European species must now be regarded as naturalized in this
country, having become thoroughly established in many localities, especially in the
Middle and South Atlantic region. A full synonymy is also given for the principal
varieties and subvarieties found in cultivation.

[2]"Silver Maple" is a very common local name erroneously applied to this form, on
account of the remote resemblance in form of its leaves to the Silver Maple (*Acer
saccharinum* L.).

Populus alba canescens (Smith) Loud.   **Silver Poplar.**

SYN.—*Populus canescens* Smith, Fl. Brit., III, 1080 (1805).
Populus alba candicans London. Arb. Frut., III, 1640 (1838),
not *P. candicans* Ait. (1789).
Populus candicans Loddiges ex Loud., l. c. (1838), not Ait.
(1789).
Populus tomentosa ex Loud., l. c. (1838).
Populus alba var. *canescens* London, Enc. Trees, 820 (1842).

Populus alba canescens umbraculifera nom. nov.
    Weeping Silver Poplar.

SYN.—*Populus pendula* Hort. ex Steudel, Nom. Bot., ed. sec., II,
381 (1841), not Burgsd. (1787), nor Hort. ex Tausch (1838).
Populus canescens a *pendula* Dippel, Handb. Laubh., zw. T.,
192 (1892), not *P. pendula* Burgsd. (1787).

Populus alba bolleana Lauche.   **Bolle Poplar.**

SYN.—*Populus Bolleana* Lauche, in Woch. Deuts. Gart., No. 32,
Aug. 10 (1878).
Populus Bolleand Hadkinson, in Ann. Rep. Nebr. Hort. Soc.
1889, 125 (1889).
Populus alba b *pyramidalis* Dippel, Handb. Laubh., zw. T.,
191 (1892), not *P. pyramid.* Salisb. (1796).

Populus alba nutans nom. nov.   **Weeping White Poplar.**

SYN.—*Populus alba pendula* Dippel, Handb. Laubh., zw. T., 191
(1892), not *P. pendula* Burgsd. (1787), nor Koch (1872).

Populus alba globosa Dippel.   **Roundtop Poplar.**

SYN.—*Populus alba* d *globosa* Dippel, Handb. Laubh., zw. T., 191
(1892).

Populus nigra Linn.   **Black Poplar.**

SYN.—*Populus nigra* Linnæus, Spec. Pl., ed. 1, II, 1034 (1753).
Populus versicolor Salisbury, Prodr., 395 (1796).
Populus rubra Hort. ex Poiret, in Lamarck, Enc. Méth. Bot.,
V, 239 (1804).
Populus flexibilis Rozier, Cours Comp. Dic., VII, 618 (1781–
1805).
Populus viminea Du Mont de Courset, Bot. Cult., ed. 2, VI,
401 (1811).
Populus Hudsonica Michaux f., Hist. Arb. Am., II, 619 (1812).
Populus betulifolia Pursh, Fl. Am. Sept., II, 619 (1814).
Populus neopolitana Tenore, Fl. Nap., V, 279 (1836).
Populus caudina Tenore, l. c., 280 (1836).
Populus nigra viridis Lindl. ex Loudon, Arb. Frut., III, 1652
(1838).

Populus nigra Linn.—Continued.

SYN.—*Populus viridis* Loddiges Cat. ed. (1836) ex Loud., l. c. (1838).
Populus nigra salicifolia Loud., l. c. (1838).
Populus salicifolia Loddiges Cat. ed. (1836) ex Loud., l. c. (1838).
Populus nigra γ betulæfolia Wesmæl, in A. de Candolle, Prodr., XVI, sect. 2, 328 (1868).
Populus vistulensis Hort. ex Loudon, Enc. Trees, 824 (1842).
Populus flexilis Koch, Dendrol., zw. Th. erst. Ab., 491 (1872).
Populus Eugenei Koch, l. c., 493 (1872).
Populus Eugenie Hort. ex Bailey, in Bull. 68, Corn. Un. Ag. Exp. Stn., 226 (1894).

Populus nigra italica Du Roi. **Lombardy Poplar.**

SYN.—*Populus nigra* var. *Italica* Du Roi, Harb. Baumz., II. 141 (1772).
Populus Italica Du Roi, l. c. (1772).
Populus fastigiata Fougeroux, in Mém. Ag. Par., 82 (1786).
Populus dilatata Aiton, Hort. Kew, ed. 1, III, 406 (1789).
Populus pyramidata Moench, Meth., 339 (1794).
Populus pyramidalis Salisbury, Prodr., 395 (1796).
Populus Pannonica Kitaibel, in Besser, En. Pl. Pod., 38 (1822).
Populus Lombardica Hort. ex Link, Handb., II, 456 (1831).
Populus croatica Waldst. & Kitaibel, in Flora, XV, II; Beibl., I, 14 (1832).
Populus nigra pyramidalis Spach, in Ann. Sc. Nat., sér. 2, XV, 31 (1841).
Populus Polonica Hort. ex Loudon, Enc. Trees, 828 (1842).

Populus[1] nigra elegans Bailey.

SYN.—*Populus nigra* var. *elegans* Bailey, in Bull. 68, Corn. Un. Ag. Exp. Stn., 227 (1894).
Populus elegans Hort. ex Bailey, l. c. (1894).

[1]The following species is of interest as a discovery lately made in Lower California by Mr. T. S. Brandegee. So far it has not been detected within the borders of the United States. As now known it attains a height of 20 to 90 feet and occurs only in mountainous regions, sometimes at 5,000 feet elevation.

Populus monticola Brandegee. **Mountain Poplar (Guerigo).**

SYN.—*Populus monticola* Brandegee, in Zoe, I, 274 (1890); in Gard. and For., IV, 24, 330, f. 56 (1891); VII, 313, f. 51 (1894).

Family BETULACEÆ.

BETULA Linn., Spec. Pl., 982 (1753).

Betula populifolia Marsh.   **White Birch.**

SYN.—*Betula lenta* Du Roi, Harbk. Baumz., I, 92 (1771), not Linn.
(1753).
Betula populifolia Marshall, Arb. Am., 19 (1785).
Betula excelsa Canadensis Wangenheim, Nordam. Holz., 86
(1787).
Betula acuminata Ehrhart, Beitr., VI, 98 (1791).
Betula (a) populifolia 2 laciniata[1] Loudon, Arb. Frut., III,
1707 (1838).
Betula laciniata Loddiges, Cat. ed. (1836) ex Loud., l. c.
(1838,) not Blom (1786).
Betula (a) populifolia 3 pendula[1] Loud., l. c. (1838).
Betula pendula Loddiges Cat. ed. (1836) ex Loud., l. c. (1838),
not Roth (1788).
Betula alba β populifolia Spach, in Ann. Sc. Nat., sér. 2, XV,
187 (1841).
Betula alba subsp. *populifolia* Regel, in Bull. Soc. Nat. Mosc.,
XXXVIII, pt. II, 399 (1865).

COMMON NAMES.

White Birch (Vt., Mass., R. I., Conn., N. Y., N. J., Pa., Del.,
Ill., Mich., Ont.).
Gray Birch (Me., R. I., Mass.).
Old Field Birch.
Poverty Birch (Me.).
Poplar-leaved Birch.
Small White Birch (Vt.).

Betula populifolia × papyrifera[2] Sargent.

SYN.—*Betula populifolia × papyrifera* Sargent, in Gard. and For.,
VIII, f. 50 (1895).

VARIETIES DISTINGUISHED IN CULTIVATION.

Betula populifolia purpurea[3] Hort. Am.

---

[1] A Variety said to be cultivated in European gardens, but apparently unknown to American nurserymen.

[2] A hybrid supposed to be from *Betula populifolia* and *B. papyrifera*, detected by Mr. Walter Faxon at Arlington, Mass. Other similar trees were found in Plymouth and Warren, N. H.

[3] It is difficult to state when this form was first distinguished as a garden variety. It appeared in the general catalogue (p. 53) of Ellwanger & Barry, Rochester, N. Y., in 1892 with a short statement of its character.

Betula papyrifera Marsh.   **Paper Birch.**

Syn.—*Betula papyrifera* Marshall, Arb. Am., 19 (1785).

Betula lenta Wangenheim, Nordam. Holz., 45 (1787), not Linn. (1753), nor Du Roi (1771).

Betula papyracea Dryander in Aiton, Hort. Kew., ed. 1, III, 337 (1789).

Betula excelsa Dryander in Ait., l. c. (1789).—?

Betulus papyracea 2 fusca Loudon, Arb. Frut., III, 1708 (1838).

Betulus fusca Bosc ex Loud., l. c. (1838), not Pall. (1776).

Betulus papyracea 3 trichoclada Hort. ex Loud., l. c. (1838).

Betulus papyracea 4 platyphylla Hort. ex Loud., l. c. (1838).

Betula alba ε papyrifera Spach, in Ann. Sc. Nat., sér. 2, XV, 188 (1841).

Betula cordifolia Regel, in Nouv. Mém. Soc. Nat. Mosc., XIII, 86, t. 12, f. 29–36 (1860).

Betula occidentalis Lyall, in Journ. Linn. Soc., VII, 134 (1864), not Hook. (1840).

Betula alba subsp. *5 β commutata* Regel, in Bull. Soc. Nat. Mosc., XXXVIII, Pt. II, 401, t. 7, f. 6–10 (1865).

Betula alba subsp. *6 α communis* Regel, l. c. (1865).

Betula alba subsp. *6 β cordifolia* Regel, l. c. (excl. t.) (1865).

Betula Ermani Rothrock, in Smithsonian Rep. 1867, 454 (1868). not Cham. (1831).

Betula alba var. *populifolia* Winchell, in Ludlow's Rep. Black Hills, Dak., 67 (1875), not Spach (1841).

Betula papyrifera β cordifolia Regel, in A. de Candolle, Prodr., XVI, sect. 2, 166 (1868).

Betula pirifolia Hort. in Koch, Dendrol., zw. Th. erst. Ab. (1872).

Betula papyracea **a** *cordifolia* Dippel, Handb. Laubh., zw. T., 177 (1892).

Betula papyracea **b** *occidentalis* Dippel, l. c., 177, f. 84 (1892), not *B. occidentalis* Hook. (1839).

COMMON NAMES.

Paper Birch (N. H., Vt., Mass., R. I., Conn., N. Y., Wis., Mich.. Minn., Ont.).

Canoe Birch (Me., Vt., N. H., R. I., Mass., N. Y., Pa., Wis., Wis., Mich., Minn., Ont.).

White Birch (Me., N. H. Vt., R. I., N. Y., N. J., Wis., Mich., Minn., Nebr., Ont.).

Silver Birch (Minn.).

Large White Birch (Vt.).

Birch.

Boleau (Quebec).

141

Betula papyrifera minor (Tuck.) Wats. & Coult.

Alpine Paper Birch.

Syn.—*Betula papyracea β minor* Tuckerman, in Am. Journ. Sci.,
XLV, 31 (1843).

Betula papyrifera var. *minor* Watson & Coulter, in Gray,
Man. Bot. N. U. S., ed. 6, 472 (1890).

Betula occidentalis Hook.

Cañon Birch.

Syn.—*Betula occidentalis* Hooker, Fl. Bor.-Am., II, 155 (1839).

Betula alba subsp. 5 *occidentalis α typica* Regel, in Bull.
Soc. Nat. Mosc., XXXVIII, Pt. II, 400, t. 7, f. 1–5 (1865).

COMMON NAMES.

Black Birch (Cal., Colo., Mont., Utah).
Cañon Birch (Utah).
Sweet Birch (Idaho).
Cherry Birch (Idaho).
Gray Birch (Mont.).
Water Birch (Colo.).
Western Birch.

Betula nigra Linn.

River Birch.

Syn.—*Betula nigra* Linnæus, Spec. Pl., ed. 1, II, 982 (1753).

Betula lanulosa Michaux, Fl. Bor.-Am., II, 181 (1803).

Betula rubra Michaux f., Hist. Arb. Am., II, 143, t. 3 (1812).

Betula angulata Loddiges Cat. ed. (1836) ex Steudel, Nom.
Bot., ed. sec., I, 201 (1840).

COMMON NAMES.

Red Birch (Mass., R. I., N. Y., N. J., Pa., Del., N. C., S. C.,
La., Mo., Ill., Wis., Kans., Nebr., Ohio).
River Birch (Mass., R. I., N. J., Del., Pa., W. Va., Ala.,
Miss., Tex., Mo., Ill., Wis., Ohio).
Water Birch (W. Va., Kans.).
Blue Birch (Ark.).
Black Birch (Fla., Tenn., Tex., Iowa).
Birch (N. C., S. C., Miss., La., Iowa).

Betula lutea Michx. f.

Yellow Birch.

Syn.—*Betula lutea* Michaux f., Hist. Arb. Am., II, 152, t. 5 (1812).

Betula excelsa Pursh, Fl. Am. Sept., II, 621 (1814), not
Dryander in Ait. (1789), nor Willd. (1805).

Betula lenta α genuina Regel, in Nouv. Mém. Soc. Nat.
Mosc., XIII, 125 (1860), in part.

Betula lenta β lutea Regel, in de Candolle, Prodr., XVI,
sect. II, 179 (1868).

Betula lutea Michx. f.—Continued.

Syn.—*Betula persicifolia* Hort. ex Koch, Dendrol., zw. Th. erst. Ab., 641 (1872).

Betula lutea a *persicifolia* Dippel, Hand. Laubh., zw. T., 185 (1892).

COMMON NAMES.

Yellow Birch (Me., N. H., Vt., Mass., Conn., R. I., N. Y., N. J., Pa., N. C., S. C., Ill., Mich., Minn., N. Dak., Ont.).
Gray Birch (Vt., R. I., Pa., Mich., Minn.).
Swamp Birch (Minn.).
Silver Birch (N. H.).
Merisier (Quebec).
Merisier Rouge (Quebec).

Betula lenta[1] Linn.   **Sweet Birch.**

Syn.—*Betula lenta* Linnæus, Spec. Pl., ed. 1, II, 983 (1753).
Betula nigra Du Roi, Obs., 30 (1771), not Linn. (1753).
Betula carpinifolia Ehrhart, Beitr., VI, 99 (1791).

COMMON NAMES.

Sweet Birch (Me., N. H., Vt., Mass., R. I., N. Y., N. J., Pa., Del., N. C., S. C., Ala., Mich., Minn.).
Birch (N. C.).
Black Birch (N. H., Vt., Mass., R. I., Conn., N. Y., N. J., Pa., W. Va., S. C., Ga., Ill., Ind., Mich., Ohio).
Cherry Birch (N. H., R. I., N. Y., Pa., Va., Del., N. C., S. C., Fla., Wis., Mich., Ont.).
River Birch (Minn.).

---

[1] The following shrubs complete the list of native species of this genus:

Betula pumila Linn.

Syn.—*Betula pumila* Linnæus, Mantiss., 124 (1767).
Betula Grayi Regel, in Bull. Soc. Nat. Mosc., XXXVIII, Pt. II, 406, t. 6, f. 9–13 (1865).

Betula pumilia × lenta Jack.

Syn.—*Betula pumila × lenta* Jack, in Gard. & For., VIII, 243, f. 36 (1895).

Betula nana Linn.

Syn.—*Betula nana* Linnæus, Spec. Pl., ed. 1, II, 983 (1753).

Betula nana flabellifolia Hook.

Syn.—*Betula nana* var. *flabellifolia* Hooker, Fl. Bor.-Am., II, 156 (1839).

Betula glandulosa Michx.

Syn.—*Betula glandulosa* Michaux, Fl. Bor.-Am., II, 180 (1803).
Betula nana Hooker, Fl. Bor.-Am., II, 156 (1839), not Linn. (1753).
Betula Littelliana Tuckerman, in Am. Journ. Sci., XLV, 31 (1843).

Betula lenta Linn.—Continued.

SYN.—Mahogany Birch (N. C., S. C.).
Mountain Mahogany (S. C.).

ALNUS Ehrhart, Beitr., III, 22 (1788).

Alnus maritima (Marsh.) Muehl.   **Seaside Alder.**

SYN.—*Betula-Alnus maritima* Marshall, Arb. Am., 20 (1785).
Alnus oblongata Regel, in Nouv. Mém. Soc. Nat. Mosc., XIII,
171, t. VI, f. 3-9 (1860), not Mill. (1768), nor Willd. (1805).
Alnus maritima α typica Regel, in Bull. Soc. Nat. Mosc.,
XXXVIII, Pt. II, 427 (1865).
Alnus maritima Muhlenberg, Mss. in Nuttall, Sylva, I, 34,
t. 10 (1842).

COMMON NAMES.

Seaside Alder (Del., Ala.).
Alder (Del.).

Alnus acuminata[1] H., B. K.   **Lanceleaf Alder.**

SYN.—*Alnus acuminata* Humboldt, Bonpland & Kunth, Nov. Gen.
Sp., II, 20 (1817).
ALNUS OBLONGIFOLIA Torrey, in Bot. Mex. Bound. Surv.,
204 (1859).
Alnus acuminata α genuina Regel, in Nouv. Mém. Soc. Nat.
Mosc., XIII, 147 (1860).
Alnus serrulata γ oblongifolia Regel, in de Candolle, Prodr.,
XVI, sect. II, 188 (1868).
Alnus rhombifolia Parry, in Bull. Cal. Acad. Sci., II, 351
(1887), in part; not Nutt. (1842).
Alnus Jorullensis var. *acuminata* Kuntze, Rev. Gen. Pl.,
Par. II, 638 (1891).

Alnus rhombifolia Nutt.   **Mountain Alder.**

SYN.—*Alnus rhombifolia* Nuttall, Sylva, I, 33 (1842).
Alnus oblongifolia Torrey ex Watson, Bot. Cal., II, 80 (1880),
in part; not Torr. (1859).

COMMON NAMES.

Alder (Cal., Oreg.).
Western or California Alder (Idaho).
Mountain Alder.

---

[1] A little-known species, occurring within the borders of the United States in New Mexico and Arizona, ranging southward into Mexico and Peru.

Alnus tenuifolia Nutt.　　　　　　　　　　**Narrowleaf Alder.**

SYN.—*Alnus incana* β Hooker, Fl. Bor.-Am., II, 157 (1839).—?
Alnus tenuifolia Nuttall, Sylva, I, 32, t. 10 (1842).
Alnus incana α *glauca* Regel, in Nouv. Mém. Soc. Nat. Mosc., XIII, 154 (1860), in part.
Alnus serrulata β *rugosa* Regel, in Bull. Soc. Nat. Mosc., XXXVIII, Pt. II, 433 (1865).
ALNUS INCANA VAR. VIRESCENS Watson, Bot. Cal., II, 81 (1880).
Alnus rhombifolia Macoun, Cat. Canad. Pl., 438 (1883), not Nutt. (1842).—?
Alnus occidentalis Dippel, Handb. Laubh., zw. T., 158, f. 78 (1892).

Alnus oregona Nutt.　　　　　　　　　　　　**Red Alder.**

SYN.—*Alnus rubra* Bongard, in Mém. Acad. Pétersburg, sér. VI, II, 162 (1833), not *Betula-Alnus rubra* Marsh. (1785).
Alnus Oregona Nuttall, Sylva, I, 28, t. 9 (1842).
Alnus incana η *rubra* Regel, in Nouv. Mém. Soc. Nat. Mosc., XIII, 157, t. 17, f. 3–4 (1860).

COMMON NAMES.

Alder (Cal., Oreg.).
Red Alder (Cal., Oreg.).
Western or Red Alder.

Alnus glutinosa[1] (Linn.) Gærtn.　　　　　**European Alder.**

SYN.—*Betula Alnus* Linnæus, Spec. Pl., ed. 1, II, 983 (1753).—?
Betula Alnus α *glutinosa* Linn., l. c. (1753).
Betula glutinosa Lamarck, Enc. Bot. Méth., I, 454 (1783).
Alnus nigra Gilbert, Exer. Phyt., II, 401 (1792).
Alnus glutinosa Gærtner, Fruct., II, 54, t. 90 (1791).
Alnus communis Nouveau Duhamel, II, 212, t. 64 (1802).
Alnus glutinosa (*vulgaris*) Persoon, Syn. Pl., II, 550 (1807).
Alnus rotundifolia Stokes, Bot. Mat. Med., IV, 369 (1812).
Alnus elliptica Requien, in Ann. Sc. Nat., sér. 1, V, 381 (1825).
Alnus macrocarpa Loddiges ex Loudon, Hort. Brit., 378 (1830).
Alnus denticulata Meyer, Verz. Pfl. Cauc., 43 (1831).
Alnus barbata Meyer, l. c. (1831).

---

[1] This alder, which is a native of Europe and northern Asia, has become perfectly naturalized in many localities in the United States, spreading and mingling with other native species in an apparently wild state. An example of this may be seen in the vicinity of Chicago, Ill.

Alnus glutinosa (Linn.) Gærtn.—Continued.

SYN.—*Alnus emarginata* Krock. ex Steudel, Nom. Bot., ed sec., I,
55 (1840).
Alnus glutinosa β subrotunda Spach, in Ann. Sc. Nat., sér. 2,
XV, 207 (1841).
Alnus nitens Koch, in Linnæa, XXII, 334 (1849).
Alnus Morisiana Bertoloni, Flor. Ital., X, 163 (1854).
Alnus glutinosa imperialis asplenifolia Verschaff., Cat. Ill.
Hort., VI, 97 (1859).
Alnus cerifera Hartig ex Regel, in de Candolle, Prodr.,
XVI, sect. 2, 187 (1868).
Alnus dubia Req. ex Regel, l. c. (1868).
Alnus Februaria Kuntze, Tasch. Flor. Leipz., 238 (1867).
Alnus prunifolia Hort. ex Koch, Dendrol., zw. Th. erst. Ab.,
630 (1872).
Alnus suaveolens Mor. ex Nyman, Consp. Flor. Eu., 671
(1878), not Req. (1825).
Alnus glutinosa var. *autumnalis* Kuntze, Revis. Gen. Pl., Par.
II, 638 (1891).
Alnus glutinosa a *denticulata*[1] Dippel, Handb. Laubh., zw.
T., 160 (1892).
Alnus intermedia Hort. ex Dippel, l. c. (1892).
Alnus oblongata Hort. ex Dippel, l. c. (1892).
Alnus glutinosa b *rubrinerva* Hort. ex Dippel, l. c. (1892).
Alnus glutinosa d *pyramidalis* (Birkiana) Hort. ex Dippel,
l. c. (1892).
Alnus glutinosa g *sorbifolia*[1] Hort. ex Dippel, l. c. (1892).
Alnus glutinosa h *laciniata α imperialis*[1] Desfossé ex Dippel,
l. c. (1892).
Alnus glutinosa imperialis Hort. ex Dippel, l. c. (1892).
Alnus imperialis Hort. ex Dippel, l. c. (1892).

PRINCIPAL VARIETIES DISTINGUISHED IN CULTIVATION.

Alnus glutinosa quercifolia Willd.

SYN.—*Alnus quercifolia* Willdenow, Berl. Baumz., ed. 1, 45 (**1796**).
Alnus glutinosa δ quercifolia Willdenow, Sp. Pl., IV, 335
(1805).

Alnus glutinosa laciniata (Ehr.) Willd.

SYN —*Alnus laciniata* Ehrhart, Beitr., III, 22 (1788).
Alnus glutinosa γ laciniata Willdenow, Sp. Pl., IV, 355
(1805).

---

[1] Cultivated in Europe as distinct garden varieties.

Alnus glutinosa laciniata (Ehr.) Willd.—Continued.

SYN.—*Alnus glutinosa incisa* Hort. ex Loudon, Arb. Frut., III,
1678 (1838), not Willd. (1805).
Alnus glutinosa pinnatifida Spach, in Ann. Sc. Nat., sér. 2,
207 (1841).

Alnus glutinosa incisa Willd.

SYN.—*Alnus glutinosa β incisa* Willdenow, Sp. Pl., IV, 335 (1805).
Alnus oxyacanthifolia Loddiges, Cat. ed. (1836)—*nomen
nudum*.
Alnus glutinosa oxyacanthifolia Loudon, Arb. Frut., IV, 1678
(1838).

Alnus glutinosa aurea (Koch) Nichol.

SYN.—*Alnus aurea* Hort. ex Koch, Dendrol., zw. Th. erst. Ab., 630
(1872).
Alnus glutinosa aurea Nicholson,[1] Dic. Gard., I, 50 (1885?);
Dippel, Handb. Laubh., zw. T., 160 (1892).

OSTRYA Scopoli, Fl. Carn., ed. 2, 243 (1772).

Ostrya virginiana (Mill.) Koch.   **Hop Hornbeam.**

SYN.—*Carpinus Ostrya* Linnæus, Spec. Pl., ed. 1, II, 998 (1753),
in part.
Carpinus Virginiana Miller, Gard. Dict., ed. 8, No. 4 (1768).
Carpinus Virginica Muenchhausen, Hausv., V, 120 (1770).
Carpinus triflora Moench, Meth., 394 (1794).
Carpinus Ostrya Americana Michaux, Fl. Bor.-Am., II, 202
(1803).
Ostrya Virginica[2] Willdenow, Sp. Pl., IV, par. 1, 469 (1805).
Ostrya vulgaris Watson, Dendrol., 4, 143 (1825), not Willd.
(1805).
Ostrya Americana Michaux, Hist. Arb. Am., III, 54 (1813).
Zugilus Virginica Rafinesque, Fl. Ludovic., 159 (1817).
Ostrya Virginica α glandulosa Spach, in Ann. Sc. Nat., sér.
2, XVI, 246 (1841).
Ostrya Virginica β eglandulosa Spach, l. c. (1841).
Ostrya Virginiana[2] Koch, Dendrol., zw. Th. zw. Ab., 6
(1873).

---

[1] The date of Nicholson's name is approximated, as Volume I of the work in which
it appeared bears no date; but Volume III bears the date 1887, leaving no doubt
but that Nicholson is the first author of this combination.

[2] By consent of most American botanists specific names having the adjective endings *cus, ca, cum,* and *anus, ana, anum,* are to be considered identical names, according
to which Willdenow (l. c.) was the first to refer the specific name *Virginiana* to *Ostrya;*
Koch, however, was the first to take up Miller's identical form *Virginiana,* and is by
most German botanists cited as the author of *O. Virginiana.* It seems advisable to
maintain the latter position in the exact statement of facts.

Ostrya virginiana (Mill.) Koch.—Continued.

SYN.—*Ostrya Ostrya*[1] (Linn.) MacMillan, Metasperm. Minn. Valley, 187 (1892).

COMMON NAMES.

Hop Hornbeam (Vt., R. I., Mass., N. Y., N. J., Pa., Del., N. C., Ala., S. C., Tex., Ark., Ill., Ind., Wis., Kans., Nebr., Minn., S. Dak., Ohio).
Ironwood (R. I., N. Y., N. J., Pa., Del., W. Va., N. C., S. C., Ala., Tex., Ark., Ky., Ind., Ill., Wis., Mich., Iowa, Minn., Nebr., Ohio, Ont., S. Dak.).
Leverwood (Vt., Mass., R. I., N. Y., Pa., Kans.).
Hornbeam (R. I., N. Y., Fla., S. C., La.).
Hardhack (Vt.).

Ostrya knowltonii[2] Coville.   **Western Hop Hornbeam.**

SYN.—*Ostrya Knowltonii* Coville, in Gard. and For., VII, 114, f. 23 (1894).

CARPINUS Linn., Spec. Pl., 998 (1753).

Carpinus caroliniana Walt.   **Blue Beech.**

SYN.—*Carpinus Betulus Virginiana*[3] Marshall, Arb. Am., 25 (1785).
Carpinus Caroliniana Walter, Fl. Caroliniana, 236 (1788).
Carpinus Americana Michaux, Fl. Bor.-Am., II, 201 (1803).
Carpinus ostryoides Rafinesque, Med. Rep., II, 333 (1811).
Carpinus Virginiana Michaux, Hist. Arb. Am., III, 57, t. 8 (1813), not Mill. (1768).

---

[1] Professor Jackson (in Index Kewensis, I, 443, 1893) cites Linnæus's name (*Carpinus Ostrya*) as a synonym of *Ostrya Virginica*, according to which our plant should be known as *Ostrya Ostrya*; but so far as can be determined from Linnæus' description (l. c.) his *Carpinus Ostrya* appears to have been founded on two distinct species—the European and American species. Without a careful study and comparison of Linnæus's type specimen it seems unsafe to take up *Ostrya* as the oldest specific name for our plant.

[2] A new, little known, and rare species, discovered on September 10, 1889, by Mr. Frank H. Knowlton on the southern slope of the canyon of the Colorado River in Arizona, 70 miles north of Flagstaff, where the post-office of Tolfree is located. It is said to occur abundantly near the trail leading to the bottom of the canyon, at 6,000 to 7,000 feet elevation. It attains a height of 20 to 30 feet and 1 to 1½ feet in diameter. As yet no other species of the genus is known to occur farther west than eastern Texas, the southwestern limit of the common Hop Hornbeam (*Ostrya virginiana*).

[3] This is the oldest name applied to the Blue Beech, but in raising it to specific rank, a combination is produced identical with Miller's *Carpinus Virginiana* published in 1768, and cited now as a synonym of *Ostrya Virginiana*, thus precluding by preoccupancy the use of this form for the Blue Beech. The next oldest name is that of Walter.

Carpinus caroliniana Walt.—Continued.

COMMON NAMES.

Syn.—Blue Beech (N. H., Vt.. R. I., N. Y.. Pa.. Miss., Tex., Ky.,
Mich., Iowa. Minn.. Ohio, Ont., Nebr.).
Water Beech (R. I.. N. Y., Pa., Del., W. Va., Ill., Ind.,
Mich., Minn., Nebr.. Kans.. Ohio).
Hornbeam (Me., N. H.. Mass.. R. I., Conn., N. Y., N. J., Pa.,
Del., N. C., S. C.. Ala., Tex.. Ky.. Ill., Kans., Minn.).
Iron Wood (Me.. Vt.. Mass.. R. I., N. Y.. N. J.. Pa., Del.,
N. C.. S. C., Fla., Ala.. La., Tex.. Mo.. Wis.. Ill., Iowa,
Kans.. Minn.. Ohio, Ont.. Nova Scotia).
O-tan-tahr-te-weh = "A lean tree" (Indians. N. Y.).

Family FAGACEÆ.
FAGUS Linn., Spec. Pl.. 997 (1753).

Fagus latifolia[1] (Muenchh.) Lond.   **Beech.**

Syn.—*Fagus Americana latifolia* Muenchhausen, Hausv.. V, 162
(1770.
Fagus sylvatica c *Americana latifolia* Du Roi. Harbk. Baumz.,
I, 269 (1771).
Fagus Sylvatica atro-punicea Marshall. Arb. Am.. 46
(1785).
Fagus sylvatica Schoepf. Mat. Med. Am., 140 (1787). not
Linn. 1753).
Fagus purpurea Desfontanes. Tabl.. ed. 1. 214 (1804).
Fagus ferruginea Aiton. Hort. Kew. ed. 1, III. 362
(1789.
Fagus sylvestris Michaux f.. Hist. Arb. Am., II. 170. t. 8
1812 .

---

[1] The synonymy of this species is difficult and appears to be involved in some
uncertainty. Professor Sargent Silva, IX, 27, 1896) takes up *Fagus Americana* Sweet
(1826), which can be deduced from *F. Americana latifolia* Muenchh. (1770), or from
F. sylvatica c *Americana latifolia* Du Roi (1771) only by inference: since these polynomials only presuppose an older *F. americana*, which is as yet unknown. But
the whole synonymy of this species now stands. it seems quite impossible to derive
F. americana from either of the above polynomials. The oldest available term
by which our plant could be designated is the *latifolia* from Muenchhausen's
F. Americana latifolia. as it is only this part of the polynomial which can be taken
to indicate the rank and definition of the plant Muenchhausen described. We can
not assume that his variety *latifolia* (although it may have been) was taken as a
form of a previously published *F. americana*: the earliest and only tangible name
which we know of as representing our plant is *F. Americana latifolia*, which gives
us *Fagus latifolia*, a name adopted by London (1838. and the one it would seem by
which our beech should be known. *F. Americana* Sweet (1826), which can be derived
only from *F. sylvatica* β *Americana* Nutt. (1818). must certainly fall before the earlier
F. Sylvatica atropunicea Marsh. (1785).

Fagus latifolia (Muenchh.) Loud.—Continued.

Syn.—*Fagus alba* Rafinesque, Fl. Ludovic., 131 (1817).
Fagus sylvatica β Americana Nuttall, Genera, II, 216 (1818).
Fagus Americana Sweet, Hort. Brit., 370 (1826).
Fagus rotundifolia Rafinesque, in Atlant. Journ., 177 (1833).
Fagus heterophylla Raf., New Fl., III, 80 (1836).
Fagus ferruginea 2 Caroliniana Loudon, Arb. Frut., III, 1980, f. 1915 (1838).
Fagus Caroliniana Lodd. Cat. ed. (1836) ex Loud., l. c. (1838).
Fagus ferruginea 3 latifolia Loud., l. c.; f. 1916 (1838).
Fagus latifolia Loud., l. c., 1980 (1838).
Fagus atropunicea (Marsh.) Sudworth, in Bull. Torr. Bot. Club, XIX, 43 (1893).

COMMON NAMES.

Beech (Me., N. H., Vt., Mass., R. I., Conn., N. Y., N. J., Pa., Del., Va., W. Va., N. C., S. C., Ga., Ala., Fla., Miss., La., Tex., Ark., Ky., Mo., Ill., Ind., Mich., Nebr., Minn., Ohio, Ont.).
Red Beech (Me., Vt., Ky., Ohio).
White Beech (Me., Ohio, Mich.).
Ridge Beech (Ark.).

CASTANOPSIS Spach, in Hist. Vég., XI, 185 (1842).

Castanopsis chrysophylla[1] (Hook.) de C. **Goldenleaf Chinquapin.**

Syn.—*Castanea chrysophylla* Hooker, Flor. Bor.-Am., II, 159 (1839).
Castanea sempervirens Kellogg, in Proc. Cal. Acad. Sci., I, 71 (1855).
Castanopsis chrysophylla A. de Candolle, in Journ. Bot., I, 182 (1863).

COMMON NAMES.

Chinquapin (Cal., Oreg.).
Chestnut (Cal.).
Western Chinquapin.

---

[1] The following low, shrubby form occurs at high elevations:
Castanopsis chrysophylla minor (Benth.) A. de C.
Syn.—*Castanea chrysophylla* var. *minor* Bentham, Pl. Hartweg., 337 (1857).
Castanopsis chrysophylla β minor A. de C., in de Candolle, Prodr., XVI, sect. 1, 110 (1864).
Castanopsis chrysophylla var. *pumila* Vasey, in Rep. Com. Ag., 1875, 175 (1876); Cat. For. Trees, 27 (1876).

CASTANEA Adanson, Fam. Pl., II, 375 (1763).

Castanea pumila [1] (Linn.) Mill.   **Chinquapin.**

SYN.—*Fagus pumila* Linnæus, Spec. Pl., ed. 1, II, 998 (1753).
Castanea pumila Miller, Gard. Dict., ed. 8, No. 2 (1768).
· *Fagus Castanea pumila* Muenchhausen, Hausv., V, 162 (1770).
/ *Castanea pumila Virginiana* Loudon, Arb. Frut., III, 2002 (1838).

COMMON NAME.

Chinquapin (Del., N. J., Pa., Va., W. Va., N. C., S. C., Ga., Ala.; Fla., Miss., La., Tex., Ark., Ky., Mo.; Mich. cult., Ohio.).

Castanea dentata (Marsh.) Borkh.   **Chestnut.**

SYN.—*Fagus Castanea* Wangenheim, Beschreib. Nordam. Holz., 90 (1781), not Linn. (1753).
Fagus-Castanea dentata Marshall, Arb. Am., 46 (1785).
Castanea dentata Borkhausen, Handb. Forstb., I, 741 (1800).
CASTANEA VESCA β AMERICANA Michaux, Fl. Bor.-Am., II, 193 (1803).
Castanea vesca Willdenow, Sp. Pl., IV, par. 1, 460 (1805), in part.
Castanea Americana Rafinesque, Fl. Ludovic., 134 (1817); New Fl., I, 82 (1836).
Castanea Americana var. *angustifolia* Raf., l. c. (1836).
Castanea Americana var. *latifolia* Raf., l. c. (1836).
CASTANEA VULGARIS γ AMERICANA A. de Candolle, in de C., Prodr., XVI, sect. 2, 114 (1864).
Castanea sativa Mill. var. *Americana* Sargent, in Gard. and For., II, 484 (1889).
Castanea Castanea (L.) var. *Americana* (Michx. f.) Sudworth, in Arboresc. Flor. of Washington, D. C., 7 (1891).
Castanea sativa Jackson, Ind. Kew., I, 455 (1893), not Mill. (1768).

---

[1] The following is a shrubby species:
Castanea alnifolia Nutt.   Dwarf Chinquapin.

SYN.— *Castanea alnifolia* Nuttall, Gen., II, 217 (1818).
Castanea nana Elliott, Sk. Bot. S. C. and Ga., II, 615 (1824).

Walter's "*Fagus pumila* var. *præcox et serrotina*" (Flor. Car., 233, 1788) possibly belongs to the latter species, but is not accompanied by a description. The *Castanea nana* Muehlenberg (Cat. Pl. Am. Sept. 86, 1813) has been taken up as the oldest name for this form, but seems not to be tenable, since it is unaccompanied by a description, the common name given, "dwarf," and the locality, "Georgia," being scarcely sufficient to establish its identity.

Castanea dentata (Marsh.) Borkh.—Continued.

SYN.—Chestnut (Me., N. H., Vt., Mass., R. I., Conn., N. Y., N. J.,
Pa., Del., Va., W. Va., N. C., Ga., Ala., Miss., Ky., Mo.,
Ill., Mich., Ont.).
O-heh-yah-tah = " Prickly Bur " (Indians, N. Y.).

QUERCUS Linn., Spec. Pl., 994 (1753).

Quercus alba Linn. **White Oak.**

SYN.—*Quercus alba* Linnæus, Spec. Pl., ed. 1, II, 996 (1753).
Quercus alba pinnatifida Michaux, Hist. Chênes Am., No. 4,
t. 5, f. 1 (1801).
Quercus alba repanda Michx., l. c., f. 2 (1801).
Quercus alba α *pinnatifido-sinuata* Hayne, Dend. Fl., 158
(1822).
Quercus alba β *sinuata* Hayne, l. c., 159 (1822).
Quercus alba γ *microcarpa* A. de Candolle, in de C., Prodr.,
XVI, sect. 2, 22 (1864).
Quercus alba b *pubescens* Dippel, Handb. Laubh., zw. T., 75
(1892), not Botero (1804).
Quercus ramosa Booth ex Dippel, l. c. (1892).
Quercus ramosa striata Booth ex Dippel, l. c. (1892).
Quercus alba c *elongata* Dippel, l. c. (1892).
Quercus paludosa Dippel, l. c. (1892), not Petzold (1864).

White Oak (Me., N. H., Vt., Mass., R. I., Conn., N. Y., N. J.,
Pa., Del., Va., W. Va., N. C., S. C., Ala., Fla., Ga., Miss.,
La., Tex., Ky., Mo., Ill., Ind., Kans., Nebr., Mich., Wis.,
Minn., S. Dak. (cult.). Iowa, Ohio, Ont.).
Stave Oak (Ark.).

Quercus alba × macrocarpa[1] Engelm.

SYN.—*Quercus alba × macrocarpa* Engelmann, in Trans. Acad. Sci.
St. Louis, III, 298 (1877).

Quercus alba × minor[2] Coulter.

SYN.—*Quercus alba × stellata* Engelmann, in Trans. Acad. Sci. St.
Louis, III, 399 (1877).

---

[1] Supposed to be a hybrid between the White and Bur Oak. Described from a single tree discovered by Mr. M. S. Bebb, near Fountaindale, Ill. Dr. Engelmann also mentions (l. c.) another similar tree found by Mr. E. Hall, near Athens, central Illinois. Mr. C. G. Pringle reported another tree near Charlotte, Vt. (Sargent, Silva, VIII, 18, t. CCCLX, 1895.)

[2] A tree discovered by Mr. M. S. Bebb, near Fountaindale, Ill., is supposed to be a hybrid between the White Oak and Post Oak, and is the first one detected.

Quercus alba × minor Coulter—Continued.

SYN.—*Quercus alba × minor* Coulter, in Mem. Torr. Bot. Club, 5, sig. 9, 121 (1894).

Quercus alba × prinus[1] Engelm.

SYN.—*Quercus alba × Prinus* Engelmann, in Trans. Acad. Sci. St. Louis, III, 399 (1877).
Quercus alba Houba, Chên. L'Am. Sept., t. (ante, p. 237) (1887), not Linn. (1753).

Quercus lobata Née.   California White Oak.

SYN.—*Quercus lobata* Née, in Anal. Cienc. Nat., III, 277 (1801).
Quercus Hindsii Bentham, Bot. Voyage Sulphur, 55 (1844).
Quercus longiglanda Torrey & Frem., in Fremont's Geogr. Mem. Upper Cal., 15, 17 (1848).
Quercus lobata var. *Hindsii* Wenzig, in Jahrb. Bot. Gart. Berl., dritt. B., 188 (1885).

COMMON NAMES.

California White Oak (Cal.).
Weeping Oak (Cal.).
Valley Oak (Cal.).
"Roble" (of Mexicans, Cal.).
White Oak (Idaho, Cal.).
Swamp Oak (Cal.).

Quercus breweri Engelm.   Shin Oak.

SYN.—*Quercus lobata* subsp. *fruticosa* Engelmann, in Trans. Acad. Sci. St. Louis, III, 389 (1877), not of previous authors.
Quercus Breweri Engelmann, in Watson, Bot. Cal., II, 96 (1880).
Quercus lobata var. *Breweri* Wenzig, in Jahrb. Bot. Gart. Berl., dritt. B., 188 (1885).
Quercus Œrstediana Greene, West. Am. Oaks, Pl. X, 19 (1889), in part (?); not R. Brown Campst. (1871).

Dr. George Vasey found a similar tree near Silver Springs, Md. (Bull. Torr. Bot. Club, X, 25 t. 29, 1883); and another was discovered by Mr. George W. Letterman, near Allentown, Mo., of evidently the same parentage (Sargent, Silva, VIII, 18, t. CCCLIX, 1895).

[1] A single tree found by Dr. George Vasey, near Soldiers' Home, Washington, D. C., in 1874, is supposed to be a hybrid between the White and Chestnut Oak; the tree is now destroyed. Another tree found by Dr. Vasey grew 2 miles north of Washington (Bull. Torr. Bot. Club, X, 26 t. 30, 1883); and still another now standing in the grounds of Mr. John Saul, near Washington, D. C., over 50 feet high (l. c., 25, t. 28). Mr. C. G. Pringle found a tree near Charlotte, Vt., in 1879, with characters suggesting similar parentage.

Quercus garryana Dougl.   **Pacific Post Oak.**

SYN.—*Quercus Garryana* Douglas, in Hooker, Fl. Bor.-Am., II, 159 (1839).
    Quercus Neæi Liebmann, in Dansk. Vidensk. Selsk. Forhandl. 1854, 173 (1854).
    Quercus Œrstediana R. Brown Campst., in Ann. Mag. Nat. Hist., ser. 4, VII, 250 (1871).—?
    Quercus Jacobi R. Brown Campst., l. c., 255 (1871).
    Quercus Gilberti Greene, West. Am. Oaks, Pt. II, 77, Pl. 37 (1890).

                COMMON NAMES.

White Oak (Cal., Oreg.).
Oregon White Oak (Cal.).
Pacific Post Oak (Oreg.).
Oregon Oak (Oreg.).
Western White Oak (Oreg.).

Quercus gambelii Nutt.   **Gambel Oak.**

SYN.—*Quercus Gambelii* Nuttall, in Journ. Phila. Acad., new ser., I, Pt. II, 179 (1848).
    Quercus alba β ? Gunnisonii Torrey, in Pacif. R. R. Rep., II, Pt. I, 130 (1855).
    Quercus stellata δ Utahensis A. de Candolle, in de C., Prodr., XVI, sect. 2, 22 (1864).
    Quercus Douglasii β Gambelii A. de C., in de C., l. c., 23 (1864).
    Quercus Douglasii γ Novo-Mexicana A. de C., l. c. (1864).
    QUERCUS UNDULATA α GAMBELII Engelmann, in Trans. Acad. Sci. St. Louis, III, 382–392 (1876).
    Quercus undulata Watson, in Am. Nat., VII, 302 (1873), in part.
    Quercus Gambelii var. *Gunnisonii* Wenzig, in Jahrb. Bot. Gart. Berl. dritt. B., 190 (1885).
    Quercus venustula Greene, West. Am. Oaks, Pt. II, 69, Pl. 32 (1890).

                COMMON NAMES.

Scrub Oak (N. Mex., Ariz., Colo., Oreg., Nev., Utah).
Rocky Mountain Scrub Oak (Nev., Oreg.).
Mountain Oak (Nev., Oreg.).
Pin Oak (Ariz.).
White Oak.
Shin Oak.

Quercus minor (Marsh.) Sargent.   **Post Oak.**

SYN.—*Quercus alba minor* Marshall, Arb. Am., 120 (1785).
    Quercus stellata Wangenheim, Nordam. Holz., 78, t. 6, f. 15 (1787).

Quercus minor (Marsh.) Sargent—Continued.

SYN.—*Quercus villosa* Walter, Fl. Caroliniana, 235 (1788).

Quercus lobulata Abbot, in Abbot & Smith, Insects Georgia, I, 47 (1797).

Quercus —— Abbot, l. c., t. 77 (1797).

QUERCUS OBTUSILOBA Michaux, Hist. Chênes Am., 1, t. 1 (1801).

Quercus Drummondii Liebmann, in Dansk. Vidensk. Selsk. Forhandl. 1854, 170 (1854).

Quercus stellata β *Floridana* A. de Candolle, in de C., Prodr., XVI, sect. 2, 22 (1864).

Quercus minor Sargent, in Gard. and For., II, 471 (1889).

COMMON NAMES.

Post Oak (Conn., R. I., N. J., Pa., Del., W. Va., N. C., S. C., Ala., Ga., Fla., Miss., La., Tex., Ark., Ky., Mo., Ill., Ind., Iowa, Kans., Nebr., Ont.).

Box White Oak (R. I.).

Iron Oak (Del., Miss., Nebr.).

Chêne etoile (Quebec).

Overcup Oak (Fla.).

White Oak (Ky., Ind.).

Box Oak (Md.).

Brash Oak (Md.).

Quercus chapmani Sargent.   **Chapman Oak.**

SYN.—*Quercus obtusiloba* var. *parvifolia* Chapman, Fl. S. States, 423 (1860), not *Q. Rob.* var. *parvifolia* Lovey & Durey (1831).

Quercus stellata Engelmann, in Trans. Acad. Sci. St. Louis, III, 389 (1877), in part.

Quercus minor var. *parvifolia* Sargent, in Gard. and For., II, 471 (1889).

Quercus Chapmani Sarg., l. c., VIII, 93 (1895).

Quercus macrocarpa Michx.   **Bur Oak.**

SYN.—*Quercus macrocarpa* Michaux, Hist. Chênes Am., 2, t. 2, 3 (1801).

Quercus olivæformis Michx. f., Hist. Arb. Am., II, 32, t. 2 (1812).

Quercus obtusiloba β *depressa* Nuttall, Genera, II, 215 (1818).

Quercus macrocarpa var. *olivæformis* Gray, Man. N. U. S., ed. 2, 404 (1856).

Quercus macrocarpa β *abbreviata* A. de Candolle, in de C., Prodr., XVI, sect. 2, 20 (1864).

Quercus macrocarpa Michx.—Continued.

SYN.—*Quercus macrocarpa* γ *minor* A. de C., in de C., l. c. (1864), not *Q. alba minor* Marsh. (1785).

    Quercus stellata γ *depressa* A. de C., in de C., l. c., 23 (1864).

    Quercus macrocarpa var. *Monstrosa* Hampton, in First Ann. Rep. O. St. For. Bur. 1885, 193 (1886).

    Quercus macrocarpa var. *muscosa* Hampton, l. c. (1886).

    Quercus macrocarpa var. *Fisherii* Hampton, l. c., 194 (1886).

    Quercus macrocarpa var. *Hamptonii* Hampton, l. c. (1886).

    Quercus macrocarpa var. *MacClarenii* Hampton, l. c. (1886).

    Quercus Hybrida Kentonii Hampton, l. c. (1886).

    Quercus pannosa Dippel, Handb. Laubh., zw. T., 80 (1892), not Bosc ex de C. (1868).

    Quercus olivæformis Hampterii Dippel, l. c. (1892).

    Quercus hybrida Dippel, l. c. (1892).

    Quercus macrocarpa unterart *olivæformis* Dippel, l. c. (1892).

COMMON NAMES.

Bur Oak (Vt., N. Y., Pa., Del., W. Va., Ala., Miss., La., Tex., Ark., Mo., Ill., Ky., Ind., Iowa, Kans., Nebr., Wis., Mich., Minn., N. Dak., S. Dak., Ohio).

Mossycup Oak (Mass., Pa., Del., S. C., Ga., Miss., La., Tex., Ark., Ill., Iowa, Nebr., Kans., Ont.).

Overcup Oak (R. I., Del., Pa., Miss., La., Ill., Minn.).

Blue Oak (Ont.).

Scrub Oak (Nebr., Minn.).

Overcup White Oak (Vt.).

Mossycup White Oak (Minn.).

Quercus lyrata Walt.   **Overcup Oak.**

SYN.—*Quercus lyrata* Walter, Fl. Caroliniana, 235 (1788).

    Quercus —— Abbot & Smith, Insects Georgia, II, t. 83 (1797).

COMMON NAMES.

Overcup Oak (N. C., S. C., Ga., Fla., Ala., Miss., La., Tex., Ark., Ill.).

Swamp Post Oak (Ala., S. C., Miss., La., Mo.).

Water White Oak (S. C., Miss.).

Oak (Ala.).

Swamp White Oak (Tex.).

Quercus prinus Linn.   **Chestnut Oak.**

SYN.—*Quercus Prinus* Linnæus, Spec. Pl., ed. 1, II, 995 (1753).

    Quercus Prinus var. *lata* Aiton, Hort. Kew., ed. 1, III, 356 (1789).

    Quercus Prinus β *oblongata* Ait., l. c. (1789).

Quercus prinus Linn.—Continued.

SYN.—*Quercus Prinus* (*Monticola*) Michaux, Hist. Chênes. Am., 5,
t. 7 (1801).
Quercus montana Willdenow, Sp. Pl., IV, Par. I, 440 (1805).
Quercus Castanea Emerson, Trees and Shr. Mass., 137, t. 5
(1846), not Née nor Muehl. (1801).
Quercus Castanea macrophylla Hampton, in First Ann. Rep.
O. St. For. Bur. 1885, 193 (1886).
Quercus Prinus c *monticola* α *parvifolia* Dippel, Handb.
Laubh., zw. T., 85, fig. 835 (1892), not *Q. Rob. parviflora*
Lov. & Dur. (1833).
Quercus caroliniana Dippel, l. c. (1892).
Quercus monticola caroliniana Dippel, l. c. (1892).
Quercus Esculus Dippel, l. c. (1892), not Linn. (1753).
Quercus Prinus caroliniana Dippel, l. c., f. 35 (1892).

COMMON NAMES.

Chestnut Oak (Mass., R. I., Conn., N. Y., N. J., Pa., Del.,
Va., W. Va., N. C., Ga., Ky.).
Rock Chestnut Oak (Mass., R. I., Pa., Del., Ala., Ill.).
Rock Oak (N. Y., Del., Pa.).
Tanbark Oak (Ala.).
. Swamp Chestnut Oak (N. C.).
Mountain Oak (Ala.).

Quercus acuminata (Michx.) Houba.   **Chinquapin Oak.**

SYN.—*Quercus Prinus* var. *acuminata* Michaux, Hist. Chênes. Am.,
5, t. 8 (1801).
Quercus Castanea Willdenow, in Neue Schrift. Gesell. Nat.
Fr. Berl., III, 396 (1801).
Quercus Mühlenbergii Engelmann, in Trans. Acad. Sci. St.
Louis, III, 391 (1877).
Quercus Acuminata Houba, Chên. L'Am. Sept., 205 (1887).
Quercus prinoides Sargent, in Tenth Cen. U. S., IX (Cat.
For. Trees N. A.), 142 (1884), in part.
Quercus acuminata Sargent, Silva, N. A., VIII, 55 t.,
CCCLXXVII (1895).

COMMON NAMES.

Chestnut Oak (Conn., Del., Ala., N. C., Miss., La., Tex.,
Ill., Mich., Kans., Nebr., Ohio).
Chinquapin Oak (Mass., R. I., Pa., Del., N. C., S. C., Ala.,
Miss., Tex., Mo., Ind., Nebr., Kans.).
Pin Oak (Kans.).
Yellow Oak (Ill., Kans., Nebr., Mich.).

Quercus acuminata (Michx.) Houba.—Continued.

SYN.—Scrub Oak (N. Y.).
Dwarf Chestnut Oak (Mass., N. C., Tenn.).
Shrub Oak (Nebr.).

Quercus prinoides[1] Willd. **Chinquapin Oak.**

SYN.—*Quercus Prinus humilis* Marshall, Arb. Am., 125 (1785), not *Q. humilis* Lam. (1783).
Quercus prinoides Willdenow, in Neue Schrift. Gesell. Nat. Fr. Berl., III, 397 (1801).
Quercus prinus (*pumila*) Michaux, Hist. Chênes Am., No. 5, t. 9, f. 1 (1801), not *Q. pumila* Walt. (1788).
Quercus Chinquapin Pursh, Fl. Am. Sep., II, 634 (1814).
Quercus Muhlenbergii Engelm. var. *humilis* Britton, in Bull. Torr. Bot. Club, XIII, 41 (1885).

Quercus platanoides (Lam.) Sudworth. **Swamp White Oak.**

SYN.—*Quercus Prinus β platanoides* Lamarck, Enc. Méth. Bot., I, 720 (1783).
Quercus alba palustris Marshall, Arb. Am., 120 (1785), not *Q. palustris* Muenchh. (1770).
Quercus Prinus tomentosa Michaux, Hist. Chênes. Am., t. 9 (1801).
QUERCUS BICOLOR Willdenow, in Muehlenberg & Willd., Neue Schrift. Gesell. Nat. Fr. Berl., III, 396 (1801).
Quercus Prinus discolor Michaux f., Hist. Arb. Am., II, 46, t. 6 (1812).
Quercus bicolor ,β mollis Nuttall, Genera, II, 215 (1818).
Quercus Prinus β bicolor Spach, Hist. Vég., XI, 158 (1842).
Quercus paludosa Petzold, Arb. Musc., 646 (1864).
Quercus bicolor β platanoides A. de Candolle, in de C., Prodr., XVI, sect. 2, 21 (1864).
Quercus Prinus lyrata Koch, Dendrol., zw. Th. zw. Ab., 48 (1873), not *Q. lyrata* Walt. (1788).
Quercus discolor var. *bicolor* Hampton, in First Ann. Rep. Ohio St. For. Bur., 1885, 195 (1886).
Quercus Prinus a palustris Dippel, Handb. Laubh., zw. T., 84 (1892), not Michx. (1801).
Quercus bicolor a angustifolia Dippel, l. c., 87 (1892).
Quercus Prinus acuminata Dippel, l. c. (1892), not Michx. (1801), nor Roxburg (1814).

---

[1] The following related species of shrubby habit (2 to 6 feet high) forms extensive thickets in southwestern Oregon and northwestern California:

Quercus sadleriana R. Brown Campst. **Sadler Oak.**

SYN.—*Quercus Sadleriana* R. Brown Campst., in Ann. Mag. Nat. Hist., ser. 2, VII 249 (1871).

Quercus platanoides (Lam.) Sudworth—Continued.

SYN.—*Quercus bicolor* **b** *lyrata* Dippel, l. c. (1892), not *Q. lyrata*
Walt. (1788).
Quercus paludosa lyrata Dippel, l. c. (1892).
Quercus bicolor **c** *cuneiformis* Dippel, l. c. (1892).
Quercus velutina Hort. ex Dippel, l. c. (1892), not Lam. (1783).
Quercus platanoides (Lam.) Sudworth, in Rep. Sec. Agric.
1892, 327 (1893).

COMMON NAMES.

Swamp White Oak (Vt., Mass., R. I., Conn., N. Y., N. J.,
Pa., Del., W. Va., Mo., Ill., Ind., Iowa, Mich., Ont.).
Swamp Oak (R. I., Pa.).

Quercus michauxii Nutt. **Cow Oak.**

SYN.—*Quercus Prinus* Walter, Fl. Caroliniana, 234 (1788), not Linn.
(1753).
Quercus Prinus (*palustris*) Michaux, Hist. Chênes Am.,
No. 5, t. 6 (1801), not *Q. palustris* Muenchh. (1770).
Quercus Michauxii Nuttall, Genera, II, 215 (1818).
Quercus Prinus var. *Michauxii* Chapman, Fl. S. States,
ed. 1, 424 (1860).
Quercus Prinus var. *discolor* Curtis, in Rep. Geolog. Surv.
N. C., 1860, III, 49 (1860), not Michx. (1812).
Quercus bicolor A. de Candolle, in de C., Prodr., XVI, sect. 2,
20 (1864), in part.
Quercus bicolor subsp. *Michauxii* Engelmann, in Trans.
Acad. Sci. St. Louis, III, 390 (1877).
Quercus Prinus **b** *tomentosa* Dippel, Handb., Laubh., zw. T.,
84 (1892), not Michx. (1801.)

COMMON NAMES.

Basket Oak (Ala., Miss., La., Tex., Ark.).
Cow Oak (Ala., Miss., Tex., Ark., Mo.).
Swamp White Oak (Del., Ala.).
Swamp Chestnut Oak (Fla.).

Quercus michauxii × macrocarpa n. hyb.[1]

---

[1] But little is known of the origin of this form, except that in October, 1888, Mr.
James Byars sent specimens to the Forestry Division Herbarium from the vicinity
of Covington, Tenn. They were among other fresh specimens of *Quercus lyrata*,
Q. michauxii, and *Q. macrocarpa*, all said to be growing upon a rich bottom. Presumably the trees were more or less near each other, but nothing definite could be
learned on this point at the time, as the trees were being cut down, and the collector did not distinguish the different species and forms.

It is a unique hybrid, in most of its leaf forms suggesting the foliage of *Q.
michauxii*, the regular rounded teeth of the latter being merely deepened so as to

Quercus breviloba (Torr.) Sargent.   **Durand Oak.**

SYN.—*Quercus obtusifolia* var.? *breviloba* Torrey, in Bot. Mex.
Bound. Surv., 206 (1858).
QUERCUS DURANDII Buckley, in Proc. Phila. Acad. Sci.,
1860, 445 (1861).
Quercus annulata Buckley, l. c. (1861).
Quercus San-sabeana Buckley, in Young, Bot. Texas, 507
(1873).
Quercus undulata Engelmann, in Trans. Acad. Sci. St. Louis,
III, 392 (1876), in part; not Torr. (1828).
Quercus brevilobata Sargent, in Gard. and For., VIII, 93
(1895), not *Q. oblongifolia* var. *brevilobata* Torr. (1874).
Quercus breviloba Sarg., Silva, VIII, 71, CCCLXXXIV
(1895).

COMMON NAMES.

White Oak (Tex.).
Texas White Oak (Ala.).
Shin Oak (Tex.).
Pin Oak (Tex.).
Bastard Oak (Ala., La., Tex.).
Basket Oak (Ala., La., Tex.).
Durand's Oak (Ala., La., Tex.).

Quercus undulata[1] Torr.   **Rocky Mountain Oak.**

SYN.—*Quercus undulata* Torrey, in Am. Lyc. N. Y., II, 248, t. 4
(1828).
Quercus Fendleri Liebmann, in Dansk. Vidensk. Selsk.
Forhandl., 170 (1854).
Quercus grisea Liebm., l. c. 171 (1854).

form shallow tooth-like lobes. The general size and outline of the hybrid leaf follows very closely that of *Q. michauxii*, with occasional remote resemblance to the typical leaf of *Q. macrocarpa*. The under surface of the hybrid leaf is pale, with a very close, invisible tomentum, much less abundant and velvety than in lower leaves of *Q. michauxii*. In general aspect the fruit resembles that of *Q. michauxii*, but slightly larger; the form of the acorn is very close to *Q. macrocarpa*. The cup is one-third deeper than *michauxii*, thinnish, with canescent scales, in form resembling those of *macrocarpa*, as also in the slightly awned upper rows of scales; the inner surface of the cups has a thick, velvety pubescence, such as is usually present in the cups of the Bur Oak (less so in the Cow Oak). The buds are large, but shorter than those of the latter oak, while the branchlets are smooth, robust, and very similar to those of *michauxii*, with no indication of the corky ridges of *Q. macrocarpa*. It is possible that *Q. lyrata* may have been one of the parents, but the larger, broader leaves, and much stouter branchlets, large, slightly mossy cups suggest the parentage of *Q. michauxii* and *Q. macrocarpa*.

[1]Heretofore this well-known southwestern species has never been found to attain the form of a tree, commonly occurring as a low shrub. But Prof. J. T. Toumey reported it (Gard. and For., VIII, 13, 1895) as a small, bushy tree in sheltered canyons of the mountains of southeastern Arizona, so that it must now be included among our arborescent species.

Quercus undulata Torr.—Continued.

Syn.—*Quercus pungens* Liebm., l. c. (1854).

Quercus oblongifolia Torrey, in Bot. Mex. Bound. Surv., 206 (1858).

Quercus undulata β *obtusifolia* A. de Candolle, in de C., Prodr., XVI, sect. 2, 23 (1864).

Quercus undulata γ *pedunculata* A. de C., in de C., l. c. (1864), not *Q. Robur.* subsp. *pedunculata* A. de C., l. c. (1864).

Quercus Emoryi Porter & Coulter, Syn. Fl. Col., 127 (1874), not Torr. (1848).

Quercus undulata γ *Jamesii* Engelmann, in Trans. Acad. Sci. St. Louis, III, 382 (1876).

Quercus undulata δ *Wrightii* Engelm., l. c. (1876).

Quercus undulata var. *pungens* Engelm., l. c., 392 (1877).

Quercus undulata var. *grisea* Engelm., l. c., 393 (1877).

Quercus turbinella Greene, West. Am. Oaks, 37 (1889); Pt. II, 59, Pl. XXVII (1890), in part.

Quercus undulata α *normalis* Kuntze, Rev. Gen. Pl., Par. II, 642 (1891).

Quercus undulata δ *heterophylla* Kuntze, l. c. (1891), not *Q. heterophylla* Michx. f. (1812).

COMMON NAMES.

Scrub Oak.
Shin Oak.

Quercus douglasii Hook. & Arn.　　　(California) **Rock Oak.**

Syn.—*Quercus Douglasii* Hooker & Arnott, in Bot. Beechey's Voyage, 391 (1841).

Quercus Ransomi Kellogg, in Proc. Cal. Acad. Sci., I, 25 (1855).

Quercus oblongifolia R. Brown Campst., in Ann. Mag. Nat. Hist., ser. 2, VII, 252 (1871), not Torr. (1858).

Quercus oblongifolia var. *brevilobata* Torrey, in Bot. Wilkes' Exped., 460 (1874).

COMMON NAMES.

Mountain White Oak (Cal.).
Rock Oak (Cal.).
White Oak (Cal.).
Blue Oak (Cal.).

Quercus engelmanni Greene.　　　　**Engelmann Oak.**

Syn.—*Quercus oblongifolia* Torrey, in Ives's Rep., Pt. IV, 28 (1861), not Torr. in Sitgreaves's Rep. (1854).

Quercus engelmanni Greene—Continued.

SYN.—*Quercus Engelmanni* Greene, West. Am. Oaks, Pt. I, 33, Pl. XV, f. 2, 3; Pl. XVII (1889).

Engelmann's Oak.
Evergreen White Oak.

Quercus oblongifolia Torr.   **Oblong-leaf Oak.**

SYN.—*Quercus oblongifolia* Torrey, in Sitgreaves's Rep., 173, t. 19 (1853).

Quercus undulata var. *oblongata* Engelmann, in Rothrock, in Wheeler's Rep., VI, 250 (1878).

Quercus undulata δ *grisea* Wenzig, in Jahrb. Bot. Gart. Berl. dritt. B., 200 (1885), in part.

    ˙ *Quercus undulata* var. *grisea* Greene, West. Am. Oaks, 29, Pl. XV, 15, f. 1 (1889), in part.

White Oak.

Quercus arizonica[1] Sargent.   **Arizona White Oak.**

SYN.—*Quercus Emoryi* Watson, in Wheeler's Rep. Geol. Expl. Surv. W. 100 Merid., 17 (1874), not Torr. (1848).

Quercus undulata var. *grisea* Engelmann, in Rothrock in Wheeler's Rep., VI, 250 (1878), not *Q. grisea* Liebm. (1854).

Quercus grisea Sargent, in Tenth Cen. U. S., IX (Cat. For. Trees N. A.), 144 (1884), not Liebm. (1854).

Quercus Arizonica Sargent, in Gard. and For., VIII, 92 (1895).

White Oak.

Quercus reticulata Humb. & Bonpl.   **Netleaf Oak.**

SYN.—*Quercus reticulata* Humboldt & Bonpland, Pl. Æquin., II, 40, t. 86 (1809).

Quercus spicata Humb. & Bonpl., l. c., 46, t. 89 (1809).

Quercus decipiens Martens & Galeotti, in Bull. Acad. Sc. Brux., X, 214 (1843).

Quercus reticulata β *Greggii* A. de Candolle, in de C., Prodr., XVI, sect. 2, 34 (1864).

---

[1] This Arizona oak has formerly been referred to Liebmann's *Q. grisea*, which proves to be only a form of Torrey's much earlier *Q. undulata* (1828), and must therefore become a synonym of the latter name.

Quercus toumeyi[1] Sargent.   **Toumey Oak.**

SYN.—*Quercus Toumeyi* Sargent, in Gard. and For., VIII, 92, f. 13, 14 (1895).

Quercus dumosa[2] Nutt.   **Scrub Oak.**

SYN.—*Quercus dumosa* Nuttall, Sylva, I, 7 (1842).
 Quercus acutidens Torrey, in Bot. Mex. Bound. Surv., 207, t. 51 (1858).
 Quercus undulata var. *pungens* Engelmann, in Watson, Bot. Cal., II, 96 (1880), in part.
 Quercus dumosa γ *acutidens* Wenzig, in Jahrb. Bot. Gart. Berl., dritt. B., 204 (1885).
 Quercus Macdonaldi Greene, West. Am. Oaks, I, 25 (1889); II, Pl., XXXIV (1890).
 Quercus MacDonaldi var. *elegantula* Greene, West. Am. Oaks, 25 (1889).
 Quercus dumosa var. *polycarpa* Greene, l. c., 36 (1889); Pt. II, 61, Pl. XXVIII (1890).
 Quercus dumosa var. *munita* Greene, l. c., 37, t. 20 (1889).
 Quercus turbinella Greene, l. c., 37 (1889); Pt. II, 59, t. 27 (1890), in part.

Quercus dumosa revoluta Sargent.   **Curl-leaf Scrub Oak.**

SYN.—*Quercus dumosa* var. *bullata* Engelmann, in Trans. Acad. Sci. St. Louis, III, 393 (1877), not *Q. tomentosa* β *bullata* A. de C. (1864).
 Quercus dumosa var. *revoluta* Sargent, in Gard. and For., VIII, 93 (1895).

Quercus virginiana[3] Mill.   **Live Oak.**

SYN.—*Quercus Virginiana* Miller, Gard. Dict., ed. 8, No. 16 (1768).

---

[1] An arborescent species, discovered by Prof. J. T. Toumey in July, 1894, on the hills above Bisbee, on the Mule Mountains, in southeastern Arizona. It is reported to be 25 to 30 feet in height and 6 to 10 inches in diameter. The trunk divides very near the ground, forming a spreading, irregular crown. It appears to resemble most nearly some of the forms of *Q. undulata*. This species is the principal one forming a narrow belt of timber on the Mule Mountains, mingling on the lower border with *Q. Emoryi*.

[2] This California oak, as it commonly occurs on the mainland, is a low shrub, and has, therefore, never been included among the arborescent species. A tree form, however, described by Professor Greene (l. c.) as *Q. Macdonaldi* occurs in the sheltered canyons on a few of the California coast islands and is believed to be the same as Nuttall's *Q. dumosa*.

[3] The following varieties are not known to be arborescent:

Quercus virginiana maritima (Michx.) Sargent.

SYN.—*Quercus Phellos* (*maritima*) Michaux, Hist. Chênes Am., No. 7, t. 13, f. 3 (1801).
 Quercus maritima Willdenow, Sp. Pl., IV, Par. I, 124 (1805).
 Quercus virens var. *maritima* Chapman, Fl. S. States, ed. 1, 421 (1860).
 Quercus Virginiana var. *maritima* Sargent, Silva, VIII, 100 (1895).

Quercus virginiana minima Sargent.

SYN.—*Quercus virens* var. *dentata* Chapman, Fl. S. States, ed. 1, 421 (1860), not *Q. dentata* Thunb. (1784).
 Quercus Virginiana var. *minima* Sargent, Silva, VIII, 101, t., CCCXCVI (1895).

Quercus virginiana Mill.—Continued.

SYN.—*Quercus Phellos* β Linnæus, Spec. Pl., ed. 1, II, 994 (1753).
    Quercus Phellos c Muenchhausen, Hausv., V, 255 (1770).
    Quercus Phellos γ *obtusifolia* Lamarck, Enc. Méth. Bot., I,
      722 (1783).
    Quercus-Phellos sempervirens Marshall, Arb. Am., 124 (1785).
    Quercus sempervirens Walter, Fl. Caroliniana, 234 (1788), not
      Mill. (1768).
    QUECUS VIRENS Aiton, Hort. Kew, ed. 1, III, 356 (1789).
    Quercus oleoides Chamisso & Schlechtendal, in Linnæa, V,
      79 (1830).
    Quercus Sagræana Nuttall, Sylva, I, 17 (1842).
    Quercus Cubana Richard, Fl. Cub., III, 230 (1853).
    Quercus retusa Liebmann, in Dansk. Vidensk. Selsk. For-
      handl. 1854, 187 (1854).

COMMON NAMES.

Live Oak (Va., N. C., S. C., Ga., Fla., Ala., Miss., La., Tex.,
    Cal.).
Chêne Vert (La.).

Quercus emoryi Torr.   **Emory Oak.**

SYN.—*Quercus Emoryi* Torrey, in Emory's Rep., 151, t. 9 (1848).
    Quercus hastata Liebmann, in Dansk. Vidensk. Selsk. For-
      handl., 1854, 171 (1854).

COMMON NAMES.

Emory's Oak (Cal.).
Black Oak (Cal., Utah).

Quercus chrysolepis Liebm.   **Cañon Live Oak.**

SYN.—*Quercus chrysolepis* Liebmann in Dansk. Vidensk. Selsk.
    Forhandl. 1854, 173 (1854).
    Quercus fulvescens[1] Kellogg, in Proc. Calif. Acad. Sci., I, 65
      (1855).
    Quercus crassipocula Torrey, in Pacif. R. R. Rep., IV, Pt. I,
      137 (1855).
    Quercus chrysophyllus Kellogg, in (reprint) Proc. Calif. Acad.
      Sci., I, 67 (1873).[1]

COMMON NAMES.

Live Oak (Cal., Oreg.).
Maul Oak (Cal.).

---

[1]Prof. E. L. Greene informs me that *Q. fulvescens*, which stands in the original
issue of Proc. Cal. Acad., was changed by Dr. Kellogg to *Q. chrysophyllus* in the 8vo
reprint of Vol. I (1873); the date of *Q. chrysophyllus* is not 1855, but that of the
publication of the reprint, 1873.

Quercus chrysolepis Liebm.—Continued.

SYN.—Iron Oak (Cal.).
Valparaiso Oak (Cal.).
Black Live Oak (Cal.).
Canyon Live Oak (Cal.).
Canyon Oak (Cal.).
Golden-cup Oak (Cal.).
Hickory Oak (Kern Co., Cal.).

Quercus chrysolepis palmeri Engelm.   **Palmer Oak.**

SYN.—*Quercus chrysolepis* subsp. *Palmeri* Engelmann, in Trans. Acad. Sci. St. Louis, III, 393 (1877).
Quercus Dunnii Kellogg, in Pacific Rural Press, June 7 (1879).
Quercus Palmeri Engelmann, in Watson, Bot. Cal., II, 97 (1880).

Quercus chrysolepis vaccinifolia[1] (Kell.) Engelm.

SYN.—*Quercus vaccinifolia* Kellogg, in Trans. Cal. Acad., I, 96 (1855).
Quercus chrysolepis subsp. *Q. vaccinifolia* Kellogg, Engelmann, in Trans. Acad. Sci. St. Louis, III, 393 (1877).

Quercus tomentella Engelm.

SYN.—*Quercus chrysolepis* Watson, (fide) Engelmann, in Proc. Am. Acad. Sci., XI, 119 (1876), not Liebm. (1854).
Quercus tomentella Engelmann, in Trans. Acad. Sci. St. Louis, III, 393 (1877).

Quercus[2] **agrifolia** Née.   **California Live Oak.**

SYN.—*Quercus agrifolia* Née, in Ann. Cienc. Nat., III, 271 (1801).

---

[1] A small, often shrubby, narrow-leaved form representing the species at high elevations.

[2] The following shrubby species occurs on the coast from North Carolina to Florida:

Quercus sericea Willd.   **Running Oak.**

SYN.—*Quercus pumila* Walter, Fl. Caroliniana, 234 (1788), not *Q. nigra pumila* Marsh. (1785).
Quercus Phellos (*pumila*) Michaux, Hist. Chênes Am., No. 7, t. 13, f. 1 (1801).
Quercus sericea Willdenow, Sp. Pl., IV. Par. I. 424 (1805).
Quercus cinerea var. *pumila* Curtis, in Rep. Geol. Surv. N. C. 1860, III, 37 (1860).
Quercus Phellos ε *nana* A. de Candolle, in de C., Prodr., XVI, sect. 2, 74 (1864), not *Q. rubra nana* Marsh (1785).
Qcurcus pumila var. *sericea* Engelmann, in Trans. Acad. Sci. St. Louis, III, 384 (1876).

Prof. C. S. Sargent has (Silva, VIII, 115, 1895) taken up for this plant Walter's *Q. pumila* (1788), a name which it seems can not stand because of the previously published *Q. nigra pumila* of Marshall (1785), which is held for another species. The next oldest available name is the *Q. sericea* of Willdenow (1805).

Quercus agrifolia Née.—Continued.

SYN.—*Quercus oxyadenia* Torrey, in Sitgreaves's Rep., 172, t. 17
(1853).
Quercus berberidifolia Liebmann, in Dansk. Vidensk. Selsk.
Forhandl. 1854, 172 (1854).
Quercus arcoglandis Kellogg, in Proc. Calif. Acad. Sci., I,
25 (1855).
Quercus agrifolia var. *frutescens* Engelmann, in Watson,
Bot. Cal., II, 98 (1880).
Quercus agrifolia γ *berberifolia* Wenzig, in Jahrb. Bot. Gart.
Berl. dritt. B., 203 (1885).

COMMON NAMES.

Coast Live Oak (Cal.).
California Live Oak (Cal.).
Encina (Cal.).
Evergreen Oak (Cal.).

Quercus hypoleuca Engelm.   **Whiteleaf Oak.**

SYN.—*Quercus confertifolia* Torrey, in Bot. Mex. Bound. Survey,
207 (1858), not H., B. & K. (1817).
Quercus hypoleuca Engelmann, Trans. Acad. Sci. St. Louis,
III, 384 (1876).
Quercus Mexicana γ *confertifolia* Wenzig, in Jahrb. Bot.
Gart. Berl., dritt. B., 209 (1883), in part.

COMMON NAMES.

Oak (Ariz.).
Mexican Oak (Cal.).
White-leaved Oak (Cal.).

Quercus wislizeni A. de C.   **Highland Live Oak.**

SYN.—*Quercus Wislizeni* A. de Candolle, in de C., Prodr., XVI, sect.
2, 67 (1864).
Quercus Wislizeni var. *frutescens* Engelmann, in Trans.
Acad. Sci. St. Louis, III, 396 (1877).
Querous parvula Greene, in Pittiona, I, 40 (1887).

COMMON NAMES.

Live Oak (Cal.).
Highland Live Oak (Cal.).

Quercus morehus Kell.   **Morehus Oak.**

SYN.—*Quercus Morehus* Kellogg, in Proc. Calif. Acad. Sci., II, 36
(1863).

Quercus morehus Kell.—Continued.

SYN.—*Quercus Wislizeni* × *Kelloggii* Curran, in Bull. Cal. Acad., I, 146 (1885).

Quercus Wislizeni × *Californica* Sargent, Silva, VIII, 120, t. CCCCVII (1895).

Quercus myrtifolia[1] Willd. **Myrtle Oak.**

SYN.—*Quercus myrtifolia* Willdenow, Sp. Pl., IV, Par. I, 424 (1805).

Quercus Phellos var. *arenaria* Chapman, Fl. S. States., ed. 1, 420 (1860).

Quercus aquatica ζ ? *myrtifolia* A. de Candolle, in de C., Prodr., XVI, sect. 2, 68 (1864).

Quercus rubra[2] Linn. **Red Oak.**

SYN.—*Quercus rubra* Linnæus, Spec. Pl., ed. 1, II, 996 (1753).

Quercus rubra β Linn., l. c. (1753).

Quercus rubra b Du Roi, Harbk. Baumz., II, 261, t. 5, f. 3 (1772).

Quercus rubra var. *latifolia* Lamarck, Enc. Méth. Bot., I, 720 (1783).

Quercus rubra γ *subserrata* Lam., l. c. (1783).

Quercus rubra maxima Marshall, Arb. Am., 122 (1785).

Quercus rubra montana Borkhausen, Handb. Forstb., I, 705 (1800), not Marsh. (1785).

Quercus ambigua Michaux f., Hist. Arb. Am., II, 120, pl. 24, as to leaf (1812).

Quercus coccinea β Spach, Hist. Vég., XI, 165 (1842).

Quercus borealis Michaux f., Sylva, I, 198 (1859), in part.

Quercus coccinea var. *ambigua* Gray, Man. Bot. N. U. S., ed. 5, 454 (1867).

Quercus rubra var. *latepinnatifida* Kuntze, Revis. Gen. Pl., Par. II, 642 (1891).

Quercus rubra var. *coccinea* Kuntze, l. c. (1891), not *Q. coccinea* Muenchh. (1770).

Quercus rubra a *viridis* Dippel, Handb. Laubh., zw. T., 118 (1892).

Quercus rubra c *Schrefeldii* Dippel, l. c. (1892).

Quercus rubra d *heterophylla* Hort. ex Dippel, l. c. (1892), not *Q. heterophylla* Nutt. (1842).

---

[1] Usually a low, much-branched shrub, sometimes reaching a height of 10 to 20 feet and 3 inches in diameter. It occurs in the coast region from South Carolina to Florida and Louisiana; abundant on islands off the coast of Alabama and Mississippi.

[2] Mr. B. F. Bush reports (Gard. and For., VIII, 33, 1895) a tree near Independence, Mo., supposed to be a hybrid from *Q. rubra* and *Q. imbricaria* (*Q. rubra* × *imbricaria* Sargent, Silva, VIII, 189 (index), 1895).

Quercus rubra Linn.—Continued.

SYN.—*Quercus rubra* e *aurea* Dippel, l. c., 119 (1892).
Quercus rubra aurea Hort. ex Dippel, l. c. (1892).
Quercus rubra Americana aurea Hort. ex Dippel, l. c. (1892).

COMMON NAMES.

Red Oak (Me., Vt., N. H., Mass., R. I., N. Y., N. J., Pa., Del., Va., W. Va., N. C., S. C., Ga., Ark., Mo., Ky., Ill., Ind., Iowa, Nebr., Kans., Mich., Minn., Ohio, Ont., S. Dak.).
Black Oak (Vt., Conn., N. Y., Wis., Iowa, Nebr., Ont., S. Dak.).
Spanish Oak (Pa., N. C.).

Quercus rubra runcinata A. de C.

SYN.—*Quercus rubra* β *runcinata* A. de Candolle, in de C., Prodr., XVI, sect. 2, 60 (1864).
Quercus runcinata Engelmann, in Trans. Acad. Sci. St. Louis, III, 542 (1877).
Quercus rubra × *digitata* Sargent, Silva, VIII, 189 (index) (1895).

Quercus texana Buckley.　　　　　　　　**Texan Oak.**

SYN.—*Quercus palustris* Torrey & Gray, in Pacif. R. R. Rep., II, Pt. III, 175 (1855), not Muenchh. (1770).
Quercus coccinea var. ? *microcarpa* Torrey, in Bot. Mex. Bound. Surv., 206 (1858), not *Q. microcarpa* La Peyr (1813), nor Liebm. (1854).
Quercus Texana Buckley, in Proc. Phila. Acad. Sci. 1860, 444 (1861).
Quercus rubra var. *Texana* Buckley, Proc. Phila. Acad. Sci. 1881, 123 (1881).

COMMON NAMES.

Red Oak (Tex.).
Spotted Oak (Tex.).
Spanish Oak (Tex.).

Quercus coccinea Muenchh.　　　　　　　**Scarlet Oak.**

SYN.—*Quercus coccinea* Muenchhausen, Hausv., V, 254 (1770).
Quercus rubra β *coccinea* Aiton, Hort. Kew., ed. 1, III, 357 (1789).
Quercus coccinea α *coccinea* A. de Candolle, in de C., Prodr., XVI, sect. 2, 61 (1864).
Quercus coccinea pendula Petzold, Arb. Musc. (1864).
Quercus coccinea undulata Petzold, l. c. (1864), not *Q. undulata* Torr. (1828).

Quercus coccinea Muenchh.—Continued.

SYN.—Scarlet Oak (Vt., Mass., R. I., Conn., N. Y., N. J., Pa., Del., N. C., Mo., Ill., Ind., Wis., Iowa, Mich., Minn., Nebr., Ont.). Red Oak (N. C., Ala., Wis., Nebr., Minn.). Black Oak (Mo., Ill., Iowa, Wis.). Spanish Oak (N. C.).

Quercus coccinea × pumila [1] Sudworth.

SYN.—*Quercus coccinea × ilicifolia* Gray, Man. Bot. N. U. S., ed. 5, 454 (1867).

Quercus ilicifolia × coccinea Engelmann, in Trans. Acad. Sci. St. Louis, III, 542 (1877).

Quercus velutina Lam.   **Yellow Oak.**

SYN.—*Quercus velutina* Lamarck, Enc. Méth. Bot., I, 721 (1783).
Quercus nigra Du Roi, Harbk. Baumz., II, 272, t. 6, f. 1 (1772), not Linn. (1753).
Quercus discolor Aiton, Hort. Kew., ed. 1, II, 358 (1789).
QUERCUS TINCTORIA [2] Bartram, Travels, ed. 1, 37 (1791)—*nomen nudum;* Michaux, Hist. Chênes Am., No. 13, t. 24, 25 (1801).
Quercus tinctoria α angulosa Michx., Fl. Bor.-Am., II, 198 (1803).
Quercus tinctoria β sinuosa Michx., l. c. (1803).
Quercus repanda Petzold, Arb. Musc., 657 (1864), not Michx. (1801).
Quercus coccinea β nigrescens A. de Candolle, in de C., Prodr., XVI, sect. 2, 61 (1864).
Quercus coccinea γ tinctoria A. de C., l. c. (1864).
Quercus nigrescens Houba, Chên. L'Am. Sept., 200, t. (1887).
Quercus hybrida Houba, l. c., 310, t. (1887).
Quercus rubra var. *tinctoria* Kuntze, Rev. Gen. Pl., Par. II, 642 (1891).
Quercus tinctoria angustifolia Hort. ex Dipple, Handb., zw. T., 121 (1892).
Quercus tinctoria a *discolor* Dippel, l. c. (1892).
Quercus tinctoria a Willdenowiana Dippel, l. c., 122 (1892).
Quercus cuneata macrophylla (Cat. Musk) ex Dippel, l. c. (1892).

---

[1] A tree 40 feet high, detected by Dr. J. W. Robbins, in 1855, near Whitinsville, Mass. Leaves similar in shape to *Q. rubra*, and the fruit almost identical with that of *Q. pumila*, but supposed to have been a hybrid between the Scarlet and Bear Oak.

[2] Bartram's *Q. tinctoria*, as pointed out (Sudworth, in Gard. and For., V, 98, 1892), is not tenable, not being founded on a description. It strictly has no place in the bibliography of this species, but is retained to avoid confusion.

Quercus velutina Lam.—Continued.

SYN.—*Quercus* a *discolor tinctoria β magnifica* Dippel, 1. c. (1892).
Quercus magnifica Hort. ex Dippel, 1. c. (1892).
Quercus americana magnifica Hort. ex Dippel, 1. c., 123 (1892).
Quercus a *discolor tinctoria γ macrophylla* Dippel, 1. c., f. 59 (1892).
Quercus macrophylla Hort. ex Dippel, 1. c. (1892), not Michx. (1801).
Quercus americana macrophylla Hort. ex Dippel, 1. c. (1892).
Quercus americana macrophylla Albertii Dippel, 1. c. (1892).
Quercus a *discolor tinctoria δ nobilis* Dippel, 1. c., 124, f. 60 (1892).
Quercus tinctoria nobilis Dippel, 1. c., f. 60 (1892).
Quercus nobilis Dippel, 1. c. (1892).
Quercus alba vera Dippel, 1. c. (1892).

COMMON NAMES.

Black Oak (Vt., Mass., R. I., N. Y., N. J., Pa., Del., Va., W. Va., N. C., S. C., Ga., Ala., Fla., Miss., La., Tex., Ill., Iowa, Kans., Nebr., Mich., Minn., Ohio, Ont., Wis.).
Quercitron Oak (Del., S. C., La., Kans., Minn.).
Yellow Oak (R. I., N. Y., Ill., Tex., Kans., Minn.).
Tanbark Oak (Ill.).
Yellow-barked Oak (Minn.).
Spotted Oak (Mo.).
Yellow Bark (R. I.).
Dyer's Oak (Tex.).

Quercus californica (Torr.) Cooper.   **California Black Oak.**

SYN.—*Quercus rubra* Bentham, Pl. Hartweg., 337 (1857), not Linn. (1753).
Quercus tinctoria var. *Californica* Torrey, in Pacif. R. R. Rep., IV, Pt. I, 138 (1856).
Quercus Kelloggii Newberry, in Pacif. R. R. Rep., VI, 28, f. 6, 89 (1857).
Quercus Californica Cooper, in Smithsonian Rep. 1858, 261 (1859).
Quercus Sonomensis Bentham ex A. de Candolle, in de C., Prodr., XVI, sect. 2, 62 (1864).

COMMON NAMES.

Black Oak (Cal., Oreg.).
Mountain Black Oak (Cal.).
Kellogg's Oak (Cal.).
California Black Oak (Cal.).

Quercus catesbæi Michx.   **Turkey Oak.**

SYN.—*Quercus rubra* Abbot & Smith, Insects Georgia, I, t. 14 (1797), not Linn. (1753).
Quercus Catesbæi Michaux, Hist. Chênes Am., No. 17, t. 29, 30 (1801).

COMMON NAMES.

Turkey Oak (Ga., Ala., Fla., Miss., La., Tex.).
Scrub Oak (Va., N. C., S. C., Fla., Miss.).
Black Jack (S. C.).
Barren Scrub Oak (Tenn.).
Forked Leaf (S. C.).
Forked-Leaf Black Jack.

Quercus sinuata[1] (Lam.) Walt.

SYN.—*Quercus nigra* γ *sinuata* Lamarck, Enc. Méth. Bot., I, 721 (1783).
Quercus sinuata Walter, Fl. Caroliniana, 235 (1788).
Quercus Catesbæi × *aquatica* Engelmann, in Trans. Acad. Sci. St. Louis, III, 400 (1877).
Quercus Catesbæi × *nigra* Sargent, Silva, VIII, 144, t. CCCCXVIII (1895).

Quercus catesbæi × brevifolia[2] nom. nov.

SYN.—*Quercus Catesbæi* × *Q. cinerea* Small, in Bull. Torr. Bot. Club, XXII, 76, Pl. 234, 235 (1895).

Quercus catesbæi × laurifolia[3] Engelm.

SYN.—*Quercus Catesbæi* × *laurifolia* Engelmann, in Trans. Acad. Sci. St. Louis, III, 539 (1877).

Quercus digitata (Marsh.) Sudworth.   **Spanish Oak.**

SYN.—*Quercus nigra digitata* Marshall, Arb. Am., 121 (1785).
Quercus rubra montana Marsh., l. c., 123 (1785).
Quercus cuneata Wangenheim, Nordam. Holz., 78, t. 5, f. 14 (1787).
Quercus rubra β *Hispanica* Castiglioni, Viag. negl. Stati Uniti, II, 347 (1790).
Quercus rubra β Abbot & Smith, Insects of Ga., I, 27, t. 14 (1797).

---

[1] A single tree, 40 feet high, was discovered many years ago by Dr. J. H. Mellichamp near Bluffton, S. C., but since destroyed. It is believed to be a hybrid.

[2] A small hybrid oak under 12 feet, recently detected by Dr. J. K. Small in Lake County, Fla.

[3] A single tree known, 40 feet high, growing in Bluffton, S. C., detected by Dr. Mellichamp.

Quercus digitata (Marsh.) Sudworth—Continued.

SYN.—*Quercus rubra* Abbot & Smith, l. c., 99, t. 50 (1797), not Linn. (1753).

Quercus elongata Willdenow, in Muehlenberg & Willd., in Neue Schrift. Gesell. Nat. Fr. Berlin, III, 400 (1801).

Quercus triloba Michaux, Hist. Chênes Am., 14, t. 26 (1801).

QUERCUS FALCATA Michx., l. c., 16, t. 28 (1801).

Quercus falcata β *triloba* Nuttall, Genera, II, 214 (1818).

Quercus falcata var. b *Pagodæfolia* Elliott, Sk. Bot. S. C. Ga., II, 605 (1824).

Quercus discolor Spach, Hist. Vég., XI, 163 (1842), not Ait. (1789).

Quercus falcata β *Ludoviciana* A. de Candolle, in de C., Prodr., XVI, sect. 2, 59 (1864).

Quercus falcata β *trinacris* Wood, Class Book, 645 (1869).

Quercus falcata var. *subintegra* Engelmann, in Trans. Acad. Sci. St. Louis, 542 (1877).

Quercus nigra var. *falcata* Kuntze (Michx.), Revis. Gen. Pl., Par. II, 642 (1891).

Quercus digitata (Marsh.) Sudworth, in Gard. and For., V, 98 (1892).

Quercus cuneata hudsonica Dippel, Handb. Laubh., zw. T., 112 (1892).

Quercus cuneata a *falcata* Dippel, l. c., f. 53, p. 113 (1892).

COMMON NAMES.

Spanish Oak (Pa., Del., Va., N. C., S. C., Ala., Fla., Miss., La., Tex., Mo., Ill.).
Red Oak (Va., Ga., Ala., Fla., Miss., La., Ind., N. C.).
Spanish Water Oak (La.).

Quercus digitata × velutina[1] n. hyb.

Quercus palustris Muenchh. **Pin Oak.**

SYN.—*Quercus palustris* Muenchhausen, Hausv., V, 253 (1770).

Quercus rubra dissecta Lamarck, Enc. Méth. Bot., I, 720 (1783).

---

[1] Specimens of this form were collected near Covington, Tenn., by Mr. James Byars, on October 9, 1888.. The collector only noted that they came from a "large forest tree," and sent them as *Quercus tinctoria* Bartr. Nothing further is known, except that in the same lot specimens of *Q. velutina* and *Q. digitata* were included, suggesting the probable association of these two species. The size and form of the hybrid leaf is identical with that of a common form of *Q. velutina*, but the under surface bears the characteristic dense coating of pale, tawny tomentum of *Q. digitata*, as also do the petioles and branchlets of the season. The acorns resemble those of *Q. digitata* so closely that they can not be distinguished. The buds are in size between those of *Q. digitata* and *Q. velutina*, but with the characteristic form of the latter.

Quercus palustris Muenchh.—Continued.

Syn.—*Quercus rubra ramosissima* Marshall. Arb. Am., 122 (1785).
Quercus palustris β cucullata Wesmael. in Bull. Féd. Soc.
Hort. Belg., 1869, 346 (1869).
Quercus rubra var. *palustris* Kuntze. Revis. Gen. Pl., Par.
II, 642 (1891).

COMMON NAMES.

Pin Oak (Mass., Conn., R. I., N. Y., Pa., Del., Va., Md., Ark.,
Mo., Ill., Wis., Iowa, Kans.).
Swamp Spanish Oak (Ark., Kans.).
Water Oak (R. I., Ill.).
Swamp Oak (Pa., Ohio, Kans.).
Water Spanish Oak (Ark.).

Quercus pumila[1] (Marsh.) nom. nov.   **Bear Oak.**

Syn.—*Quercus nigra pumila* Marshall, Arb. Am., 122 (1785).
Quercus rubra nana Marsh., l. c., 123 (1785).
QUERCUS ILICIFOLIA Wangenheim, Nordam. Holz. Forst.,
79. t. 6. f. 17 (1787).
Quercus Banisteri Michaux. Hist. Chênes Am., No. 15, t. 27
(1801).
Quercus discolor γ Banisteri Spach. Hist. Vég., XI, 164
(1842).
Quercus nigra var. *ilicifolia* Kuntze (Wang.). Revis. Gen.
Pl., Par. II, 642 (1891).
Quercus ilicifolia arborescens Dippel. Handb. Laubh., zw.
T., 113 (1892).—?
Quercus nana Sargent. in Gard. and For., VIII, 93 (1895),
not Willd. (1805), nor *Q. Phello ε nana* A. de C. (1864).

COMMON NAMES.

Bear Oak.
Barren Oak (Md.).
Dwarf Black Oak.

---

[1] Formerly not included among the arborescent flora. As it usually occurs on poor, dry soil, it is a low, much-branched shrub. Professor Sargent (Silva, VIII, 155, 1895) records its occurrence as a small tree 18 or 20 feet high in Huntingdon County, Pa., and recently I have detected (Montgomery County, Md.) several individuals of this species growing in deep rich soil on the border of an oak forest, where it has reached a height of 12 to 15 feet with a single trunk 3 to 4 inches in diameter.

Prof. C. S. Sargent (l. c.) has proposed for this species the varietal name *nana*, from Marshall's *Q. rubra nana* (l. c.), a name which seems to be well identified with this tree; but I can see no good reason for passing over the same author's *Q. nigra pumila* which occurs on a preceding page (l. c.), where the characters mentioned could certainly apply to no other of the black oaks growing in Maryland or Virginia.

Quercus georgiana[1] Curtis. **Georgia Oak.**

SYN.—*Quercus Georgiana* Curtis, in Am. Journ. Sci., ser. 2, VII, 406 (1849).

Quercus georgiana × marilandica[2] Sargent.

SYN.—*Quercus Georgiana × Q. nigra* Small, in Bull. Torr. Bot. Club, XXII, 75, Pl. 233 (1895).
Quercus Georgiana × Marilandica Sargent, Silva, VIII, 159 (1895).

Quercus marilandica[3] Muenchh. **Black Jack.**

SYN.—*Quercus nigra* β Linnæus, Spec. Pl., ed. 1, II, 996 (1753).
Quercus Marilandica Muenchhausen, Hausv., V, 253 (1770).
Quercus Marylandica Du Roi, Harbk. Baumz., II, 274 (1772).
Quercus nigra Wangenheim, Besch. Nordam. Holz., 133 (1781), not Linn. (1753).
Quercus nigra β *latifolia* Lamarck, Enc. Méth. Bot., I, 721 (1783).
Quercus nigra integrifolia Marshall, Arb. Am., 121 (1785).
Quercus ferruginea Michaux f., Hist. Arb. Am., II, 92, t. XVIII (1812).
Quercus nigra β *quinqueloba* A. de Candolle, in de C., Prodr., XVI, sect. 2, 64 (1864).
Quercus quinqueloba Engelmann, in Trans. Acad. Sci. St. Louis, III, 542 (1877).
Quercus nigra var. *triloba* Kuntze, Rev. Gen. Pl., Par. II, 641 (1891).
Quercus ferruginea a *hybrida* Dippel, Handb. Laubh., zw. T., 111 (1892), not *Q. laurifolia hybrida* Michx. (1801), nor *Q. aquatica* var. *hybrida* Chapm. (1860), nor *Q. hybrida* Dippel (1892).

COMMON NAMES.

Black Jack (Pa., Del., W. Va., N. C., S. C., Ga., Ala., Miss., La., Tex., Ark., Mo., Ill., Ind., Kans., Nebr., Mich., Minn., Iowa, S. Dak.).
Jack Oak (N. Y., W. Va., Miss., Tex., Mo., Ill., Kans., Nebr., S. Dak., Ohio).

---

[1] Formerly known only as low shrub and only from Stone Mountain, Dekalb County, Ga. Recently Dr. J. K. Small (1893) discovered it growing on Little Stone Mountain, 9 miles south of Stone Mountain; on a granite hill 18 miles east of Stone Mountain, and also on another granite hill 12 miles east of the first. It occasionally reaches 20 to 30 feet in height, with a diameter of 6 to 14 inches.

[2] A small hybrid oak, under 30 feet in height, found by Dr. J. K. Small, on the north slope of Stone Mountain, Dekalb County, Ga., in 1894.

[3] The discovery that Linnæus' *Quercus nigra*, so long applied to this species, belongs to the Water Oak, makes it necessary to take up the next oldest name for the Black Jack, which is the *Quercus marilandica* of Muenchhausen.

Quercus marilandica Muenchh.—Continued.

Syn.—Iron Oak (Tenn.).
    Black Oak (Ark., Wis.).
    Barren Oak (Kans., Tenn.).
    Barrens Oak (Fla.).
    Scrub Oak (S. C.).

Quercus brittoni[1] Davis.   **Britton Oak.**

Syn.—*Quercus Brittoni* Davis, in Sc. Am., LXVII, 145 (1892); in
    Bull. Torr. Bot. Club, XIX, 301 (1892).
    Quercus nigra × *ilicifolia* Coulter, in Mem. Torr. Bot. Club,
    5, 133 (1894).
    Quercus Marilandica × *nana* Sargeut, Silva, VIII, 189 (index)
    (1895).

Quercus marilandica × velutina[2] Bush.

Syn.—*Quercus Marilandica* × *velutina* Bush, in Gard. and For.,
    VIII, 464 (1895).

Quercus nigra[3] Linn.   **Water Oak.**

Syn.—*Quercus nigra* Linnæus, Spec. Pl., ed. 1, II, 995 (1753).
    Quercus nigra α *aquatica* Lamarck, Enc. Méth. Bot., I, 721
    (1783).
    Quercus nigra trifida Marshall, Arb. Am., 121 (1785).—?
    Quercus uliginosa Wangenheim, Nordam. Holz., 80, t. 6, f. 18
    (1787).
    Quercus aquatica Walter, Fl. Caroliniana, 234 (1788).
    Quercus hemisphærica Willdenow, Sp. Pl., IV, Par. I, 443
    (1805).
    Quercus nana Willd., l. c. (1805).
    Quercus aquatica α *cuneata* Aiton, Hort. Kew., ed. 2, V, 290
    (1813), not *Q. cuneata* Wang. (1787).
    Quercus aquatica γ *elongata* Ait., l. c. (1813).
    Quercus aquatica δ *indivisa* Ait., l. c. (1813).

---

[1] Mr. W. T. Davis describes (l. c.) a group of shrubby trees found by him at
Watchogue, Staten Island, N. Y., in 1892. He supposed them to be a hybrid between
the Black Jack and Bear Oak.

[2] A new hybrid discovered by Mr. B. F. Bush (on September 20, 1895) near Sapula,
Ind. Ter. It is described as a slender tree, in its trunk and branches resembling *Q.
marilandica*, and in branchlets and buds like *Q. velutina*. Found growing in a rocky
hollow associated with *Q. texana*, *Q. marilandica*, *Q. minor*, and *Q. velutina*.

[3] Linnæus' *Quercus nigra* has been long supposed to apply to the Black Jack, to
which his short description "*foliis cuneiformibus obsolete trilobis*" seems to point.
His additional citation, however, of Catesby's plate (Nat. Hist. Car. Fla. and Bahama Isl., 20, t. 20, 1731), as representing the plant he had in hand, leaves no doubt
whatever but that he named our Water Oak *Quercus nigra*, and not the Black Jack.
The oak figured by Catesby being none other than the Water Oak, which must now
bear the name *Quercus nigra*.

Quercus nigra Linn.—Continued.

SYN.—*Quercus ε attenuata* Ait., l. c. (1813).
Quercus hemisphærica β nana Nuttall, Gen., II, 214 (1818).
Quercus aquatica var. *hybrida* Chapman, Fl. S. States, ed. 1,
421 (1860).
Quercus nigra Koch, Dendrol., zw. Th. erst. Ab., 61 (1873), in
part.

COMMON NAMES.

Water Oak (Del., N. C., S. C., Ala., Fla., Miss., La., Tex.,
Ark., Mo.).
Spotted Oak (Tex., Ala.).
Duck Oak.
Possum Oak.
Punk Oak.

Quercus laurifolia Michx.   **Laurel Oak.**

SYN.—*Quercus laurifolia* Michaux, Hist. Chênes Am., No. 10, t. 17
(1801).
Quercus laurifolia hybrida Michx., l. c., t. 18 (1801).
Quercus laurifolia α acuta Willdenow, Sp. Pl., IV, Par. I,
428 (1805).
Quercus laurifolia β obtusa Willd., l. c. (1805).
Quercus obtusa Pursh, Fl. Am. Sept., II, 627 (1814).
Quercus Phellos var. *laurifolia* Chapman, Fl. S. States, ed. 1,
420 (1860).
Quercus aquatica β laurifolia A. de Candolle, in de C., Prodr.,
XVI, sect. 2, 68 (1864).

COMMON NAMES.

Laurel Oak (N. C., S. C., Ala., Fla.).
Swamp Laurel Oak (Tenn.).
Darlington Oak (S. C.).
Willow Oak (Fla., S. C.).
Water Oak (Ga.).

Quercus brevifolia (Lam.) Sargent.   **Blue Jack.**

SYN.—*Quercus Phellos β brevifolia* Lamarck, Enc. Méth. Bot., I,
722 (1783).
Quercus humilis Walter, Fl. Caroliniana, 234 (1788), not Lam.
(1783), nor Marsh. (1785).
Quercus Phellos β sericea Aiton, Hort. Kew., ed. 1, III, 354
(1789).
Quercus Phellos β latifolia Castiglioni, Viag. negli Stati
Uniti, II, 345 (1790), not Marsh. (1785).
Quercus Phellos β Smith, in Abbot & Smith, Insects of Ga.,
II, 103, t. 52 (1797).

Quercus brevifolia (Lam.) Sargent—Continued.

Syn.—Quercus cinerea Michaux, Hist. Chênes Am., No. 8, t. 14 (1801).

Quercus Phellos β humilis Pursh., Fl. Am. Sept., II, 625 (1814).

Quercus cinerea β dentato-lobata A. de Candolle, in de C. Prodr., XVI, sect. 2, 73 (1864).

Quercus cinerea γ humilis A de C., l. c., 74 (1864).

COMMON NAMES.

Upland Willow Oak (N. C., Ala., Tex.).
Blue Jack (N. C., Fla., Tex., Ga.).
Sand Jack (Tex.).
High-ground Willow Oak (S. C.).
Turkey Oak (S. C., Ga.).
Shin Oak (Tex.).
Cinnamon Oak (Fla.)

Quercus imbricaria Michx.   **Shingle Oak.**

Syn.—*Quercus imbricaria* Michaux, Hist. Chênes. Am., No. 9, t. 15, 16 (1801).

Quercus Phellos β imbricaria Spach, Hist. Vég., XI, 160 (1842).

Quercus imbricaria β spinulosa A. de Candolle, in de C. Prodr., XVI, sect. 2, 63 (1864).

Quercus imbricaria var. *inæqualifolia* Kuntze, Revis. Gen. Pl., Par. II, 641 (1891).

Quercus imbricaria macrophylla Dippel, Handb. Laubh., zw. T., 105 (1892).

Quercus imbricaria diversifolia Dippel, l. c. (1892).

COMMON NAMES.

Shingle Oak (Del., N. C., S. C., Ky., Mo., Ind., Ill., Kans., Iowa, Nebr.).
Laurel Oak (Pa., Del., S. C., Ky., Ill., Nebr.).
Jack Oak (Ill.).
Water Oak (N. C.).

Quercus tridentata[1] (de C.) Engelm.

Syn.—*Quercus nigra γ tridentata* de Candolle, in de C., Prodr., XVI, sect. 2, 64 (1864).

---

[1]A small tree discovered by Dr. Engelmann in 1849, near St. Louis, Mo., but since destroyed, supposed to be a hybrid between *Q. imbricaria* and *Q. marilandica*. Trees with similar foliage have since been detected near Allentown, Mo., by Mr. George W. Letterman. There are also two small trees with foliage of the same general characters standing in the edge of a forest 6 miles south of Ann Arbor, Mich. *Q. marilandica* and *Q. imbricaria* are growing with them.

Quercus tridentata (de C.) Engelm.—Continued.

SYN.—*Quercus imbricaria* × *nigra* Engelmann, in Trans. Acad. Sci.
St. Louis, III, 539 (1877).
Quercus tridentata Engelm. (in Hb.), l. c. (1877).
Quercus imbricaria × *Marilandica* Sargent, Silva, VIII, 176,
t. CCCCXXXIII (1895).

Quercus leana[1] Nutt.   **Lea Oak.**

SYN.—*Quercus Leana* Nuttall, Sylva, I, 13, t. V (bis) (1842).
Quercus nigra var. ? *leana* Cooper, in Smithsonian Rep. 1858,
255 (1859)—*nomen nudum*.
Quercus imbricaria × *coccinea* Engelmann, in Trans. Acad.
Sci. St. Louis, III, 540 (1877).
Quercus imbricaria × *velutina* Sargent, Silva, VIII, 176, t.
CCCCXXXIV (1895).

Quercus imbricaria × **palustris**[2] Engelm.

SYN.—*Quercus imbricaria* × *palustris* Engelmann, in Trans. Acad.
Sci. St. Louis, III, 539 (1877).

Quercus phellos Linn.   **Willow Oak.**

SYN.—*Quercus Phellos* Linnæus, Spec. Pl., ed. 1, II, 994 (1753).
Quercus Phellos α *longifolia* Lamarck, Enc. Méth. Bot., I,
722 (1783).
Quercus Phellos δ *subrepanda* Lam., l. c. (1783).
Quercus Phellos ε *sublobata* Lam., l. c (1783).
Quercus Phellos angustifolia Marshall, Arb. Am., 124 (1785).
Quercus Phellos latifolia Marsh., l. c. (1785).
Quercus Phellos α *viridis* Aiton, Hort. Kew., ed. 1, III, 354
(1789).
Quercus Phellos (*sylvatica*) Michaux, Hist. Chênes Am., No.
7, t. 12 (1801).

COMMON NAMES.

Willow Oak (R. I., N. Y., Pa., Del., Ala., N. C., S. C., Fla.,
Miss., La., Tex., Ark., Mo.).
Peach Oak (N. J., Del., Ohio).

---

[1]Supposed to be a hybrid between the Shingle and Scarlet or the Shingle and
Black Oak. The first individual was discovered nearly seventy years ago by Mr. T.
G. Lea, near Cincinnati, Ohio. Since that time single individuals have been found
from the District of Columbia and western North Carolina to southern Michigan,
central and northern Illinois, and southeastern Missouri. Usually 40 to 60 feet high,
with a diameter of 8 to 14 inches.

[2]A small tree discovered in 1870 by Dr. Engelmann, 8 miles west of St. Louis, Mo.,
since destroyed, believed to be a hybrid between the Shingle and Pin Oak. He states
(l. c.) that acorns planted by Thomas Meehan, of Germantown, Pa., produced plants
with foliage like the parent.

Quercus phellos Linn.—Continued.

Syn.—Water Oak (S. C.).
Swamp Willow Oak (Tex.).

Quercus phellos × digitata[1] Small.

Syn.—*Quercus Phellos × Q. digitata* Small, in Bull. Torr. Bot. Club,
XXII, 74, Pl. 232 (1895).

Quercus phellos × pumila[2] nom. nov.

Syn.—*Quercus Phellos × ilicifolia* Peters, in Bull. Torr. Bot. Club,
XX. 295 (1893).

Quercus heterophylla[3] Michx. f.   **Bartram Oak.**

Syn.—*Quercus heterophylla* Michaux f., Hist. Arb. Am., II, 87, t.
16 (1812).
Quercus aquatica β heterophylla Aiton. Hort. Kew., ed. 2, V,
290 (1813).
Quercus nigra var. ? *heterophylla* Cooper. in Smithsonian Rep.
1858, 255 (1859).
Quercus Phellos × tinctoria Gray, Manual N. U. S., ed. 4,
406 (1856).
Quercus Phellos var. Gray, l. c., ed. 5, 453 (1867).
Quercus Phellos × coccinea Engelmann, in Trans. Acad. Sci.
St. Louis, III, 541 (1876).
Quercus Phellos × rubra Hollick, in Bull. Torr. Bot. Club,
XV, Pl. 84, 85, 309 (1888).
Quercus Phellos latifolia Loddiges (Cat. ed. 1836) ex Dippel,
Handb. Laubh., zw. T., 106, f. 48 (1892), not Marsh. (1785).
Quercus sonchifolia Booth ex Dippel, l. c. (1892).
Quercus longifolia Hort. ex. Dippel, l. c. (1892).

COMMON NAMES.

Bartram's Oak (Pa., Del., Miss., La., Ala.. Mo.).
Burriers Oak (Lit.).

---

[1] A large hybrid oak 75 to 100 feet high, discovered recently by Dr. J. K. Small on
the "Hills west of the Falls of the Yadkin River (Stanley County), North Carolina."

[2] Specimens of foliage from this supposed hybrid were first collected by Mr. J. C.
Gifford and Rev. J. E. Peters at Mays Landing. N. J., in July, 1890. Rev. Peters
states that a "clump" of low trees, 6 to 8 feet high, was found only in one locality,
the trees much resembling *Q. pumila*, which together with *Q. phellos* and *Q. marilandica* grew very near the hybrid.

[3] The first individual known was described by the younger Michaux (l. c.) in 1812.
The tree grew in a field belonging to John Bartram, near Philadelphia, Pa., but has
long since been destroyed. Groups and single individuals have since been detected
near Camden, N. J.; Wilmington, Del.; Staten Island, New York; Alexandria, Va.;
District of Columbia; western North Carolina; near Falkville, Ala.; Houston, Tex.
The trees are from 25 to 40 feet in height and 8 to 20 inches in diameter.

Quercus subimbricaria[1] (de C.) nom. nov.

SYN.—*Quercus phellos β subimbricaria* de Candolle, in de C., Prodr., XVI, sect. 2, 63 (1864).
Quercus Phellos × *nigra* Britton, in Bull. Torr. Bot. Club, IX, 13, Pl. 10–12, f. 1–7, 9, 10 (1882).
Quercus Rudkini Britton, l. c., 14, Pl. 10–12, f. 1–7, 9, 10 (1882).
Quercus Phellos × *nigra* Watson & Coulter, in Gray, Man. N. States, ed. 6, 479 (1889).
Quercus Phellos × *Marilandica* Sargent, Silva, VIII, 181, t. CCCCXXXVII (1895).

Quercus densiflora Hook. & Arn.   **Tanbark Oak.**

SYN.—*Quercus densiflora* Hooker & Arnott, in Bot. Beechey's Voyage, 391 (1841).
Quercus echinacea Torrey, in Pacif. R. R. Rep., IV, Pt. I, 137, t. 14 (1856).
. *Pasania densiflora* Örsted, in Vidensk. Medd. fra Nat. For. Kjöb. 1866, 83 (1866).

COMMON NAMES.

Tanbark Oak (Cal.).
Chestnut Oak (Cal., Nev.).
California Chestnut Oak (Oreg.).
Peach Oak (Oreg.).
Live Oak (Oreg.).

Quercus densiflora echinoides[2] (R. Br. Campst.) Sargent.

SYN.—*Quercus echinoides* R. Brown Campst., in Ann. Mag. Nat. Hist., ser. 4, VII, 251 (1871).
Quercus densiflora var. *echinoides* Sargent, Silva, VIII, 183, t. CCCCXXXVIII, f. 9 (1895).
Quercus densiflora Greene, West. Am. Oaks, Pt. I, Pl. 24 (1889).

---

[1] Supposed to be a hybrid between the Willow and Black Jack Oak. The probable parentage was first indicated in 1881, when Messrs. W. H. Rudkin and W. Brown discovered about ten individuals between Keyport and South Amboy, N. J., for which Dr. Britton (l. c.) proposed the name of *Quercus rudkini*, indicating at the same time its origin as a hybrid between *Q. phellos* and *Q. marilandica*. It is now thought to be more or less common on Staten Island, New York, and trees of probably similar parentage have been detected by Dr. J. K. Small at the Falls of the Yadkin River, North Carolina (1892), and by Mr. Ravenel near Aiken, S. C.

[2] A low, shrubby form confined to high altitudes in southern Oregon and northern California, apparently connected with the larger-leaved species (*Q. densiflora*) by small trees with foliage intermediate in form and character.

Family ULMACEÆ.

ULMUS Linn., Spec. Pl., 225 (1753).

Ulmus crassifolia Nutt.   **Cedar Elm.**

SYN.—*Ulmus crassifolia* Nuttall, in Trans. Am. Phil. Soc., n. ser.,
V, 169 (1837).
Ulmus opaca Nutt., Sylva, I, 35, t. 11 (1842).
Ulmus americana opaca Browne, Trees Am., 503 (1846).

COMMON NAMES.

Cedar Elm (Tex.).
Red Elm (Tex.).
Basket Elm (Ark.).

Ulmus pubescens Walter.   **Slippery Elm.**

SYN.—*Ulmus pubescens* Walter, Fl. Caroliniana, 112 (1788).
Ulmus Americana α rubra Aiton, Hort. Kew., I, 319 (1789).
ULMUS FULVA Michaux, Fl. Bor.-Am., I, 172 (1803).
Ulmus crispa Willdenow, Enum., 295 (1809).
Ulmus rubra Michaux f., Hist. Arb. Am., III, 278, t. 6 (1813).
Ulmus pinguis Rafinesque, Fl. Ludovic., 115 (1817).
Ulmus americana fulva Browne, Trees Am., 501 (1846).

COMMON NAMES.

Slippery Elm (Vt., N. H., Mass., R. I., N. Y., N. J., Pa., Del.,
Va., W. Va., N. C., S. C.. Fla., Ala., Ga., Miss., La., Tex.,
Ky., Mo., Kans., Nebr., Ill., Ind., Wis., Mich., Ohio, Ont.,
Iowa, Minn.).
Red Elm (Vt., Mass., N. Y.. Del.. Pa.. W. Va., S. C., Ala., Miss.,
La., Tex., Ark.. Ky., Mo., Ill., Kans., Nebr., Iowa, Ohio,
Ont., Wis., Mich., Minn.).
Red-wooded Elm (Tenn.).
Rock Elm (Tenn.).
Orme-gras (La.).
Moose Elm.
Oo-hoosk-ah="It slips" (Indians, N. Y.).

Ulmus americana Linn.   **White Elm.**

SYN.—*Ulmus Americana* Linnæus, Spec. Pl., ed. 1, I, 226 (1753).
Ulmus mollifolia Marshall, Arb. Am., 156 (1785).
Ulmus Americana β alba Aiton, Hort. Kew., ed. 1, I, 320
(1789).
Ulmus oborata Rafinesque, New Fl., 3d Pt., 38 (1836).
Ulmus dentata Raf., l. c. (1836).

Ulmus americana Linn.—Continued.

Syn.—*Ulmus sessilis* Raf., l. c. (1836).
    Ulmus Americana β scabra Spach, in Ann. Sc. Nat., sér. 2,
      XV, 364 (1841), not *U. scabra* Mill. (1768).
    Ulmus Americana subsessilifolia Browne, Trees Am., 500
      (1846).
    Ulmus grandidentata Browne, l. c. (1846).
    Ulmus Americana obovata Browne, l. c. (1846).
    Ulmus Americana α glabra Walpers, Ann., III, 424 (1852).
    Ulmus Americana γ ? Bartramii Walp., l. c. (1852).
    Ulmus Americana var. *? aspera* Chapman, Fl. S. States, 416
      (1865).
    Ulmus Floridana Chap., l. c. (1865).
    Ulmus Americana a *nigricans* Dippel, Handb. Laubh., zw. T.,
      33 (1892).

COMMON NAMES.

American Elm (Vt., Mass., R. I., N. Y., Del., Pa., N. C., Miss.,
    Tex., Ill., Ohio, Kans., Nebr., Mich., Minn., Ont.).
White Elm (Me., N. H., Vt., Mass., R. I., N. Y., Pa., N. J.,
    Del., Va., W. Va., N. C., S. C., Ala., Fla., Miss., La., Tex.,
    Ark., Ky., Mo., Ill., Ind., Kans., Nebr., Ohio, Ont., Iowa,
    Mich., Minn., N. Dak., S. Dak.).
Water Elm (Miss., Tex., Ark., Mo., Ill., Iowa, Mich., Ohio,
    Minn., Nebr.).
Elm (Mass., R. I., Conn., N. J., Pa., N. C., S. C., Iowa, Wis.).
Orme Maigre (La.).
Swamp Elm.
Rock Elm.

VARIETY DISTINGUISHED IN CULTIVATION.

Ulmus americana pendula Ait.   **Weeping American Elm**.
Syn.—*Ulmus Americana γ pendula* Aiton, Hort. Kew., ed. 1, 320
      (1789).
    Ulmus pendula Willdenow, Berl. Baumz., zw. Ausg., 519
      (1811).
    Ulmus alba Rafinesque, Fl. Ludovic., 115 (1817).

Ulmus racemosa Thomas.   **Cork Elm**.
Syn.—*Ulmus racemosa* Thomas, in Am. Journ. Sci., XIX, 170
      (1831).
    Ulmus Americana Planchon, in de Candolle, Prodr., XVII,
      155 (1873), in part; not Linn. (1753).

COMMON NAMES.

Cork Elm (Vt., Mass., R. I., N. Y., N. J., Ark., Ky., Mo., Wis.,
    Mich., Ohio, Iowa).

Ulmus racemosa Thomas—Continued.

SYN.—Rock Elm (R. I., W. Va., Ark., Ky., Mo., Ill., Wis., Iowa,
  Mich., Nebr., Ont.).
  Hickory Elm (Ark., Mo., Ill., Ind., Iowa).
  White Elm (Ont.).
  Thomas Elm (Tenn.).
  Northern Cork-barked Elm (Tenn.).
  Corkbark Elm (N. Y.).
  Northern Cork Elm (Vt.).
  Wahoo (Ohio).
  Cliff Elm (Wis.).
  Corky White Elm.

Ulmus alata Michx.   **Wing Elm.**

SYN.—*Ulmus pumila* Walter, Fl. Caroliniana, 111 (1788), not Linn.
  (1753).
  Ulmus alata Michaux, Fl. Bor.-Am., I, 173 (1803).
  Ulmus longifolia Rafinesque, New Fl., 3d Pt., 38 (1836).
  Ulmus Americana γ *alata* Spach, in Ann. Sc. Nat., sér. 2,
    XV, 364 (1841).
  Ulmus dimidiata Raf., l. c., 39 (1836).—?
  Ulmus americana longifolia Browne, Trees Am., 502 (1846).
  Ulmus americana dimidiata Browne, l. c., 503 (1846).—?

COMMON NAMES.

Winged-Elm (N. C., S. C., Ark., Tex., Ill., Ind.).
Wahoo (W. Va., N. C., S. C., La., Tex., Ky., Mo.).
Wahoo Elm (Mo.).
Witch Elm (W. Va.).
Elm (W. Va.).
Cork Elm (Fla., S. C., Tex.).
Water Elm (Ala.).
Small-leaved Elm (N. C.).
Red Elm (Fla., Ark.).
Whahoo (S. C.).
Corky Elm (Tex.).
Mountain Elm (Ark.).

PLANERA Gmelin, Syst. Nat., II, 150 (1791).

Planera aquatica (Walt.) Gmel.   **Planertree.**

SYN.—*Anonymos aquatica* Walter, Fl. Caroliniana, 230 (1788).
  Planera aquatica Gmelin, Syst. Nat., II, Pt. I, 150 (1791).
  Planera ulmifolia Michaux f., Hist. Arb. Am., III, 283, t. 7
    (1813).

Planera aquatica (Walt.) Gmel.—Continued.

SYN.—*Ulmus aquatica* Rafinesque, Fl. Ludovic., 165 (1817).
Planera americana Hort. ex Steudel, Nom. Bot., ed. sec., II, 347 (1841).
Planera Richardi Torrey & Gray, in Pac. R. R. Rep., II, 175 (1855), not Michx. (1803).

COMMON NAMES.

American Plane Tree (Ala.).
Planer Tree (N. C., S. C., Fla., La., Tex., Ark., Tenn.).
Water Elm (Fla.).
Sycamore (N. C.).
Plene (La.).

CELTIS[1] Linn., Spec. Pl., 1043 (1753).

Celtis occidentalis Linn. **Hackberry.**

SYN.—*Celtis occidentalis* Linnæus, Spec. Pl., ed. 1, II, 1044 (1753).
Celtis obliqua Mœnch, Meth., 344 (1794).
Celtis procera Salisbury, Prodr., 175 (1796).
Celtis crassifolia Lamarck, Enc. Méth. Bot., IV, 138 (1797).
Celtis cordifolia L'Heritier, in Nouv. Duham., II, 37 (1802).
Celtis occidentalis var. *scabriuscula* Willdenow, Sp. Pl., IV, Par. II, 995 (1805).

---

[1] The following varieties and species of shrubs are included to complete the list found in the United States:

Celtis occidentalis pumila (Pursh) Gray. **Shrubby Hackberry.**
SYN.—*Celtis pumila* Pursh, Fl. Am. Sept., I, 200 (1814).
Celtis tenuifolia Nuttall, Gen., I, 202 (1818).
Celtis occidentalis var. *pumila* Gray, Man. N. States, ed. 2, 373 (1856).

Celtis tala pallida (Torr.) Planch. **Granjeno.**
SYN.—*Celtis pallida* Torrey, in Mex. Bound. Surv., 203, t. 50 (1859).
Celtis Tala ε pallida Planchon, in de C., Prodr., XVII, 191 (1873).

Celtis iguanæus (Jacq.) Sargent.
SYN.—*Rhamnus iguanæus* Jacquin, Enum. Pl. Carib., 16 (1760).
Celtis aculeata Swartz, Prodr., 53 (1788).
Zizyphus iguanea Lamarck, Enc. Méth. Bot., III, 318 (1789).
Celtis rhamnoides Willdenow, Sp. Pl., IV, Par. II, 998 (1805).
Mertensia zizyphoides Humboldt, Boupland & Kunth, Nov. Gen. Sp., II, 31 (1817).
Zizyphus commutata Roemer & Schultes, Syst., V, 336 (1819).
Mertensia rhamnoides Roem. & Schult., l. c., VI, 313 (1820).
Momisia Ehrenbergiana Klotzsch, in Linnæa, XX, 538 (1847).
Momisia aculeata Klotzsch, l. c., 539 (1847).
Celtis Ehrenbergiana Liebmann, in Dansk. Vid. Selsk. Skrift., ser. 5, II, 339 (1851).
Celtis iguanæus Sargent, Silva, VII, 64 (1895).

184

Celtis occidentalis Linn.—Continued.

Syn.—*Celtis occidentalis* var. *tenuifolia* Persoon, Syn.. I, 292 (1805).
Celtis cordata Pers., l. c. (1805).
Celtis occidentalis var. *cordata* Willdenow. Berl. Baumz., zw.
Ausg.. 82 (1811).
Ulmus occidentalis cordata Willd., l. c. (1811).
Celtis alba Rafinesque. Fl. Ludovic., 25 (1817).
Celtis canina Raf.. in Am. Monthly Mag. and Crit. Rev.. II,
43 (1817).
Celtis maritima Raf.. l. c.. 44 (1817).
Celtis grandidentata Tenore, Ind. Sem. Hort.. 15 (1833).
Celtis urticifolia Rafinesque, New. Flor. 3d. pt.. 32 (1836).
Celtis morifolia Raf.. l. c., 34 (1836).
Celtis heterophylla Raf., l. c.. 37 (1836).
Celtis patula Raf., l. c. (1836).
Celtis Floridana Raf., l. c. (1836).
Celtis parvifolia Raf.. l. c.. 36 (1836).
Celtis crassifolia var. *tiliaefolia* Spach. in Ann. Sc. Nat..
sér. 2, XVI, 39 (1841).
Celtis crassifolia var. *morifolia* Spach, l. c. (1841).
Celtis crassifolia var. *eucalyptifolia* Spach. l. c.. 40 (1841).
Celtis occidentalis var. *grandidentata* Spach, l. c. (1840).
Celtis occidentalis var. *serrulata* Spach. l. c.. 41 (1841).
Celtis Audibertiana Spach. l. c. (1841).
Celtis Audibertiana var. *orata* Spach. l. c. (1841).
Celtis Audibertiana var. *oblongata* Spach, l. c. (1841).
Celtis Douglasii Planchon, in Ann. Sc. Nat.. sér. 3, X, 293
(1848).
Celtis occidentalis var. *crassifolia* Gray. Man. N. States, ed.
2, 397 (1856).
Celtis reticulata Cooper. in Am. Nat.. III. 407 (1869), not *C.
occid.* var. *reticulata* Torr. (1828).
Celtis aspera Koch. Dendrol., zw. T. erst. Ab.. 431 (1872).
Celtis scabra Koch. l. c. (1872).
Celtis occidentalis a *Audibertiana* Koch. l. c. (1872).
Celtis occidentalis b *aspera* Dippel. Handb. Laubh.. zw. T..
44 (1892).

COMMON NAMES.

Hackberry (N. H.. Vt.. R. I.. N. Y.. N. J.. Del., Pa.. W. Va..
N. C.. S. C.. Ala.. Fla.. Miss.. La.. Tex.. Ariz.. Ark.. Ky..
Mo., Ill.. Ind.. Wis.. Iowa, Kans.. Nebr.. Mich.. Minn..
S. Dak.. Ohio. Ont.).
Sugarberry (N. Y.. Pa.. Del., N. C.. S. C.. Minn.).
Nettle Tree (R. I.. Mass.. Del.. Mich.).
American Nettle Tree (Tenn.).

Celtis occidentalis Linn.—Continued.

SYN.—Hoop Ash (Vt.).
 Ope Berry (R. I.).
 Hack Tree (Minn.).
 Juniper Tree (N. J.).
 Bastard Elm (N. J.).

Celtis occidentalis reticulata (Torr.) Sargent.   **Palo Blanco.**

SYN.—*Celtis reticulata* Torrey, in Ann. Lyc. N. Y., II, 247 (1828).
 Celtis brevipes Watson, in Proc. Am. Acad. Sci., 3 ser., XVI, 297 (1879).
 Celtis occidentalis var. *reticulata* Sargent, in Rep. Tenth Census U. S., IX, (Cat. For. Trees N. A.) 126 (1884).
 Celtis Mississippiensis var. *reticulata* Sargent, Silva, VII, 72, t. CCCXIX (1895).

COMMON NAMES.

 Hackberry (Tex.).
 Palo Blanco (Tex.).

Celtis mississippiensis Bosc.   **Sugarberry.**

SYN.—*Celtis Mississippiensis* Bosc, Enc. Méth. Agric., nouv. éd., 41 (1810).
 Celtis lævigata Willdenow, Berl. Baumz., zw. Ausg., 81 (1811).
 Celtis occidentalis β *integrifolia* Nuttall, Gen., I, 202 (1818).
 Celtis longifolia Rafinesque, in Atl. Journ., 177 (1833).
 Celtis salicifolia Raf., l. c., 34 (1836).
 Celtis fuscata Raf., New Fl. and Bot., 3d Pt., 33 (1836).
 Celtis integrifolia Nuttall, in Trans. Am. Phil. Soc., new ser., V, 169 (1837), not Lam. (1797).
 Celtis Berlandieri Klotzsch, in Linnæa, XX, 541 (1847).
 Celtis Texana Scheele, in Linnæa, XXII, 146 (1849).
 Celtis Lindheimeri Engelmann ex Koch, Dendrol., zw. Th. erst. Ab., 434 (1872).
 Celtis Americana Hort. ex Planchon, in de C., Prodr., XVII, 176 (1873), not Mill. (1768).
 Celtis occidentalis Sargent, in Tenth Cen. U. S., IX (Cat. For. Trees N. A.), 125 (1884), in part; not Linn. (1753).

COMMON NAMES.

 Sugar Berry (Fla., Ala., Miss.).
 Connu (La.).
 Bois inconnu (La.).

Family MORACEÆ.

MORUS Linn., Spec. Pl., 986 (1753).

Morus rubra Linn.   **Red Mulberry.**

SYN.—*Morus rubra* Linnæus, Spec. Pl., ed. 1, II, 986 (1753).
Morus canadensis Poiret. in Lamarck, Enc. Méth. Bot., IV, 380 (1797).
Morus scabra Willdenow, Enum., 967 (1809).
Morus tomentosa Rafinesque, Fl. Ludovic., 113 (1817).
Morus Pennsylvanica Nois. ex Loudon, Hort. Brit., 378 (1830).
Morus riparia Rafinesque, New Fl. and Bot., 3d Pt., 46 (1836).
Morus rubra var. *pallida* Raf., l. c. (1836).
Morus parvifolia Raf., l. c., 47 (1836).
Morus rubra var. *heterophylla* Raf., l. c. (1836).
Morus rubra var. *canadensis* Loudon, Arb. Frut., III, 1360 (1838).
Morus reticulata Rafinesque. Am. Man. Mulberry Trees, 28 (1839).
Morus rubra var. *purpurea* Raf., l. c. (1839).
Morus Missouriensis Audibert ex Moretti, in Giorn. Lomb. Sc., I, 181 (1841).
Morus Caroliniana Hort. ex Moretti, l. c. (1841).
Morus rubra β *tomentosa* Bureau, in de Candolle, Prodr., XVII, 246 (1873).
Morus rubra γ *incisa* Bureau, in de C., l. c. (1873).

COMMON NAMES.

Red Mulberry (Me., Vt., Mass., R. I., N. Y., N. J., Pa., Del., Va., W. Va., N. C., Fla., Ala., Ga., Miss., La., Tex., Ark., Ky., Mo., Ill., Ind., Kans., Nebr., Mich., Minn., Iowa, Ohio, Ont.).
Mulberry (Pa., N. C., S. C., Fla., Ala., Ark., Tex., Ky., Mo., Ill., Iowa, Nebr., Ohio).
Black Mulberry (N. J., Pa., W. Va.).
Virginia Mulberry Tree (Tenn.).
Murier Sauvage (La.).

Morus celtidifolia H., B. & K.   **Mexican Mulberry.**

SYN.—*Morus celtidifolia* Humboldt. Bonpland & Kunth. Nov. Gen. Sp., II, 33 (1817).
Morus mexicana Bentham, Pl. Hartweg., 71 (1839).

Morus celtidifolia H., B. & K.—Continued.

SYN.—*Morus microphylla* Buckley, in Proc. Phila. Acad. Nat. Sci. 1862, 8 (1862).

COMMON NAME.

Mexican Mulberry (Tex.).

Morus alba[1] Linn.   **White Mulberry.**

SYN.—*Morus alba* Linnæus, Spec. Pl., ed. 1, II, 986 (1753).

Morus rubra Laureiro, Fl. Coch., 555 (1790), not Linn. (1753).

Morus italica Poiret ex Lamarck, Enc. Méth. Bot., IV, 377 (1797).

Morus pumila Balbis, Cat. Hort. Taur. App., 52 (1813).

Morus atropurpurea Roxburgh, Hort. Beng., 67 (1814).

Morus serrata Wallich, Cat. No. 4648 (1828), not Roxb. (1814).

Morus bullata Balbis ex Loudon, Arb. Frut., III, 1348 (1838).

Morus alba nervosa Loddiges Cat. ed. (1836) ex Loud., l. c., 1349 (1838).

Morus nervosa Bon Jard. (1836) ex Loud., l. c. (1838).

Morus subalba nervosa Loddiges Cat. ed. (1836) ex Loud., l. c. (1838).

Morus alba italica Hort. ex Loud., l. c. (1838).

Morus hispanica Hort. ex. Loud., l. c. (1838).

Morus alba membranacea Lodd., l. c. (1838).

Morus heterophylla Loud., l. c., 1361 (1838).

Morus Integrifolia Rafinesque, Am. Man. Mulberry Trees, 16 (1839).

Morus Lobata Raf., l. c. (1839).

Morus Columbiana Raf., l. c. (1839).

Morus Rubella Raf., l. c. (1839).

Morus Spherica Raf., l. c. (1839).

Morus Cinerea Raf., l. c., 17 (1839).

Morus Hispanica prolifera Raf., l. c. (1839).

Morus Syriaca Raf., l. c. (1839).

Morus Lactiflua Raf., l. c., 18 (1839).

Morus Mariettii Hort. ex Steudel, Nom. Bot., ed. sec., II, 162 (1841).

Morus membranacea Hort. ex. Steudel, l. c. (1841).

Morus subalba Hort. ex Steudel, l. c. (1841).

Morus furcata Hort. ex Steudel, l. c. (1841).

---

[1] The wide introduction of this exotic species in the United States has resulted in its becoming perfectly naturalized in many localities, notably in the south Atlantic region and in the central Western States, where it is running wild.

Morus alba Linn.—Continued.

Syn.—*Morus Guzziola* Hort. ex Steudel, l. c. (1841).
Morus Venassaini Hort. ex Steudel, l. c. (1841).
Morus tortuosa Audibert ex Moretti, in Giorn. Lomb. Sc.,
   I, 181 (1841).
Morus patavina Hort. ex Spach, l. c. (1842).
Morus romana Loddiges ex Spach, l. c., 45 (1842).
Morus stylosa Seringe. Descr. Mûr., 225 (1855).
Morus Moretti Audibert ex Bureau, in de Candolle, Prodr.,
   XVII, 39 (1873).
Morus serotina Mart. ex Bur., l. c. (1873).

VARIETIES DISTINGUISHED IN CULTIVATION.

Morus alba tatarica (Linn.) Loud.   **Russian Mulberry.**

Syn.—*Morus tatarica* Linnæus, Spec. Pl., ed. 1, II. 986 (1753).
Morus alba tatarica Loudon, Arb. Frut., III. 1358, f. 1225
   (1838).

Morus alba rosea Loud.

Syn.—*Morus alba rosea* Hort. ex Loudon, Arb. Frut., III, 1349
   (1838).
Morus lucida Hort. ex Lond., l. c., 1350 (1838).
Morus rotundifolia Rafinesque, Am. Man. Mulberry Trees,
   16 (1839).
Morus alba α vulgaris, s. v. rosea Bureau, in A. de Candolle,
   Prodr., XVII, 239 (1873).
Morus alba lucida Audib. ex Dippel. Handb. Laubh., zw. T.,
   9 (1892).
Morus rosacea Hort. ex Dippel, l. c. (1892).

Morus alba macrophylla (Moretti) Loud.   **Largeleaf Mulberry.**

Syn.—*Morus macrophylla* Moretti. Del. Sem. Hort. (1829).
Morus alba macrophylla, Loddiges Cat. ed. (1836) ex London,
   Arb. Frut., III. 1349 (1838).
Morus chinensis Hort. Lodd., l. c., ex Loud., l. c., 1350 (1838).
Morus alba Morettiana Lodd., l. c. (1838).
Morus alba romana Lodd., l. c. (1838).
Morus alba oratifolia ex Loud., Arb. Frut., III, 1350 (1838).
Morus alba α vulgaris, s. v. macrophylla Bureau, in A. de
   Candolle, Prodr., XVII. 239 (1873).
Morus alba latifolia Hort. ex Bur., l. c. (1873), not Spach
   (1842).
Morus latifolia Hort. ex Spach, Hist. Vég., XI, 42 (1842),
   not Poir. in Lam. (1797).

Morus alba vulgaris tokwa (Sieb.) Bureau.

SYN.—*Morus Tokwa* Siebold (ined.), Cat. Rais. Pl. Jap., 5 (1856);
Koch, Dendrol., zw. Th. erst. Ab., 447 (1872).
Morus alba α vulgaris, s. v. *Tokwa* Bureau, in A. de Candolle, Prodr., XVII, 240 (1873).
Morus japonica Audibert ex Seringe, Desc. Mûr., 226 (1855).

Morus alba pyramidalis Seringe.

SYN.—*Morus alba pyramidalis* Seringe, Desc. Mûr., 212 (1855).
Morus alba (untera.) *vulgaris* e *pyramidalis* Dippel, Handb.
Laubh., zw. T., 10 (1892).
Morus alba fastigiata Hort. ex Dippel, l. c. (1892).
Morus fastigiata Hort. ex Dippel, l. c. (1892).

Morus alba pendula (Dipp.) nom. nov.

SYN.—*Morus alba* (untera.) *vulgaris*, f *pendula* Hort. ex Dippel,
Handb. Laubh., zw. T., 10 (1892).

Morus alba constantinopolitana (Poir.) Loud.

SYN.—*Morus Constantinopolitana* Poiret, in Lamarck, Enc. Méth.
Bot., IV, 381 (1797).
Morus Byzantina Sieber, Herb. Fl. Cret. (1820).
Morus alba Constantinopolitana Loudon, Arb. Frut., III,
1358 (1838).
Morus alba (untera.) *vulgaris* Dippel, Handb. Laubh., zw. T.,
10 (1892).

Morus alba multicaulis (Parrot.) Loud.

SYN.—*Morus multicaulis* Parrottet, in Ann. Soc. Linn. Par., III, 129
(1825).
Morus sinensis Hort. ex Loudon, Hort. Brit., 378 (1830).
Morus cucullata Bonafous, Cult. Mûr., 7 (1831).
Morus intermedia Parrottet, in Arch. Bot., I, 234 (1833).
Morus alba multicaulis Loudon, Arb. Frut., III, 1348 (1838).
Morus alba sinensis Hort. ex Loud., l. c., 1350 (1838).
Morus alba latifolia Hort. ex Spach, Hist. Vég., XI, 45 (1842).
Morus alba Lhou Seringe, Desc. Mûr., 208 (1855).
Morus alba (untera.) *multicaulis* Dippel, Handb. Laubh.,
zw. T., 11 (1892).

Morus alba venosa Delile.

SYN.—*Morus alba venosa* Delile, in Bull. Soc. d'Ag. d'Her., XIII,
328 (1826).
Morus nervosa Delile, in Spach, Hist. Vég., II, 33 (1834).
Morus alba nervosa Loddiges, Cat. ed. (1836)—*nomen nudum.*
Morus venosa Delile ex Spach, Hist. Vég., XI, 43 (1842).

Morus alba venosa Delile—Continued.

SYN.—*Morus alba fibrosa* Seringe, Desc. Mûr., 212 (1855).
Morus alba Urticæfolia Koch, Dendrol.,zw. Th. erst. Ab., 443 (1872).
Morus urticæfolia Hort. ex Dippel. Handb. Laubh.,zw. T..12 (1892).
Morus alba (untera.) *renosa* Dippel. l. c. (1892).

TOXYLON[1] Raf.. in Am. Month. Mag.. II. 118 (1817).

Toxylon pomiferum[2] Raf.   **Osage Orange.**

SYN.—*Ioxylon pomiferum* Rafinesque. in Am. Month. Mag., II. 118 (1817).
MACLURA AURANTIACA Nuttall, Gen., II, 234 (1818).
Broussonetia tinctoria Sprengel. Syst. Nat., II, 901 (1825), not H., B. & K. (1817).
Toxylon Maclura Rafinesque. New Fl. 3d Pt., 43 (1836).

COMMON NAMES.

Osage Orange (Mass., R. I., N. Y., N. J., Pa., Del., Va., W. Va.. N. C., S. C., Ga.. Ala., Miss., La.. Tex., Ky.. Mo., Ill., Kaus., Nebr., Iowa, Mich., Ohio).
Bois D'Arc (La., Tex., Mo.).
Bodock (Kans.).
Mock Orange (La.).
Bow Wood (Ala.).
Osage Apple Tree (Tenn.).
Yellow Wood (Tenn.).
Hedge (Ill.).
Hedge Plant (Iowa, Nebr.).
Osage (Iowa).

VARIETY DISTINGUISHED IN CULTIVATION.

Toxylon pomiferum inerme[3] André.   **Thornless Osage Orange.**

SYN.—*Maclura aurantiaca inerme* André, in Rev. Hort. 1896, 33, t. 10 (1896).
Toxylon pomiferum inerme André, l. c. (1896).

---

[1] The first appearance of *Toxylon* is recorded in Rafinesque's New Flora and Botany, Part I, 43 (1836); the author states (l. c., 12). however, that *Ioxylon* Raf., which appeared much earlier (Am. Month. Mag., II, 118, 1817) is a misprint for *Toxylon,* a printer's error and poor proof reading allowing an I to stand for T.

[2] Supposed to be indigenous in southwestern Arkansas, southeastern Indian Territory, and northern Texas. Widely distributed by cultivation.

[3] A thornless variety originated in cultivation.

FICUS Linn., Spec. Pl., 1059 (1753).

Ficus aurea Nutt.   **Golden Fig.**

SYN.—*Ficus aurea* Nuttall, Sylva, II, 4, t. 43 (1849).
Ficus aurea var. *latifolia* Nutt., Sylva, II, 4 (1849).

COMMON NAMES.

Wild Fig (Fla.).
India Rubber Tree (Fla.).
Wild Rubber Tree (Fla.).
Rubber Tree (Fla.).

Ficus populnea Willd.   **Poplarleaf Fig.**

SYN.—*Ficus populnea* Willdenow, Sp. Pl., IV, Par. II, 1141 (1805).
Urostigma pedunculatum Miquel, in Hooker, in Lond. Journ.
Bot., VI, 537, t. 21 A. (1847).
FICUS PEDUNCULATA Nuttall, Sylva, ed. 1, II, t. 41 (1849),
not Dryander in Ait. (1789), nor Miq. in Hook. (1848).
Ficus brevifolia Nutt., l. c., 3, t. 42 (1849).

COMMON NAMES.

Wild Fig (Fla.).
India-Rubber Tree (Fla.).

Family POLYGONACEÆ.

COCCOLOBIS[1] Browne, Nat. Hist. Jam., 209 (1756).

Coccolobis uvifera (Linn.) Sargent.   **Sea Grape.**

SYN.—*Polygonum Uvifera* Linnæus, Spec. Pl., ed. 1, I, 365 (1753).
COCCOLOBA UVIFERA Jacquin, Enum. Pl. Carib., 19 (1760).
Coccoloba Leoganensis Jacq., l. c. (1760).
Coccoloba Uvifera β Leoganensis Willdenow, Sp. Pl., II, Par.
I, 457 (1799).
Coccoloba Uvifera var. *ovalifolia* Meisner, in de Candolle
Prodr., XIV, 152 (1857).
Uvifera Leoganensis Kuntze, Rev. Gen. Pl., Par. II, 561
(1891).
Coccolobis Uvifera Sargent, Silva, VI, 115, t. CCXCVIII,
CCXCIX (1894).

COMMON NAMES.

Sea Grape (Fla.).
Seaside Plum.

---

[1] The original spelling was changed by Linnæus to *Coccoloba* (Syst. Nat., ed. 10, 1007, 1759), and was followed by subsequent authors.

Coccolobis laurifolia (Jacq.) Sargent.   **Pigeon Plum.**

Syn.—*Coccoloba laurifolia* Jacquin. Hort. Schœnbr., III, 9, t. 267
(1798).
Coccoloba parcifolia Nuttall. Sylva, ed. 1, III, 25, t. 89
(1849), not Poir. (1804).
Coccoloba Floridana Meisner. in de Candolle. Prodr.,
XIV, 165 (1857).
Coccoloba tenuifolia Eggers, in Vidensk. Medd. For. Kjöb.
1876, 142 (1876), not Linn. (1759), nor Griseb. (1864).
Coccoloba Leoganensis Eggers, l. c. (1876), not Jacq. (1760).
Coccoloba Curtissii Lindau. in Bot. Jahrb. dreizehnt. B., 159
(1891).
Urifera Curtissii (Lindau) Kuntze, Revis. Gen. Pl., Par. II,
561 (1891).
Urifera laurifolia (Jacq.) Kuntze, l. c. (1891).
Coccolobis laurifolia Sargent, Silva, VI, 119, t. CCC (1894).

COMMON NAME.

Pigeon Plum (Fla.).

Family NYCTAGINACEÆ.

PISONIA Linn.. Spec Pl., 1026 (1753).

Pisonia obtusata Jacq.   **Blolly.**

Syn.—*Pisonia obtusata* Jacquin. Hort. Schœnbr., III, 35, t. 314
(1798).
Pisonia cuneifolia Schlechtendal, in Linnæa, XXIII, 571
(1850).

COMMON NAMES.

Pigeon Wood (Fla.).
Beef Wood (Fla.).
Cork Wood (Fla.).
Pork Wood (Fla.).
Blolly (Fla.).

Family MAGNOLIACEÆ.

MAGNOLIA Linn.. Spec. Pl.. 535 (1753).

Magnolia fœtida (Linn.) Sargent.   **Magnolia.**

Syn.—*Magnolia Virginiana β fœtida* Linnæus. Spec. Pl., ed. 1, 1, 536
(1753).

193

Magnolia fœtida (Linn.) Sargent—Continued.

SYN.—MAGNOLIA GRANDIFLORA Linn., Syst. Nat., ed. 10, II, 1082 (1759.)
Magnolia grandiflora α elliptica Aiton, Hort. Kew., ed. 1,III, 329 (1789).
Magnolia grandiflora β obovata Ait., 1. c. (1789), not *M. obovata* Thunb. (1794).
Magnolia grandifolia var. *α rotundifolia* Sweet, Hort. Brit., 11 (1827), ex Don, Hist. Dichl. Pl., I, 82 (1831).
Magnolia obovata Aiton ex Link, Handb., II, 375 (1831), not Thunb. (1794).
Magnolia grandiflora macrantha Leyroy (in Cat.) ex Koch, Dendrol., erst. Th., 368 (1869).
Magnolia grandiflora salicifolia ex Koch, 1. c. (1869).
Magnolia Hartwegi (in Cat.) ex Koch, 1. c. (1869).
Magnolia fœtida Sargent, in Gard. and For., II, 615 (1889).

COMMON NAMES.

Magnolia (N. C., S. C., Ala., Fla., Miss., La., Tex., Ky.).
Big Laurel (N. C., S. C., Miss., La.).
Bull Bay (Ala., Ga., Miss.).
Great Laurel Magnolia (Ala.).
Laurel-leaved Magnolia (Tenn.).
Large-flowered Evergreen Magnolia (Tenn.).
Bat Tree (Tenn.).
Laurel Bay (Tenn.).
Laurel (S. C.).

VARIETIES DISTINGUISHED IN CULTIVATION.

Magnolia fœtida lanceolata (Ait.) nom. nov.   **Exmouth Magnolia.**

SYN.—*Magnolia grandiflora γ lanceolata* Aiton, Hort. Kew., ed. 1, II, 251 (1789).
Magnolia grandiflora α elliptica Sims, in Bot. Mag., 1952 (1818). •
Magnolia exoniensis Loddiges, Bot. Cab., t. 824 (1824).
Magnolia elliptica Link, Handb., II, 375 (1831).—?
Magnolia lanceolata Link, 1. c. (1831).—?
Magnolia grandifolia 4 exoniensis Hort. ex Loudon, Arb. Frut., I, 261 (1838).
Magnolia stricta Lond., 1. c. (1838).
Magnolia fœtida var. *Exoniensis* Sargent, Silva N. A., I, 4 (1891).

Magnolia fœtida præcox (Loud.) Sargent.

SYN.—*Magnolia grandiflora 8 præcox* Hort. ex Loudon, Arb. Frut., I, 261 (1838).
Magnolia fœtida var. *præcox* Sargent, Silva, N. A., I, 4 (1891).

18158—No. 14——13

Magnolia fœtida angustifolia (Loud.) Sargent.

Syn.—*Magnolia grandiflora 8 angustifolia* Hort. ex Loudon, Arb.
Frut., I, 261 (1838).
Magnolia Hartwicus (in Hort. Ital.) ex Nicholson, in Gard.
and For., II, 532 (1889).
Magnolia fœtida var. *angustifolia* Sargent, Silva, N. A., I, 4
(1891).
"*Magnolia Hartwegus*" (in Hort. Ital.) ex Sarg., l. c. (1891).

Magnolia fœtida ferruginea (Sims) nom. nov.

Syn.—*Magnolia grandiflora* (δ) *ferruginea* Sims, in Bot. Mag., 1952
(1818).

Magnolia glauca[1] Linn.   **Sweet Magnolia.**

Syn.—*Magnolia virginiana* Linnæus, Spec. Pl., ed. 1, I, 535 (1753),
in part.—?
Magnolia Virginiana α glauca Linn., l. c. (1753).
Magnolia Virginiana γ grisea Linn., l. c., 536 (1753).
Magnolia glauca Linn., Syst. Nat., ed. 10, II, 1082 (1759).
Magnolia glauca α latifolia Aiton. Hort. Kew., ed. 1, II, 251
(1789).
Magnolia fragrans Salisbury, Prodr., 379 (1796).
Magnolia glauca γ Argentea Pursh! in Herb. Lamb. ex de
Candolle, Syst. Nat., Vol. Prim., 452 (1818).
Magnolia glauca var. *pumila* Nuttall, in Am. Journ. Sci., ser.
1, V, 295 (1822).
Magnolia glauca var. δ *Gordoniana* (Hort.) ex Don, Hist.
Dichl. Pl., I, 82 (1831).
Magnolia glauca var. ε *Burchelliana* (Hort.) ex Don, l. c.
(1831).
Magnolia Burchelliana Hort. ex Steudel, Nom. Bot., ed. sec.,
II, 89 (1841).
Magnolia Gordoniana Hort. ex Steudel, l. c., 90 (1841).
Magnolia glauca arborea Browne, Trees of Am., 8 (1846).
Magnolia glauca sempervirens Browne, l. c. (1846).

COMMON NAMES.

Sweet Bay (Mass., R. I., Pa., N. J., N. C., S. C., Ala., Fla.,
Miss., La., Ark., Mo.).

---

[1] The late Thomas Morong, in his revision of this genus for the List Pteridophyta
and Spermatophyta in Northeastern North America (155, 1894), takes up Linnæus'
M. virginiana (Sp. Pl., ed. 1, I, 535, 1753) for *M. virginiana α glauca*, a name which occurs
immediately after *M. virginiana*, on the same page. The identity of the latter name
is very uncertain, however, as its foundation ("*Magnolia foliis ovato-lanceolatis*")
may apply in part as well to two others of our native magnolias, *M. tripetala* and
M. acuminata. The identity of *M. virginiana α glauca* is perfectly clear, and seems to
be the earliest name that can be safely maintained.

Magnolia glauca Linn.—Continued.

SYN.—White Bay (N. C., S. C., Ala., Fla., Miss., La.).
Swamp Laurel (Mass., N. C., Ga., Miss.).
Swamp Sassafras (Del., Pa., Tenn.).
Swamp Magnolia (N. J., Pa., Tenn.).
Magnolia (N. J., Del., Pa.).
White Laurel (Del., Miss., La.).
Beaver Tree (Del., S. C., Miss.).
Bay (S. C.).

VARIETIES DISTINGUISHED IN CULTIVATION.

Magnolia glauca longifolia Ait.

SYN.—*Magnolia glauca* var. *longifolia* Aiton, Hort. Kew., ed. 1, II,
251 (1789).
Magnolia longifolia Sweet, Hort. Brit., 11, (1827).—?

Magnolia glauca major[1] Sims. **Thompson Magnolia.**

SYN.—*Magnolia glauca* γ *major* Sims, in Bot. Mag., XLVII, t. 2164
(1822).
Magnolia glauca var. γ *Thompsoniana* Don, Hist. Dichl. Pl.,
I, 82 (1831).
Magnolia Thompsoniana Hort. ex Steudel, Nom. Bot., ed.
sec., II, 90 (1841).
Magnolia Thompsoniana × Sargent, in Gard. and For., I, 268,
f. 43 (1888).

Magnolia acuminata Linn. **Cucumber-tree.**

SYN.—*Magnolia virginiana* Linnæus, Spec. Pl., ed. 1, I, 535 (1753),
in part.—?
Magnolia Virginiana ε *acuminata* Linn., l. c., 536 (1753).
Magnolia acuminata Linn., Syst. Nat., ed. 10, II, 1082
(1759).
Magnolia rustica Hort. ex de Candolle, Syst. Nat., Vol. Prim.,
453 (1818).
Magnolia Pennsylvanica Hort. ex de C., l. c. (1818).
Magnolia de Candollii Savi, in Bibl. Ital., XV, 1, 224, t.
(1819).
Magnolia acuminata β *Candolli* de Candolle, Prodr., I, 80
(1824).
Magnolia maxima Loddiges ex Don, in Loudon, Hort. Brit.,
226 (1830).
Magnolia Candollii Link, Handb., II, 375 (1831).

---

[1] A supposed hybrid from *Magnolia tripetala* and *M. glauca* originated in England
from seed of *M. glauca*.

Magnolia acuminata Linn.—Continued.

Syn.—*Magnolia acuminata 3 maxima* Loddiges ex Loudon, Arb. Frut., I, 273 (1838).
Tulipastrum Americanum Spach, Hist. Vég., VII, 483 (1839).
Magnolia excelsa Loddiges ex Koch. Dendrol.. erst. Th.. 372 (1869).

COMMON NAMES.

Cucumber Tree (R. I.. Mass.. N. Y.. Pa.. D. C.. N. C.. S. C.,
    Ala.. Miss.. La.. Ark.. Ky.. W. Va.. Ohio. Ind.. Ill.).
Mountain Magnolia (Miss.. Ky.).
Cucumber (W. Va.).
Black Lin (W. Va.).
Magnolia (Ark.).
Pointed-leaved Magnolia (lit.).

Magnolia acuminata cordata[1] (Michx.) Loud.

Yellow-flower Cucumber-tree.

Syn.—*Magnolia cordata* Michaux, Flor. Bor.-Am., I. 328 (1803).
Tulipastrum Americanum var. *subcordatum* Spach, Hist.Vég..
    VII. 483 (1839).
Magnolia (? acu.) cordata Loudon, Arb. Frut.. I, 275 (1838).
Magnolia acuminata var. *cordata* Sargent, in Am. Journ. Sci..
    ser. 3. XXXII. 473 (1886).

COMMON NAMES.

Cucumber Tree (Va.. N. C.. Miss.. La.).
Yellow-flowered Magnolia (Ala.. La.).
Yellow-flowered Cucumber Tree (Ala.).
Yellow Cucumber Tree (Ala.).
Heart-leaved Cucumber Tree (N. C.).

Magnolia macrophylla Michx.   **Largeleaf Umbrella.**

Syn.—*Magnolia macrophylla* Michaux. Fl. Bor.-Am.. I. 327 (1803).

COMMON NAMES.

Large-leaved Cucumber Tree (Ala.. Miss., La.).
Great-leaved Magnolia (N. C.. Miss.. La.).
Large-leaved Umbrella Tree (N. C.. Tenn.).
Cucumber (Ky.).
Cucumber Tree (Fla.).
Long-leaved Magnolia (S. C.).

---

[1] The most perfect type of this variety is found in cultivation: only occasional wild forms approach it.

Magnolia tripetala Linn.   **Umbrella-tree.**

SYN.—*Magnolia Virginiana* Linnæus, Spec. Pl., ed. 1, I, 535 (1753),
   in part.—?
   Magnolia Virginiana δ tripetala Linn., l. c., 536 (1753).
   Magnolia tripetala Linn., Syst. Nat., ed. 10, II, 1082 (1759).
   MAGNOLIA UMBRELLA Desrousseaux in Lamarck, Enc.
   Méth. Bot., III, 673 (1789).
   Magnolia frondosa Salisbury, Prodr., 379 (1796).

COMMON NAMES.

Umbrella Tree (Pa., W. Va., N. C., S. C., Ala., Miss., La.).
Cucumber (Ky.).
Magnolia (W. Va.).
Elk Wood.

Magnolia fraseri Walter.   **Fraser Umbrella.**

SYN.—*Magnolia Fraseri* Walter, Fl. Caroliniana, 159 and t. (1788).
   Magnolia auriculata Desrousseaux in Lamarck, Enc. Méth
   Bot., III, 673 (1789).
   Magnolia pyramidata Bartram ex Pursh, Fl. Am. Sept., II,
   382 (1814).
   Magnolia auricularis Salisbury, Parad. Lond., I, t. 43
   (1807).
   Magnolia (? aur.) pyramidata Bartr. ex Loudon, Arb. Frut.,
   I, 276 (1838).

COMMON NAMES.

Longleaved Cucumber Tree (N. C., S. C.).
Earleaved Umbrella Tree (N. C., S. C., Miss.).
Earleaved Cucumber Tree (N. C., Fla.).
Indian Physic (N. C., Tenn.).
Indian Bitters (N. C.).
Cucumber (Ky.).
North Carolina Bay Tree (W. Va.).
Cucumber Tree (Fla.).
Water Lily Tree.
Mountain Magnolia.
Whahoo.

LIRIODENDRON Linn., Spec. Pl., 535 (1753).

Liriodendron tulipifera Linn.   **Tulip-tree.**

SYN.—*Liriodendron Tulipifera* Linnæus, Spec. Pl., ed. 1, I, 535
   (1753).
   Liriodendron procera Salisbury, Prodr., 379 (1796).

Liriodendron tulipifera Linn.—Continued.

SYN.—*Lyriodendron flavum* Hort. ex de Candolle, Syst. Nat., Vol.
Prim., 462 (1818).
Liriodendron Tulipifera acutiloba Michaux, Fl. Bor.-Am., I.
326 (1803).
Liriodendron Tulipifera acutifolia Michaux ex Loudon, Arb.
Frut., I, 285 (1838).
Liriodendron Tulipifera flava Hort. ex Loud., l. c. (1838).
Liriodendron lutea ex Koch, Dendrol., erst. Th., 331 (1869).
Liriodendron leucantha ex Koch, l. c. (1869).
Liriodendron heterophylla ex Koch, l. c. (1869).
Liriodendron crispum ex Koch, l. c. (1869).
Liriodendron tulipiflora St. Lag., in Ann. Soc. Bot. Lyon.
VII, 129 (1880).
Hiriodendron Tulipifermus Hadkinson, in Ann. Rep. Nebr.
Hort. Soc. 1889, 124 (1889).
Tulipifera Liriodendron Miller ex Kuntze, Rev. Gen. Par.,
I, 7 (1891).

COMMON NAMES.

Tulip Tree (Vt., Mass., R. I., Conn., N. Y., N. J., Del., Pa.,
Va., W. Va., D. C., N. C., S. C., Ga., Ark., Ky., Ohio, Ind.,
Ill., Ont.).
White Wood (Vt., Mass., R. I., Conn., N. Y., N. J., Del.,
S. C., Ky., Ohio, Ill., Mich., Ont.).
Yellow Poplar (N. Y., N. J., Pa., Del., Va., W. Va., N. C.,
S. C., Ala., Ark., Ky., Ohio, Ind., Mo.).
Tulip Poplar (Del., Pa., S. C., Ill.).
Poplar (R. I., Del., N. C., S. C., Fla., Ohio).
White Poplar (Pa., Ky., Ind.).
Blue Poplar (Del., W. Va.).
Hickory Poplar (Va., N. C., W. Va.).
Popple (R. I.).
Cucumber Tree (N. Y.).
Canoe Wood (Tenn.).
Old-Wife's-Shirt Tree (Tenn.).
Ko-yen-ta-ka-ah-ta = "White Tree" (Onondaga Indians
N. Y.).
Basswood (Ohio).

VARIETIES DISTINGUISHED IN CULTIVATION.

Liriodendron tulipifera obtusiloba Michx.

SYN.—*Liriodendron Tulipifera obtusiloba* Michaux, Fl. Bor.-Am., I,
326 (1803).
Liriodendron truncatifolia Stokes, Bot. Mat. Med., III,
233 (1812).

Liriodendron tulipifera obtusiloba Michx.—Continued.

SYN.—*Lyriodendron integrifolium* Hort, ex de Candolle, Syst. Nat.,
Vol. Prim., 462 (1818).
Liriodendron Tulipifera integrifolia Hort. ex Loudon, Arb.
Frut., I, 285 (1838).
Liriodendron integrifolia Koch, Dendrol., erst. Th., 381 (1869).

Liriodendron tulipifera pyramidalis Dippel.

Pyramidal Tulip-tree.

SYN.—*Liriodendron Tulipifera* c *pyramidalis* Dippel, Handb.
Laubh., dritt. T., 155 (1893).
Liriodendron Tulipifera fastigiata Hort. ex Dippel, l. c.
(1893).

Liriodendron tulipifera penache Ellw. & B.

Variegated Tulip-tree.

SYN.—*Liriodendron tulipifera* var. *penache* Ellwanger & Barry,
Cat., 45 (1890).
Liriodendron tulipifera d *rariegata* Dippel, Handb. Laubh.,
dritt. T., 155 (1893).

Liriodendron tulipifera aureo-maculata[1] (Arb. Kew.) nom. nov.

SYN.—*Liriodendron tulipifera aureo-maculatum* ex Hand-list Trees
and Shr. Arb. Kew[2], Pt. I, 19 (1895)—*nomen nudum.*

Family ANNONACEÆ.

ASIMINA Adanson, Fam. Pl., II, 365 (1763).

Asimina triloba (Linn.) Dunal. **Papaw.**

SYN.—*Annona triloba* Linnæus, Spec. Pl., ed. 1, I, 537 (1753).
Anona pendula Salisbury, Prodr., 380 (1796).
Anona palustris Abbot, Insects of Ga., I, t. 4 (1797), not
Linn. (1762).
Orchidocarpum arietinum Michaux, Fl. Bor.-Am., I, 329 (1803).
Porcelia triloba Persoon, Syn., II, 95 (1807.)
Asimina triloba Dunal, Monograph Anon., 83 (1817).
Asimina grandiflora Hort. ex Spach, Hist. Vég., V, 530
(1836).
Uvaria triloba Torrey & Gray, Fl. N. A., I, 45 (1838).
Anona campaniflora Spach, Hist. Vég., VII, 528 (1839).
Asimina conoidea Spach, l. c., 530 (1839).
Asimina glabra Hort, ex Koch, Dendrol., erst. Th., 384 (1869).

---

[1] This form is distinguished by the golden-yellow areas on its leaves.
[2] Published anonymously.

Asimina triloba (Linn.) Dunal—Continued.

SYN.—Papaw (R. I., Del., N. Y., N. J., D. C., Va., W. Va., N. C., S. C.,
  Ga., Miss., La., Ky., Ohio, Ill., Ind., Mo., Iowa, Kans., Nebr.).
  Custard Apple (Ont., Del., Pa., Ohio, S. C., Miss.).
  Banana (Ark.).
  False Banana (Ill.).
  Jasmine (La.).
  Jasminier (La.).
  Fetid Shrub (N. C.).

ANNONA Linn., Spec. Pl., 536 (1753).

Annona glabra Linn.   **Pond Apple.**

SYN.—*Annona glabra* Linnæus, Spec. Pl., ed. 1, I, 537 (1753).
  Anona laurifolia Dunal, Mon. Anon., 65 (1817).
  Anona Audubon, Birds Amer., II, t. 162 (1833), not Persoon
    (1807).

Pond Apple (Fla.).
Custard Apple (Fla.).

Family LAURACEÆ.

PERSEA Gaertner f., Frut., III, 222 (1805).

Persea borbonia (Linn.) Spreng.   **Red Bay.**

SYN.—*Laurus Borbonia* Linnæus, Spec. Pl., ed. 1, I, 370 (1753).
  Laurus Carolinensis Catesby, Carol., ed. 2, I, 63, t. 63 (1754).
  Laurus elongata Salisbury, Prodr., 344 (1796).
  Laurus Caroliniana Poiret, in Lamarck, Enc. Méth. Bot.
    Suppl., III, 323 (1813).
  Larus Carolinensis α glabra Pursh, Fl. Am. Sept., I, 276
    (1814).
  Laurus Carolinensis γ obtusa Pursh, l. c. (1814).
  Persea Borbonia Sprengel, Syst. Vég., II, 268 (1825).
  PERSEA CAROLINENSIS Nees, Syst. Laurin., 150 (1836).
  Persea fœtens Willdenow ex Nees, l. c., 151 (1836), not Ait.
    (1786).
  Tamala borbonia Rafinesque, Sylva Tellur., 136 (1838).
  Persea Carolinensis var. *glabriuscula* Meisner, in de Can-
    dolle, Prodr., XV, sect. 1, 51 (1864).
  Notaphœbe Borbonia Pax, in Engler & Prantl, Nat. Pfl., III,
    Abt. II, 116 (1889).

Persea borbonia (Linn.) Spreng.—Continued.

SYN.—*Persea Carolinensis* forma *glabriuscula* Garcke & Urban, Jahrb. Kon. Bot. Gart., V, 176 (1889).

COMMON NAMES.

Red Bay (N. C., S. C., Ga., Fla., Ala., Miss., La.).
Bay Galls (Tenn.).
Laurel Tree (La.).
Laurier Petit Magnolia (La.).
Florida Mahogany (Fla.).
Sweet Bay (Fla.).
False Mahogany (lit.)
Tisswood.

Persea pubescens (Pursh) Sargent.   **Swamp Bay.**

SYN.—*Laurus Carolinensis* Michaux f., Hist. Arb. Am., III. t. 2 (1813), not Michx. (1803).
Laurus Carolinensis β *pubescens* Pursh, Fl. Am. Sept., I, 276 (1814).
Persea Carolinensis α Nees, Syst. Laurin., 150 (1836).
Tamala palustris Rafinesque, Sylva Tellur., 137 (1838).
PERSEA CAROLINENSIS var. PALUSTRIS Chapman, Fl. S. States, 393 (1865).
Persea Carolinensis β *pubescens* Meisner, in de Candolle, Prodr., XV, sect. 1, 51 (1864).
Persea Carolinensis forma *pubescens* Garcke & Urban, Jahrb. Kon. Bot. Gart., V, 176 (1889).
Persea pubescens Sargent, Silva, VII, 7 (1895).
Persea palustris Sarg., l. c., t. CCCII (1895).

COMMON NAME.

Swamp Red Bay (Fla.).

OCOTEA Aublet, Pl. Guian., II, 780 (1775).

Ocotea catesbyana (Michx.) Sargent.   **Lancewood.**

SYN.—*Laurus Catesbyana* Michaux, Fl. Bor.-Am., I, 244 (1803).
Laurus Catesbæi Persoon, Syn. Pl., I, 449 (1805).
NECTANDRA WILLDENOVIANA Meisner, in de Candolle, Prodr., XV., sect. 1, 165 (1864), not Nees (1836).
Gymnobalanus Catesbyanus Nees, Syst. Laurin., 483 (1836).
Nectandra anonyma Steudel, Nom. Bot., ed. sec., II, 187 (1841).
Nectandra Bredemeyeriana Nees, in Linnæa, XXI, 505 (1848).
Nectandra Cigua Richard, Fl. Cub. Phan., II, 187 (1853).

Ocotea catesbyana (Michx.) Sargent—Continued.

SYN.—*Persea Catesbyana* Chapman, Fl. S. States, 393 (1865).
Nectandra coriacea Mez, in Jahrb. Kon. Bot. Gart., V, 459 (1889), not Griseb. (1864).
Nectandra Catesbiana Sargent, in Gard. and For., II, 448 (1889).
Nectandra sanguinea Hitchcock, in Rep. Mo. Bot. Gard., IV, 125 (1893), not Rollander ex Rottb. (1778).
Ocotea Catesbyana Sargent, Silva, VII, 11, t. CCCIII (1895).

COMMON NAMES.

Lance Wood (Fla.).
Sweetwood (Jamaica).
Cigua (Cuba).
Avispillo (Puerto Rico).
Canela (Puerto Rico).

SASSAFRAS Nees & Eberm., Handb. Med. Pharm. Bot., I, 418 (1830).

Sassafras sassafras (Linn.) Karsten.   **Sassafras.**

SYN.—*Laurus Sassafras* Linnæus, Spec. Pl., ed. 1, I, 371 (1753).
Laurus variifolia Salisbury, Prodr., 344 (1796).
Laurus diversifolia Stokes, Bot. Mat. Med., II, 426 (1812).
Laurus albida Nuttall, Gen., I, 259 (1818).
Tetranthera albida Sprengel, Syst. Vég., II, 267 (1825).
Persea Sassafras Sprengel, l. c., 270 (1825).
SASSAFRAS OFFICINALE Nees & Ebermaier, Handb. Med. Pharm. Bot., I, 418 (1830).
Sassafras albidum Nees, Syst. Laurin., 490 (1836).
Sassafras Sassafras Karsten, Deutsch. Fl., 505 (1882).
Sassafras variifolium Kuntze, Rev. Gen. Pl., Par. II, 574 (1891).

COMMON NAMES.

Sassafras (Vt., N. H., Mass., Conn., R. I., N. Y., N. J., Pa., Del., Md., Va., W. Va., N. C., S. C., Ga., Fla., Miss., La., Tex., Ark., Ky., Mo., Ill., Ind., Kans., Nebr., Minn., Mich., Ohio, Ont.).
Saxifrax (Fla.).
Saxifrax Tree (Tenn.).
Sassafac (W. Va.)
Sassafrac (Del.).
Gumbo file (La., negro dialect).
Wah-eh-nah-kas = "Smelling stick" (Onondaga Indians, N.Y.).

UMBELLULARIA Nutt., Sylva, I, 87 (1842).

Umbellularia californica (Hook. & Arn.) Nutt.   California Laurel.

SYN.—*Tetranthera ? Californica* Hooker & Arnot, Bot. Beechey's
Voyage, 159 (1833).
Laurus regia Douglas, in Companion Bot. Mag., II, 127
(1836).
Oreodaphne Californica Nees, Syst. Laurin., 463 (1836).
Drimophyllum pauciflorum Nuttall, Sylva, I, 85, t. 22 (1842).
Umbellularia Californica Nutt., l. c., 87 (1842).
Ocotea californica ex Nicholson, Dic. Gard., IV, 121 (1889).

COMMON NAMES.

California Laurel (Cal., Nev.).
Mountain Laurel (Cal., Nev.).
California Bay Tree (Cal., Nev.).
Myrtle Tree (Oreg.).
Spice Tree (Oreg.).
Cagiput (Oreg.).
California Olive (Oreg.).
Myrtle (Oreg.).
Spice Tree (Cal., Nev., Oreg.).
Laurel (Cal.).
Bay Tree (Cal.).
Oreodaphne (Cal.).
Californian Sassafras.

Family CAPPARIDACEÆ.[1]

CAPPARIS Linn., Spec. Pl., 503 (1753).

Capparis jamaicensis Jacq.   Florida Caper.

SYN.—*Capparis siliquosa* Linnæus, Syst. Nat., ed. 10, II, 1071 (1759).
Capparis Jamaicensis Jacquin, Enum. Pl. Carib., 23 (1760).
Capparis torulosa Swartz, Prodr. Veg. Ind. Occ., 81 (1788).
Capparis uncinata Loddiges ex Eichler, in Martin's Fl. Bras.,
I, 270 (1840).

---

[1] *Forchhammeria Watsoni* Rose (in Contr. U. S. Nat. Herb., I, No. 9, 302 and t., 1895)
is a new arborescent species of this family discovered in Guaymas, Mexico, and in
the cape region of Lower California by Dr. Edward Palmer in 1887, 1890, and 1891,
and by Mr. T. S. Brandegee in 1889 and 1892. It is said to be about 15 feet in height,
with a diameter of 1 to 5 feet. The pear-shaped, purple-red fruit, is said to be much
eaten by birds. It is a close ally of the Mexican *F. pallida* Liebmann (in Kjoeb
Vidensk. Med., 1853, 94, 1854), which is a tree 15 to 20 feet high and 6 to 10 inches in
diameter.

Capparis jamaicensis (Linn.) Jacq.—Continued.

Syn.—*Capparis emarginata* Richard. Fl. Cub., 78, t. 9 (1853), not
Presl (1836).
Capparis Jamaicensis var. *emarginata* Grisebach. Fl. Brit.
W. Ind., 18 (1864).

COMMON NAME.

Caper Tree (Fla.).

Family SAXIFRAGACEÆ.

LYONOTHAMNUS Gray. in Proc. Am. Acad. Sci., ser. 2. XII. 291
(1877).

Lyonothamnus floribundus Gray. **Santa Cruz Ironwood.**

Syn.—*Lyonothamnus floribundus* Gray. Proc. Am. Acad. Sci., ser. 2,
XII, 292 (1877).
Lyonothamnus asplenifolius Greene. in Bull. Cal. Acad. Sci ,
I, 187 (1885): II, 149, t. 6 (1886).
Lyonothamnus floribundus var. *asplenifolius* Brandegee, in
Zoe, I, 136 (1890).

Family HAMAMELIDACEÆ.

HAMAMELIS Linn.. Spec. Pl.. 124 (1753).

Hamamelis virginiana Linn. **Witch Hazel.**

Syn.—*Hamamelis Virginiana* Linnæus. Spec. Pl.. ed. 1. I. 124 (1753).
Hamamelis dioica Walter. Fl. Caroliniana. 255 (1788).
Hamamelis androgyna Walter, l. c. (1788).
Hamamelis corylifolia Moench. Meth.. 273 (1794).
Hamamelis macrophylla Pursh. Fl. Am. Sept.. I. 116 (1814).
Hamamelis Virginiana γ *parvifolia* Nuttall. Genera. I. 107
(1818).
Hamamelis caroliniana Walt. ex Steudel. Nom. Bot.. ed. 1.
388 (1821).
Hamamelis virginiana Rafinesque. Med. Bot.. I. t. 45 (1828).
Trilopus Virginica Raf.. New Fl.. III. 15 (1836).
Trilopus riparia Raf.. l. c.. 16 (1836).
Hamamelis riparia Raf.. l. c. (1836).
Trilopus hyemalis Raf.. l. c. (1836).
Hamamelis hyemalis Raf.. l. c. (1836).
Trilopus nigra Raf.. l. c. (1836).

Hamamelis virginiana Linn.—Continued.

SYN.—*Hamamelis nigra* Raf., l. c. (1836).
Trilopus estivalis Raf., l. c. (1836).
Hamamelis estivalis Raf., l. c. (1836).
Trilopus rotundifolia Raf., l. c. (1836).
Hamamelis rotundifolia Raf., l. c. (1836).
Trilopus parvifolia Raf., l. c., 17 (1836).
Trilopus dentata Raf., l. c. (1836).
Hamamelis dentata Raf., l. c. (1836).

COMMON NAMES.

Witch Hazel (Me., Vt., Mass., R. I., Conn., N. Y., N. J., Del.,
Pa., Va., W. Va., N. C., S. C., Ga., Ala., Fla., Miss., La.,
Tex., Ky., Mo., Ill., Wis., Iowa, Ohio, Mich., Nebr.).
Oe-eh-nah-kwe-ha-he = " Spotted stick" (Onondaga Indians
N. Y.).
Winter Bloom, }
Snapping Hazel, } (Lit. of domestic medicine).
Spotted Alder. }

LIQUIDAMBAR Linn., Spec. Pl., 999 (1753).

Liquidambar styraciflua Linn.   **Sweet Gum.**

SYN.—*Liquidambar styraciflua* Linnæus, Spec. Pl., ed. 1, II, 999
(1753.)
Liquidambar gummifera Salisbury, Prodr., 893 (1796).
Liquidambar barbata Stokes, Bot. Mat. Med., IV, 332 (1812).
Liquidambar macrophylla Örsted, Am. Cent., XVI, t. 10
(1863).
Liquidambar Syraciflua var. *Mexicana* Örsted, l. c., t. 11
(1863).

COMMON NAMES.

Sweet Gum (Mass., R. I., N. Y., N. J., Pa., Del., Va., W. Va.,
N. C., S. C., Ga., Ala., Fla., Miss., La., Tex., Ark., Ky., Mo.,
Ill., Ind., Ohio).
Liquidamber (R. I., N. Y., Del., N. J., Pa., La., Tex., Ohio,
Ill.).
Red Gum (Va., Ala., Miss., Tex., La.).
Gum (Va.).
Gum Tree (S. C., La.).
Alligator Wood (N. J.).
Bilsted (N. J.).
Star-leaved Gum.
Satin Walnut.

Family PLATANACEÆ.

PLATANUS Linn., Spec. Pl., 999 (1753).

Platanus occidentalis Linn.   Sycamore.

SYN.—*Platanus occidentalis* Linnæus, Spec. Pl., ed. 1, II, 999 (1753).
Platanus lobata Moench, Meth., 358 (1794).
Platanus excelsa Salisbury, Prodr., 393 (1796).
Platanus hybridus Brotero, Fl. Lus., II, 487 (1804).
Platanus vulgaris ε *angulosa* Spach, in Ann. Sc. Nat., sér. 2, XV, 293 (1841).
Platanus occidentalis var. *Hispanica* Wesmael, in Mém. Soc. Sc. Hainaut, sér. 3. I, 12, f. 5 (1867).
Platanus occidentalis β *lobata* Bommer, Les Platan. Cult., 17, f. 5, 6 (1869).
Platanus macrophylla ex Koch, Dendrol., zw. Th. erst. Ab., 469 (1872), not Cree ex Baxt. in Loud. (1832).
Platanus integrifolia ex Koch, l. c. (1872).
Platanus pyramidalis ex Koch, l. c. (1872).
Platanus orientalis var. *occidentalis* Kuntze (L.), Rev. Gen. Pl., Par. II, 636 (1891).

COMMON NAMES.

Sycamore (Vt., N. H., Mass., Conn., R. I., N. Y., N. J., Pa., Del., Va., W. Va., N. C., S. C., Ga., Fla., Ala., Miss., La., Tex., Ky., Ark., Mo., Ill., Ind., Iowa, Kans., Nebr., Mich., Wis., Ohio, Ont.).
Button Wood (Vt., N. H., R. I., Mass., N. Y., N. J., Pa., Del., S. C., Ala., Miss., La., Tex., Ark., Mo., Ill., Nebr., Mich., Minn., Ohio, Ont.).
Buttonball Tree (Mass., R. I., Conn., N. Y., N. J., Pa., Del., Miss., La., Mo., Ill., Iowa, Mich., Nebr., Ohio).
Buttonball (R. I., N. Y., Pa., Fla.).
Plane Tree (R. I., Del., S. C., Kans., Nebr., Iowa.)
Water Beech (Del.).
Platane (La.).
Cotonier (La.).
Bois puant (La.).
Oo-da-te-cha-wun-nes=" Big stockings" (Indians, N. Y.).

Platanus racemosa Nutt.   California Sycamore.

SYN.—*Platanus occidentalis* Hooker & Arnott, Bot. Beechey's Voyage, 160, 390 (1833), not Linn. (1753).
Platanus racemosa Nuttall, Sylva, I, 47, t. 15 (1842).
Platanus Californica Bentham, Bot. Voyage Sulphur, 54 (1844).

Platanus racemosa Nutt.—Continued.

SYN.—*Platanus Mexicana* Torrey, in Sitgreaves's Rep., 172 (1853), not Moricand (1830), nor Torr. (1848).
Platanus orientalis var. *racemosa* Kuntze (Nutt.), Rev. Gen. Pl., Par. II, 636 (1891).

COMMON NAMES.
Sycamore (Cal.).
Buttonwood.
Buttonball Tree (Cal.).
Buttonball (Cal.).

Platanus wrightii Watson.   **Arizona Sycamore.**

SYN.—*Platanus Mexicana* Torrey, in Emory's Rep., 151 (1848), not Moricand (1830).
Platanus racemosa Watson, Pl. Wheeler, 16 (1874), not Nutt. (1842).
Platanus Wrightii Watson, in Proc. Am. Acad. Sci., X, 349 (1875).

COMMON NAMES.
Sycamore (Ariz.).
Arizona Sycamore (Cal.).

Family ROSACEÆ.

VAUQUELINIA Correa ex Humboldt, Bonpland & Kunth, Pl. Aeq.. I, 140 (1808).

Vauquelinia californica (Torr.) Sargent.   **Vauquelinia.**

SYN.—*Spiræa Californica* Torrey, in Emory's Rep., 140 (1848).
Vauquelinia corymbosa Torr., in Bot. Mex. Bound. Survey, 64 (1859) not Correa (1815).
VAUQUELINIA TORREYI Watson, in Proc. Am. Acad. Sci., XI, 147 (1876).
Vauquelinia Californica Sargent, in Gard. and For., II, 400 (1889).

CERCOCARPUS[1] Humboldt, Bonpland & Kunth, Nov. Gen. Sp., VI, 232 (1823).

Cercocarpus ledifolius Nutt.   **Mountain Mahogany.**

SYN.—*Cercocarpus ledifolius* Nuttall, mss. in Torrey & Gray, Fl. N. A., I, 427 (1840).

---

[1] The following variety is a shrub:
Cercocarpus ledifolius intricatus (Wats.) Jones.   **Utah Mountain Mahogany.**
SYN.—*Cercocarpus breviflorus* Watson, in King's Rep., V, 83 (1871), not Gray (1871).
Cercocarpus intricatus Watson, in Proc. Am. Acad. Sci., X, 346 (1875).
Cercocarpus Arizonicus Jones, in Zoe, II, 14 (1891).
Cercocarpus ledifolius var. *intricatus* Jones, l. c., 14, 244 (1891).

Syn.—Mountain Mahogany (Cal.. N. Mex.. Utah, Idaho, Mont.).

Cercocarpus parvifolius Nutt. Valley Mahogany.

Syn.—*Cercocarpus fothergilloides* Torrey. Ann. Lyc. N. Y.. II, 198
(1828), not H., B. K. (1823).
Cercocarpus parvifolius Nuttall. in Hooker & Arnott. Bot.
Beechey's Voyage, 337 (1841).

COMMON NAMES.

Mountain Mahogany (Cal.. N. Mex.. Utah, Idaho, Colo.).
Valley Mahogany (Cal.).
Feather Tree (Cal.).

Cercocarpus parvifolius betuloides (Nutt.) Sargent.
Birch-leaf Mahogany.

Syn.—*Cercocarpus betuloides* Nuttall, mss. in Torrey & Gray. Fl.
N. A., I, 427 (1840).
Cercocarpus betulæfolius Nuttall. MSS. in Hooker. Icon.. IV,
t. 322 (1840).
Cercocarpus parvifolius var. *glaber* Watson. in Brewer &
Watson. Bot. Cal., I, 175 (1880).
Cercocarpus parvifolius var. *betuloides* Sargent, Silva. IV, 66
(1892).

Cercocarpus parvifolius breviflorus (Gr.) Jones.
Short-leaf Mahogany.

Syn.—*Cercocarpus breviflorus* Gray, Pl., Wright., II, 54 (1853).
Cercocarpus parvifolius var. *breviflorus* Jones, in Zoe, II, 245
(1891).
Cercocarpus parvifolius var. *brevifolius* Jones ex Sargent,
Silva, iv, 66 (1892).

Cercocarpus parvifolius paucidentatus Watson.
Entire-leaf Mahogany.

Syn.—*Cercocarpus parvifolius* var. *paucidentatus* Watson. in Proc.
Am. Acad. Sci.. XVII. 353 (1882).

PYRUS Linn.. Spec. Pl.. 479 (1753).

Pyrus coronaria Linn. Sweet Crab.

Syn.—*Pyrus coronaria* Linnæus. Spec. Pl.. ed. 1. I, 480 (1753).
Malus coronaria Miller. Gard. Dict. ed. 8, No. 2 (1768).
Crataegus coronaria Salisbury. Prodr.. 357 (1796).
Malus microcarpa coronaria Carrière. in Rev. Hort. 1884,
104. f. 24 (1884).

Pyrus coronaria Linn.—Continued.

COMMON NAMES.

SYN.—American Crab (R. I., N. J., Del., Pa., Ala., Miss., La., Ill.,
Ohio, Ont., Kans., Nebr., Mich., Minn.).
Sweet-scented Crab (Mass., Del., Pa., N. C., S. C., Miss., Ill.,
Ohio, Iowa).
Crab Apple (S. C., La., Ky., Mo., Ill., Ohio, Iowa, Kans.,
Nebr.).
Wild Crab (N. Y., Ill., Ind., Wis., Iowa, Kans., Minn., Mo.,
Ark.).
Crab (W. Va., N. C., Ga., Miss., Wis.).
American Crab Apple (Nebr.).
Fragrant Crab.

Pyrus ioensis (Wood) Bailey.   **Iowa Crab.**

SYN.—*Pyrus coronaria β Ioensis* Wood, Class-book Bot., rev. ed.,
333 (1869).
Pyrus Ioensis Bailey, in Am. Gard., XII, 473, f. 7, 8 (1891).

Pyrus soulardi Bailey.   **Soulard Apple.**

SYN.—*Pyrus Soulardi* Bailey, in Am. Gard., XII, 472 (1891).

Pyrus malus [1] Linn.   **Wild Apple.**

SYN.—*Pyrus Malus* Linnæus, Spec. Pl., ed. 1, I, 479 (1753).
Malus communis Desfontaines, Hist. Arb., II, 140 (1809).

Pyrus angustifolia Ait.   **Narrowleaf Crab.**

SYN.—*Pyrus angustifolia* Aiton, Hort. Kew., ed. 1, II, 176 (1789).
Pyrus coronaria Wangenheim, Nordam. Holz., 61, t. 21, f.
47 (1787), not Linn. (1753).
Cratægus pyrifolia Lamarck, Enc. Méth. Bot., I, 84 (1783).
Malus angustifolia Michaux, Fl. Bor.-Am., I, 292 (1803).
Malus sempervirens Desfontaines, Hist. Arb., II, 141 (1809).
Pyrus coronaria var. *angustifolia* Wenzig, in Linnæa,
XXXVIII, 41 (1871).
Chloromeles sempervirens Decaisne, in Fl. des Serres, XXIII,
156 (1883).
Malus microcarpa sempervirens Carrière, Pom. Micr., 136, f.
1, 18 (1884).

COMMON NAMES.

Southern Crab Apple (Del., N. C., Ala., Miss., La.).
American Crab Apple (R. I., N. J., Miss.).
Narrowleaf Crab (Ala.).

---

[1] Naturalized.

Pyrus angustifolia Ait.—Continued.

Syn.—Crab Apple (N. C., S. C., Ga.).
Wild Crab Apple.
Narrowleaved Crab Apple (N. C.).
Narrowleaved Crab (S. C.).
Crabtree (Va., Fla.).

Pyrus rivularis Dougl.   **Oregon Crab.**

Syn.—*Pyrus fusca* [1] Rafinesque, Med. Fl., II, 254 (1830).—?
Pyrus rivularis Douglas. MSS. in Hooker. Fl. Bor.-Am., I,
203, t. 68 (1833).
Pyrus diversifolia Bongard. in Mém. Acad. Sci. St. Petersb.,
sér. 6, II, 133 (1834).
Pyrus subcordata Ledebour. Fl. Ross., II, 95 (1846).
Malus subcordata Roemer. Fam. Nat. Syn., III, 192 (1847).
Malus rivularis Roem., l. c., 215 (1847).
Malus diversifolia Roem., l. c. (1847.)
Pyrus rivularis β *leripes* Nuttall. Sylva, II, 24 (1849).

COMMON NAMES.

Oregon Crab Apple (Cal., Wash., Oreg.).
Crab or Wild Apple.

Pyrus americana (Marsh.) de C.   **Mountain Ash.**

Syn.—*Sorbus Americana* Marshall, Arb. Am., 145 (1785).
Sorbus aucuparia Poiret, in Lamarck. Enc. Méth. Bot., VII,
234 (1806). in part.
Sorbus aucuparia var. *Americana* Persoon. Syn. Pl., II, 38
(1807).
Pyrus Americana de Candolle. Prodr., II, 637 (1825).
Pyrus aucuparia Meyer. Pl. Lab., 81 (1830). in part.
Sorbus humifusa Raf. ex Watson, Ind. N. A. Bot., 290 (1878).
Pyrus Sorbus polonica Hort., ex Hand-list Trees and Shr.
Arb. Kew. Pt. I, 189 (1894).
Pyrus americana var. *nana*,[2] ex Hand-list, l. c. (1894).

COMMON NAMES.

Mountain Ash (Vt., N. H., Mass., R. I., N. Y., N. J., Pa.,
Va., W. Va., N. C., Ky., Mich., Ont.

---

[1] Rafinesque's *Pyrus fusca* (printed "*P. fusea*") is placed here on very meager
evidence, insufficient, it is believed, to warrant its replacing the later *P. rivularis*.

[2] A form cultivated as a distinct variety in the Royal Kew Gardens, but the name
was published without description, so that nothing is known of its true character.
It appears to be unknown in American gardens.

Pyrus americana (Marsh.) de C.—Continued.

SYN.—American Mountain Ash (Pa.).
Mountain Sumach (N. C., S. C.).
Wine Tree (N. C.).
Round Wood (Me.).
Life of Man (N. Y.).
Rowan Berry (Ont.).

Pyrus americana microcarpa (Pursh) Torr. & Gr.
Smallfruit Mountain Ash

SYN.—*Sorbus aucuparia* var. *α* Michaux, Fl. Bor.-Am., I, 290 (1803),
Sorbus microcarpa Pursh, Fl. Am. Sept., I, 341 (1814).
Pyrus microcarpa Sprengel, Syst. Veg., II, 511 (1825).
Sorbus riparia Rafinesque, New Fl., 3d Pt., 15 (1836).
Pyrus Americana var. *microcarpa* Torrey & Gray, Fl. N. A.,
I, 472 (1840).
Sorbus Americana var. *microcarpa* Wenzig, in Linnæa,
XXXVIII, 73 (1874).

Pyrus sambucifolia Cham. & Schlecht.   **Elderleaf Mountain Ash.**

SYN.—*Sorbus aucuparia* var. *β* Michaux, Fl. Bor.-Am., I, 290 (1803).
Pyrus sambucifolia Chamisso & Schlechtendal, in Linnæa,
II, 36 (1827).
Pyrus aucuparia Meyer, Pl. Labrador, 81 (1830), in part.
Sorbus aucuparia Schrank, Pfl. Lab., 25 (1839), in part; not
Linn. (1753).
Sorbus sambucifolia Roemer, Fam. Nat. Syn., III, 139 (1847).
Sorbus Sitchensis Roem., l. c. (1847).
Pyrus Americana Newberry, in Pac. R. R. Rep., VI, 73
(1857), not de C. (1825).

COMMON NAMES.

Mountain Ash (Vt., Ont., Mont., Minn.).
Elder-leaved Mountain Ash (Minn.).

AMELANCHIER Medic., Phil. Bot., I, 155 (1789).

Amelanchier canadensis (Linn.) Medic.   **Service-tree.**

SYN.—*Mespilus Canadensis* Linnæus, Spec. Pl., ed. 1, I, 478 (1753).
Pyrus Botryapium Linn. f., System. Veg., ed. 13, Suppl., 255
(1781).
Cratægus racemosa Lamarck, Enc. Méth. Bot., I, 84 (1783).
Mespilus nivea Marshall, Arb. Am., 90 (1785).
Amelanchier Canadensis var. *prunifolia* Castiglioni, Viag.
Stati Uniti, II, 293 (1790).
Amelanchier Canadensis Medicus, Gesch. Bot., 79 (1793).

Amelanchier canadensis (Linn.) Medic.—Continued.

SYN.—*Mespilus Canadensis* var. *cordata* Michaux, Fl. Bor.-Am., I.
291 (1803).
Mespilus Amelanchier Walter, Fl. Caroliniana, 148 (1788),
not Linn. (1753).
Cratægus amœna Salisbury, Prodr., 357 (1796).
Aronia Botryapium Persoon, Syn. Pl., II, 39 (1807).
Mespilus arborea Michaux f., Hist. Arb. Am., III, 68, t. 11
(1813).
Amelanchier Botryapium Borkhausen, Handb. Forstb., II,
1260 (1803).
Aronia arborea Barton, Compend. Fl., Philadelph., I, 228
(1818).
Amelanchier sanguinea Lindley, in Bot. Reg., XIV, t. 1171
(1828), not de C. (1825).
Aronia cordata Rafinesque, Med. Bot., II, 196 (1830).
Amelanchier ovalis Hooker, Fl. Bor.-Am., I, 202 (1833), in
part.
Pyrus Bartramiana Tausch, Reg. Fl., II, 715 (1838).
Pyrus Wangenheimiana Tausch, l. c. (1838).
Amelanchier Canadensis var. *Botryapium* Torrey & Gray,
Flor. N. A., I, 473 (1840).
Amelanchier Bartramiana Roemer, Fam. Nat. Syn., III, 145
(1847).
Amelanchier Wangenheimiana Roem., l. c., 146 (1847).

COMMON NAMES.

June Berry (Mass., N. Y., Pa., Del., Ill., Kans., S. Dak.).
Shad Bush (Vt., Mass., R. I., N. Y., Pa., Del., S. C.).
Service Berry (Ill., Del., Ark., Fla., S. Dak.).
Service Tree (Pa., Del.).
May Cherry (Pa., N. C.).
Indian Cherry (Pa.).
Wild Indian Pear (Newfoundland).
Currant Tree (Fla., Ala.).
Shad Berry (Fla.).

Amelanchier canadensis obovalis (Michx.) B. S. P.
Longleaf Service-tree.

SYN.—*Mespilus Canadensis* var. *obovalis* Michaux, Fl. Bor.-Am., I,
291 (1803).
Pyrus sanguinea Pursh. Fl. Am. Sept., 1, 340 (1814). in part.
Pyrus ovalis Bigelow, Fl. Bost.. ed. 2. 195 (1824), not Willd.
(1796).
Amelanchier intermedia Spach. Hist. Vég., II, 85 (1834).
Aronia ovalis Torrey, Fl. U. S., 479 (1840), not Pers. (1807).

Amelanchier canadensis obovalis (Michx.) B. S. P.—Continued.

SYN.—*Amelanchier Canadensis* var. *oblongifolia* Torrey & Gray, Fl. N. A., I, 473 (1840).
Amelanchier oblongifolia Roemer, Fam. Nat. Syn., III, 147 (1847).
Amelanchier spicata Decaisne, in Nouv. Arch. Mus., X, 135, t. 9, f. 5 (1875).
Amelanchier Canadensis var. *oboralis* (Michx.) B. S. P., in Britton, Cat. Pl. N. J., 100 (1889).

Amelanchier canadensis spicata (Lam.) Sargent.

SYN.—*Cratægus spicata* Lamarck, Enc. Méth. Bot., I, 84 (1783).
Pyrus ovalis Willdenow, Berl. Baumz., ed. 1, 259 (1796).
Mespilus Canadensis var. *rotundifolia* Michaux, Flor. Bor.-Am., I, 291 (1803).
Amelanchier ovalis Borkhausen, Handb. Forstb., II, 1259 (1800).
Aronia ovalis Persoon, Syn. Pl., II, 40 (1807).
Amelanchier Canadensis γ *rotundifolia* Torrey & Gray, Fl. N. A., I, 473 (1840).
Amelanchier rotundifolia Roemer, Fam. Nat. Syn., III, 146 (1847), not Dum. Cour. (1811).
Amelanchier Canadensis var. *spicata* Sargent, Silva, IV, 129 (1892).

Amelanchier alnifolia[1][2] Nutt.   **Western Service-tree.**

SYN.—*Pyrus sanguinea* Pursh, Fl. Am. Sept., I, 340 (1814), in part.
Aronia alnifolia Nuttall, Gen., I, 306 (1818).
Pyrus alnifolia Sprengel, Syst. Veg., II, 509 (1825).
Amelanchier florida Lindley, in Bot. Reg., XIX, t. 1589 (1829).
Amelanchier ovalis var. *semiintegrifolia* Hooker, Fl. Bor.-Am., I, 202 (1833).
Amelanchier alnifolia Nuttall, in Journ. Phila. Acad. Sci., VII, 22 (1834).

---

[1] Prof. E. L. Greene describes the following shrubby variety from the "north sides of hills near Humboldt Wells, in eastern Nevada, July, 1893:"

Amelanchier alnifolia arguta (Greene) nom. nov.

SYN.—*Amelanchier pallida* Greene, var. *arguta* Greene, in Erythea, I, 221 (1893).

[2] The following is a recently described variety peculiar to poor rocky and gravelly soils in Arizona and Utah at elevations from "4,000°" to "6,000°" altitude. Mr. Jones does not state (l. c.) whether this plant is a shrub or tree.

Amelanchier alnifolia utahensis (Koehne) Jones.

SYN.—*Amelanchier Utahensis* Koehne, Gattung. Poma. Berl. Ost., 32, t. 2 (1890).
Amelanchier alnifolia var. *Utahensis* (Koehne, l. c.) Jones, in Proc. Cal. Acad. Sci., 2d ser., V, pt. 1, 679 (1895).

Amelanchier alnifolia Nutt.—Continued.

SYN.—*Amelanchier Canadensis* var. *alnifolia* Torrey & Gray, Fl.
N. A., I, 473 (1840).
Amelanchier Canadensis var. *pumila* Torr. & Gr., l. c., 474
(1840).
Amelanchier pumila Roemer, Fam. Nat. Syn., III, 145 (1847).
Amelanchier Canadensis var. *oblongifolia* Bentham, Pl. Hart-
weg., 309 (1849), not Torr. & Gr. (1840).
Amelanchier diversifolia var. *alnifolia* (Torr. & Gr.) Torrey,
in Fremont's Rep., 89 (1845).
Amelanchier Canadensis Anderson, Cat. Pl. Nev., 120 (1870),
not Medic. (1793).
Amelanchier glabra Greene, Fl. Franciscana, I, 52 (1891).
Amelanchier pallida Greene. l. c., 53 (1891).

COMMON NAMES.

Pigeon Berry (So. Oreg.).
Service Berry.
Western Service Berry.

CRATÆGUS[1] Linn., Spec. Pl., 475 (1753).

Cratægus douglasii Lindl.   **Western Haw.**

SYN.—*Cratægus punctata* β *brevispina* Douglas, in Hooker, Fl. Bor.
Am., I, 201 (1834)—*nomen nudum*.
Cratægus Douglasii Lindley, in Bot. Reg., XXI, t. 1810
(1835).
Cratægus brevispina Dougl. ex Steudel, Nom. Bot., ed. sec.,
I, 432 (1840).
Cratægus sanguinea var. *Douglasii* Torrey & Gray, Fl. N.
A., I, 464 (1840).
Anthomeles Douglasii Roemer, Fam. Nat. Syn., III, 140
(1847).
Cratægus sanguinea Nuttall, Sylva, II, 6, t. 44 (1849), not
Pall. (1784).
Cratægus rivularis Brewer & Watson, Bot. Cal., I, 189
(1876), not Nutt. in Torr. & Gr. (1840).

---

[1] The following closely related species is a shrub:
Cratægus rivularis Nutt.   **Narrowleaf Haw.**

SYN.—*Cratægus rivularis* Nuttall, mss. in Torrey & Gray, Fl. N. A., I, 464 (1840).
Mespilus quitensis Koch, in Wochenschr., V, 364 (1862).
Mephilus rivularis Koch, l. c., 372 (1862).
Cratægus Douglasii var. *rivularis* Sargent, in Gard. and For., II, 400 (1889).
Cratægus macnabiana Hort., ex Hand-list Trees and Shr. Arb. Kew, Pt. I,
199 (1896).

Cratægus douglasii Lindl.—Continued.

SYN.—Thorn Apple (Cal., Utah, Wash., Idaho, Nev.).
Hawthorn (Cal.).
Black Haw (Mont.).
Western Haw (Oreg.).
Thorn (N. Mex., Mont., Idaho).
Haw (Oreg.).
Black Thorn (Idaho, Utah, Wash.).
Western Hawthorn (Utah).
River Hawthorn (Utah).
Wild Hawthorn (Utah).
Thorntree.
Wild Thorn (Oreg.).

Cratægus brachyacantha Sarg. & Engelm.   **Hog Haw.**

SYN.—*Cratægus spathulata* Hooker, in Companion Bot. Mag., I, 25
   (1835), not Michx. (1803).
Cratægus brachyacantha Sargent & Engelmann, ex Engel-
   mann, in Bot. Gaz., VII, 128 (1882).
Cratægus brachyacantha var. *maxima* Kuntze, Rev. Gen.
   Pl., Par. I, 215 (1891).

Hog's Haw (La.).
Red Haw (La.).
Pomette Bleue.

Cratægus crus-galli Linn.   **Cockspur.**

SYN.—*Cratægus Crus-galli* Linnæus, Spec. Pl., ed. 1, I, 476 (1753).
Cratægus lucida Miller, Gard. Dic., ed. 8, No. 6 (1768).
Mespilus Crus-galli Marshall, Arb. Am., 88 (1785).
Mespilus lucida Ehrhart, Beitr., IV, 17 (1789).
Mespilus cuneifolia Moench, Meth., 684 (1794).
Cratægus Crus-galli var. *splendens* Aiton, Hort. Kew., ed. 2,
   III, 202 (1811).
Mespilus nana Dumont de Courset, Bot. Cult., ed. 2, VII,
   286 (1814).
Mespilus racemiflora Desfontaines, Cat. Hort. Par., ed. 3, 409
   (1829).
Cratægus laurifolia Medicus, Gesch., 84 (1829).
Cratægus horrida Medic., l. c. (1829).
Cratægus viridis Medic., l. c., 85 (1829), not Linn. (1753).
Mespilus Poiretiana Sweet, Hort. Brit., ed. 2, 176 (1830).
Mespilus Watsoniana Spach, Hist. Vég., II, 57 (1834).

Cratægus crus-galli Linn.—Continued.

SYN.—*Cratægus pyracanthifolia* Don, in Sweet, Hort. Brit., ed. 3, 209 (1839).

Cratægus Watsoniana Roemer. Fam. Nat. Syn., III, 117 (1847).

Cratægus Carrierei Carrière, in Rev. Hort., 1883, 108, t. (1883).

Cratægus Lavallei Hort.   Par. ex Sargent, Silva, IV, 91 (1892).

Cratægus Crus-galli var. *arbutifolia,* ex Hand-list Trees and Shr. Arb. Kew, Pt. I, 197 (1894), not *C. arbutifolia* Lam. (1783).

Cratægus Crus-galli var. *pruinosa,* ex Hand-list, l. c., 199 (1896), not *C. pruinosa* Wendl. (1822).

COMMON NAMES.

Cockspur Thorn (Vt., N. H., R. I., N. Y., N. J., Pa., Del., W. Va., N. C., S. C., Ala., Fla., Miss., Mo., Ill., Kans., Ont.).

Red Haw (Tex., Ill., Mich., Miss.).

Newcastle Thorn (Del., Miss.).

Thorn Apple (N. Y., W. Va.).

Thorn Bush (Pa.).

Thorn (Pa., Ky.).

Pin Thorn (W. Va.),

Thorn Plum (Me.).

Cockspur Hawthorn (Pa.)

Hawthorn (Pa.).

Haw (S. C.)

Cratægus crus-galli salicifolia (Medic.) Ait.

Willowleaf Cockspur.

SYN.—*Cratægus salicifolia* Medicus, Bot. Beobacht., II, 345 (1782).

Cratægus Crus-galli var. *salicifolia* Aiton, Hort. Kew., ed. 1, II, 170 (1789).

Cratægus Crus-galli var. *pyracanthifolia* Ait., l. c. (1789).

Mespilus Crus-galli var. *salicifolia* Hayne, Dend. Fl., 80 (1822).

Mespilus Crus-galli var. *pyracanthifolia* Hayne, l. c. (1822).

Cratægus pyricanthifolia Sweet. Hort. Brit., ed. 2, 175 (1830).

Cratægus Coursetiana Roemer. Fam. Nat. Syn., III, 117 (1847).

Mespilus salicifolia ex Koch, Dendrol., erst. Th., 144 (1869).

Cratægus alpestris ex Koch, l. c. (1869).

Cratægus crus-galli prunellifolia (Poir.) nom. nov.

Broadleaf Cockspur.

SYN.—*Mespilus prunellifolia* Bosc ex Poiret, in Lamarck, Enc. Méth. Bot., Suppl., IV, 72 (1816).

Cratægus crus-galli prunellifolia (Poir.) nom. nov.—Continued.

SYN.—*Mespilus ovalifolia* Hornemann, Hort. Hafn., Suppl., 52 (1819).
Cratægus ovalifolia Hornem., l. c., 52 (1819).
Cratægus Crus-galli var. *ovalifolia* Lindley, in Bot. Reg., XXII, t. 1860 (1836).
Cratægus prunellifolia Bosc, in de Candolle, Prodr., II, 627 (1825).
Mespilus elliptica Guimpel, in Otto & Hayne, Abb. Holz., 170, t. 144 (1830), not Lam. (1797).

Cratægus crus-galli angustifolia (Ehr.) nom. nov.

Narrowleaf Cockspur.

SYN.—*Mespilus lucida* var. *angustifolia* Ehrhart, Beitr., IV, 18 (1789).
Cratægus angustifolia Borckhausen, in Roemer, Arch. Bot., I, 3, 86 (1798).
Cratægus linearis Persoon, Syn. Pl., II, 37 (1807).
Mespilus linearis Desfontaines, Hist. Arb., II, 156 (1809).
Cratægus Crus-galli δ *linearis* de Candolle, Prodr., II, 626 (1825).
Cratægus Crus-galli b *salicifolia* α *linearis* Dippel, Handb. Laubh., dritt. T., 442 (1893).
Cratægus liniarifolia Hort., ex Hand-list Trees and Shr. Arb. Kew, Pt. I, 197 (1896).
Cratægus stipulacea microphylla· Hort., ex Hand-list, l. c. (1896).

Cratægus crus-galli prunifolia (Marsh.) Torr. & Gr.

Plumleaf Cockspur.

SYN.—*Mespilus prunifolia* Marshall, Arb. Am., 90 (1785).
Cratægus prunifolia Persoon, Syn. Pl., II, 37 (1807).
Cratægus (*C.*) *prunifolia 2 ingestria* Loudon, Arb. Frut., II, 821 (1838).
Cratægus ingestria Lodd., ex Loud., l. c. (1838).
Cratægus Crus-galli var. *prunifolia* Torrey & Gray, Fl. N. A., I, 464 (1840).
Cratægus caroliniana Lodd., ex Hand-list Trees and Shr., Arb. Kew, Pt. I, 199 (1896), not Pers. (1807).

Cratægus crus-galli fontanesiana (Spach) Wenzig.

SYN.—*Mespilus Fontanesiana* Spach, Hist. Vég., II, 58, t. 10, f. k. (1834).
Mespilus Bosciana Spach, l. c., 58 (1834).
Cratægus Fontanesiana Steudel, Nom. Bot., ed. sec., I, 432 (1840).

Cratægus crus-galli fontanesiana (Spach.) Wenzig—Continued.

SYN.—*Cratægus Bosciana* Roemer, Fam. Nat. Syn., III, 118 (1847).
Cratægus Crus-galli var. *Fontanesiana* Wenzig, in Linnæa,
XXXVIII, 141 (1874).

Cratægus crus-galli berberifolia (Torr. & Gr.) Sargent.
Barberryleaf Cockspur.

SYN.—*Cratægus berberifolia* Torrey & Gray, Fl. N. A., I, 469
(1840).
Mespilus berberifolia Koch, in Wochenschr., V, 383 (1862);
Wenzig, in Linnæa, XXXVIII, 125 (1874).
Cratægus Crus-galli var. *berberifolia* Sargent, in Gard. and
For., II, 464 (1889).

Cratægus coccinea Linn. **Scarlet Haw.**

SYN.—*Cratægus coccinea* Linnæus, Spec. Pl., ed. 1, I, 476 (1753).
Cratægus rotundifolia Moench, Verzeich. Bäume, 29, t. 1
(1785).
Mespilus coccinea Marshall, Arb. Am., 87 (1785).
Mespilus rotundifolia Ehrhart, Beitr., III, 20 (1788).
Mespilus coccinea var. *viridis* Castiglioni, Viag. Stati Uniti,
II, 293 (1794).
Cratægus glandulosa Willdenow, Sp. Pl., II, 1002 (1799),
in part; not Moench (1785), nor Solander in Ait. (1797),
nor Georgi (1800).
Mespilus maxima Dumont de Courset, Bot. Cult., ed. 2, V,
451 (1811).
Cratægus viridis Elliott, Sk. Bot. S. C. and Ga., I, 551 (1821),
not Linn. (1753).
Mespilus odorata Wendland, Regend. Fl., 1823, 700 (1823).
Mespilus viridis Sweet, Hort. Brit., ed. 1, 134 (1827).
Mespilus Wendlandii Opiz, in Rengensb. Fl., 1834, 590
(1834).
Mespilus flabellata Spach, Hist. Vég., II, 63 (1834).
Cratægus coccinea corallina Loudon, Arb. Frut., II, 817
(1838).
Cratægus coccinea indentata Loud., l. c., f. 566 (1838).
Cratægus indentata Loddiges Cat. ex Loud., l. c. (1838).
Cratægus coccinea 4 maxima Lodd., Cat. ex Loud., l. c. (1838).
Cratægus coccinea spinosa Godefroy ex Loud., l. c. (1838).
Cratægus coccinea β viridis Torrey & Gray, Fl. N. A., I, 465
(1840).
Cratægus coccinea δ oligandra Torr. & Gr., l. c. (1840).
Halmia flabellata Roemer, Fam. Nat. Syn., III, 136 (1847).
Anthomeles rotundifolia Roem., l. c., 140 (1847).
Phænopyrum coccineum Roem., l. c., 156 (1847).

Cratægus coccinea Linn.—Continued.

SYN.—*Phœnopyrum Wendlandii* Roem., l. c. (1847).
    Mespilus pyriformis Roem., l. c. (1847).
    Cratægus chlorocarpa Koch, Ind. Sem. Hort. Ber., 17 (1855).
    Cratægus glandulosa β rotundifolia Regel, in Act. Hort.
        Petrop., I, 120 (1871).
    Cratægus coccinea c *flabellata* Dippel, Handb. Laubh., dritt.
        T., 435 (1893).
    Cratægus rotundifolia a *minor* Dippel, l. c., 440 (1893).
    Cratægus rotundifolia b *succulenta* Dippel, l. c. (1893).
    Cratægus accrifolia Hort., ex Hand-list Trees and Shr. Arb.
        Kew, Pt. I, 195 (1896), not Moench (1785).

COMMON NAMES.

Scarlet Haw (N. H., Mass., N. Y., N. J., Pa., N. C., S. C.,
    Miss., Ark., Mo., Ill., Nebr., Iowa, Minn.).
Red Haw (R. I., N. Y., W. Va., S. C., Ga., Miss., La., Tex.,
    Mo., Ill., Nebr., Ohio, Iowa, Minn., S. Dak.).
White Thorn (Vt., R. I., Del., Miss., Iowa, Ill., Kans., Minn.,
    Ont.).
Scarlet Thorn (Vt., Mass., R. I., N. J., Del., Ont.).
Scarlet-fruited Thorn (Mich., Minn.).
Red Thorn (Ky.).
Hawthorn (Pa., Iowa).
Thorn (Vt., N. Y., Ky., Mont.).
Thorn Bush (R. I., Pa.).
Thorn Apple (Vt., Mont.).
Thorn Apple Tree (Minn.).
Thorn Plum (Me., Vt., N. Y.).
Haw Bush (Mont.).
Scarlet Thorn-Haw (Fla.).
Hedge Thorn (Mont.).
Red Thorn Bush (Ky., Ind.).

Cratægus coccinea populifolia Torr. & Gr.   **Poplarleaf Haw.**

SYN.—*Mespilus populifolia* Poiret, in Lamarck, Enc. Méth. Bot.,
    IV, 447 (1797).
    Cratægus populifolia Elliott, Sk. Bot. S. C. and Ga., I, 553
        (1821), not Walt. (1788).
    Cratægus coccinea var. *typica* Regel, in Act. Hort. Petrop.,
        I, 121 (1871).
    Cratægus coccinea γ populifolia Torrey & Gray, Fl. N. A., I,
        465 (1840).
    Phœnopyrum populifolium Roemer, Fam. Nat. Syn., III, 153
        (1847).

Cratægus macracantha (Lindl.) Loud.   **Largeflower Haw.**

SYN.—*Cratægus glandulosa* Moench, Verzeich. Bäume, 31 (1785).—?
    Pyrus glandulosa Moench, Meth. 680 (1794).—?
    Cratægus glandulosa Willdenow, Berl. Baumz., ed. 1, 84
      (1796), not Ait. (1789).
    Mespilus glandulosa Willd., Enum. Pl., 523 (1809).
    Mespilus sanguinea[1] Du Mont de Courset, Bot. Cult., ed. 2,
      V, 452 (1811).
    Cratægus glandulosa β macracantha Lindley, in Bot. Reg., t.
      1912 (1836).
    Cratægus macracantha Loddiges ex Loudon, Arb. Frut., II,
      819 (1838).
    Cratægus sanguinea Torrey & Gray, Fl. N. A., I, 464 (1840),
      not Pall. (1789).
    Cratægus coccinea Brandegee, in Rep. Chief Eng. U. S. A.,
      Append. S., 1841 (1875), not Linn. (1753).
    Cratægus coccinea Var. *viridis* Torrey, in Pac. R. R. Rep.,
      IV, 86 (1857), not Torr. & Gr. (1840).
    Cratægus Douglasii Macoun, Cat. Canad. Pl., Pt. III, 522
      (1886), not Lindl. (1835).
    Cratægus coccinea var. *macracantha* Dudley, in Bull. Corn.
      Un., II, 33 (1886).
    Cratægus coccinea d *pruinosa* Dippel, Handb. Laubh., dritt.
      T., 436 (1893).

Cratægus mollis (Torr. & Gr.) Scheele.   **Downy Haw.**

SYN.—*Mespilus coccinea* Schmidt, Oestr. Baumz., IV, 30, t. 210
    (1822) not Marsh. (1785).
    Mespilus pubescens Wendland, Regensb. Fl., 1823, 700 (1823),
      not Presl. (1822), nor H., B. K. (1823).
    Mespilus coccinea β pubescens Tausch, Regensb. Fl., 1838,
      Pt. II, 718 (1838).
    Cratægus coccinea var. *mollis* Torrey & Gray, Fl. N. A., I,
      465 (1840).
    Cratægus subvillosa Schrad. ex Torr. & Gr., l. c. (1840).
    Cratægus tomentosa Emerson, Trees & Shr. Mass., ed. 1, 435
      (1846), not Linn. (1753).
    Phænopyrum subvillosum Roemer, Fam. Nat. Syn., III, 154
      (1847).
    Cratægus mollis Scheele, in Linnæa, XXI, 569 (1848).
    Cratægus Texana Buckley, in Proc. Phila. Acad. Sci., 1861,
      454 (1861).
    CRATÆGUS TOMENTOSA var. MOLLIS Gray, Man. N. States,
      ed. 5, 160 (1867).

---

[1]This is an older specific name than Lindley's *C. gl. β macracantha*, but if taken
up under *Cratægus*, a combination is produced identical with that of Pallas (1789),
applied to a distinct species.

Cratægus mollis (Torr. & Gr.) Scheele—Continued.

SYN.—*Mespilus tiliæfolia* Koch, in Wochenschr., V, 375 (1862).
    Cratægus coccinea glandulosa Hort., ex Hand-list Trees and
      Shr. Arb. Kew, Pt. I, 203 (1896), not *C. glandulosa* Moench
      (1785), nor Ait. (1789), nor Georgi (1802).
    Cratægus coccinea Kelmanni Hort., ex Hand-list, l. c. (1896).
    Cratægus mespilifolia Hort., ex Hand-list, l. c. (1896).
    Cratægus urasina Hort., ex Hand-list., l. c. (1896).

COMMON NAMES.

Scarlet Haw.
Red Thorn Apple (Mich.)

Cratægus oxyacantha[1] Linn.   **English Hawthorn.**

SYN.—*Cratægus Oxyacantha* Linnæus, Spec. Pl., ed.1, I, 477 (1753).
    Mespilus oxyacantha Crantz, Stirp. Austr., ed. 1, II, 39
      (1763).
    Cratægus monopyrena Pallas, Reise, I, 349 (1771).
    Cratægus spinosa Gilibert, Fl. Lith., II, 231 (1781).
    Cratægus maura Linnæus f., Syst. Veg., ed. 13, Suppl., 253
      (1781).
    Mespilus Cratægus Borkhausen, Beschr. Hess., 190 (1790).
    Mespilus lævigata Poiret, in Lamarck. Enc. Méth. Bot., IV,
      439 (1797).
    Cratægus glandulosa Georgi, Beschr. Russ. Nachtr., III, 273
      (1800), not Moench (1785).
    Cratægus pauciflora Persoon, Syn. Pl., II, 37 (1807).
    Cratægus maroccana Pers., l. c. (1807).
    Mespilus nigra Willdenow, Enum. Pl., 524 (1809).
    Mespilus maura Poiret, in Lamarck, Enc. Méth. Bot., Suppl.,
      IV, 72 (1816).
    Mespilus maroccana Poir., in Lam., l. c., 74 (1816).
    Mespilus digyna Gray, Nat. Arr. Brit., II, 565 (1821).
    Cratægus lævigata de C., l. c., 630 (1825).
    Cratægus Kyrtostyla Fing. ex Schlectendal, in Linnæa, IV,
      372 (1829).
    Mespilus polycantha Gussone, Fl. Sic. Prodr., Suppl., 154
      (1843).
    Cratægus lagenaria Fisch. & Mey., in Bull. Soc. Nat. Mosc.,
      367 (1838).
    Cratægus nigra Pall. ex Steudel, Nom. Bot., ed. sec., I, 433
      (1840).

---

[1] This European species is widely introduced in the United States by cultivation, and in many localities in the Eastern States has escaped and become thoroughly naturalized.

Cratægus oxyacantha Linn.—Continued.

SYN.—*Cratægus ribesius* Bertoloni, Misc. Bot., XXII, 14, t. 2 (1863).

Cratægus digyna Pallas in Ledebour, Fl. Ross., II, 88 (1846).

Oxyacanthus kyrtostyla Roemer, Syn. Fl. Rositl., III, 106 (1847).

Oxyacanthus macrocarpa Roem., l. c., 107 (1847).

Oxyacanthus vulgaris Roem., l. c., 109 (1847).

Mespilus caucasica Koch, in Wochenschr., V, 405 (1862).

Cratægus hirsuta Schur, Enum. Pl. Transsil., 206 (1866).

Cratægus rosæformis Janka, in Oestr. Bot. Zeitsch., XX, 250 (1870).

Cratægus spectabilis Carrière, in Rev. Hort., 471 (1873).

Cratægus polyacantha Jan. ex Nyman, Consp., 244 (1878.)

Cratægus calycina Peterm. ex Nyman, l. c. (1878).

Cratægus Oxyacantha subsp. *monogyna* var. *crispa pendula*[1] Hort., ex Hand-list Trees and Shr. Arb. Kew, Pt. I, 203 (1894).

Cratægus Oxyacantha subsp. *monogyna* var. *fusca*[1] Hort., ex Hand-list, l. c., 205 (1894).

Cratægus Oxyacantha subsp. *monogyna* var. *fuscata*[1] Hort., ex Hand-list, l. c. (1894).

Cratægus Oxyacantha subsp. *monogyna* var. *Gumperi versicolor*[1] Hort., ex Hand-list, l. c. (1894).

Cratægus Oxyacantha subsp. *monogyna* var. *oxyphylla*,[1] ex Hand-list, l. c. (1894).

Cratægus Oxyacantha subsp. *monogyna* var. *rubrinervis*,[1] ex Hand-list, l. c. (1894).

Cratægus Oxyacantha subsp. *monogyna* var. *salisburifolia*,[1] ex Hand-list, l. c. (1894).

Cratægus Oxyacantha subsp. *monogyna* var. *sesteriana*[1] Hort., ex Hand-list, l. c. (1894).

Cratægus Oxyacantha subsp. *oxyacanthoides* var. *atrofusca*[1] Hort., ex Hand-list, l. c., 207 (1894).

VARIETIES DISTINGUISHED IN CULTIVATION.

Cratægus oxyacantha oxyacanthoides (Thuill.) Reich.

SYN.—*Mespilus Oxyacanthoides* Thuillier, Fl. Par. Distr., ed. 1, 245 (1790).

Mespilus oxyacanthoides de Candolle, Fl. Fran., IV, 433 (1805).

---

[1] Forms cultivated as distinct varieties in the Royal Kew Gardens. Nothing is known of their distinctive features, as their catalogue names are unaccompanied by description; moreover, they are not known to have been described elsewhere.

Cratægus oxyacantha oxyacanthoides (Thuill.) Reich.—Continued.

SYN.—*Cratægus integrifolia* Wallroth, Sched. Crit., 219 (1822).
Mespilus oxyacantha integrifolia Wallr., 1. c. (1822).
Cratægus Oxyacantha α obtusata de Candolle, Prodr., II, 628 (1825),
Cratægus Oxyacantha β oxyacanthoides Reichenbach, Fl. Germ., II, 628 (1832).
Cratægus Oxyacantha glomerata Hort., ex Hand-list Trees and Shr. Arb. Kew, Pt. I, 207 (1894).

Cratægus oxyacantha monogyna (Jacq.) Loud.

SYN.—*Cratægus monogyna* Jacquin, Fl. Austr., III, 50, t. 292, f. 1 (1775).
Mespilus monogyna Allioni, Fl. Pedem., II, 141 (1785).
Mespilus triloba Poiret, Voy. Barb., II, 171 (1789).
Cratægus elegans Poiret, in Lamarck, Enc. Méth. Bot., IV, 439 (1797).
Mespilus apiifolia Medicus, Gesch., 82 (1829), not Poir. in Lam. (1816).
Cratægus apiifolia Med., 1. c., 83 (1829), not Michx. (1803).
Mespilus elegans Poiret, in Lamarck, Enc. Méth. Bot., Suppl., IV, 72 (1816).
Cratægus Oxyacantha δ monostyla de Candolle, Prodr., II, 628 (1825).
Cratægus Oxyacantha 3 sibirica Loudon, Arb. Frut., II, 830, f. 555 (1838).
Cratægus sibirica Lodd., Cat: ex Loud., 1. c. (1838).
Cratægus Oxyacantha 4 transylvanica Hort. ex Loud., 1. c. (1838).
Cratægus Oxyacantha 26 monogyna Loud., 1. c., 834 (1838).
Oxyacanthus monogyna Roemer, Syn. Fl., III, 106 (1847).
Cratægus transylvanica Hort. ex Roem., 1. c., 107 (1847).
Oxyacanthus elegans Roem., 1. c., 108 (1847).
Oxyacanthus apiifolia Roem., 1. c. (1847).
Oxyacantha media Roem., 1. c. (1847).

Cratægus oxyacantha laciniata (Borkh.) de C.

SYN.—*Cratægus laciniata* Borkhausen, Handb. Forstb., II, 1355 (1803).
Cratægus dissecta Borkh., in Roemer, Arch. I, III, 86 (1805).
Cratægus fissa Bosc (ined.) ex de Candolle, Prodr., II, 628 (1825).
Cratægus Oxyacantha γ laciniata de C., 1. c. (1825).
Mespilus dissecta Du Mont de Courset, Bot. Cult., ed 2, V, 454 (1811).
Mespilus fissa Poiret, in Lamarck, Enc. Méth. Bot., Suppl., IV, 72 (1816).

Cratægus oxyacantha laciniata (Borkh.) de C.—Continued.

SYN.—*Cratægus Oxyacantha 5 quercifolia* Hort. ex Loudon. Arb.
Frut.. II. 830, f. 608 (1838).
Cratægus Oxyacantha 7 pteridifolia Loudon. l. c.. 831. f. 607
(1838).
Cratægus pterifolia Lodd.. Cat. ex Loud.. l. c. (1838).
Mespilus monogyna laciniata ex Koch. Dendrol.. erst. Th.,
160 (1869).
Cratægus oxyacantha filicifolia Nicholson, Dic. Gard., I, 394
(1885).—?
Cratægus oxyacantha dissecta Hort. ex Dippel. Handb.
Laubh., dritt. T., 459 (1893).
Cratægus oxyacantha fissa Hort. ex Dippel. l. c. (1893).
Cratægus oxyacantha pectinata Hort. ex Dippel. l. c. (1893),
not *C. pectinata* Bosc ex de C. (1825).
Cratægus Oxyacantha subsp. *monogyna* var. *pteridifolia* Loud.,
ex Hand-list Trees and Shr. Arb. Kew, Pt. I, 205 (1894).

Cratægus oxyacantha diversifolia (Poir.) nom. nov.

SYN.—*Mespilus diversifolia* Poiret, in Lamarck. Enc. Méth. Bot.,
IV. Suppl.. 72 (1816).
Cratægus diversifolia Steudel. Nom. Bot. ed. 1, 234 (1821).
Cratægus Azarella Grisebach, Spicil. Fl. Rum., I, 88 (1843).
Oxyacantha Azarella Roemer. Syn. Fl.. III, 106 (1847).
Cratægus heterophylla Steven. Verz. Taur.. 147 (1857), not
Fluegg. (1808).
Cratægus monogyna a *heterophylla* Dippel. Handb. Laubh.,
dritt. T., 458 (1893).

Cratægus oxyacantha incisa Regel.

SYN.—*Cratægus monogyna* Pallas, Fl. Ross., t. 12 (1784). not Jacq.
(1775).
Cratægus Oxyacantha β incisa Regel, in Act. Hort. Petrop.,
I. 117 (1871).
Cratægus oxyacantha b *incisa* Dippel. Handb. Laubh., dritt.
T.. 456 (1893).
Cratægus Oxyacantha subsp. *oxyacanthoides* var. *incisa*, ex
Hand-list Trees and Shr. Arb. Kew, Pt. I, 207 (1894).

Cratægus oxyacantha auriculata Dippel.

SYN.—*Cratægus oxyacantha* c *auriculata* Dippel. Handb. Laubh.,
dritt. T., 457 (1893).

Cratægus oxyacantha sorbifolia (Desf.) Dippel.

SYN.—*Cratægus sorbifolia* Desfontaines, Cat. Hort. Par., ed. III,
408 (1829).

Cratægus oxyacantha sorbifolia (Desf.) Dippel.—Continued.

SYN.—*Cratægus oxyacantha* d *sorbifolia* Dippel, Handb. Laubh., dritt. T., 457 (1893).

Cratægus oxyacantha pinnatiloba (Lange) nom. nov.

SYN.—*Cratægus pinnatiloba* Lange, in Bot. Tidsskr., XIII, 22, t. 3 (1883).
Cratægus monogyna c *pinnatiloba* Dippel, Handb. Laubh., dritt. T., 458 (1893).
Cratægus Oxycantha subsp. *monogyna* var. *pinnatiloba*, ex Hand-list Trees and Shr. Arb. Kew, Pt. I, 205 (1894).

Cratægus oxycantha ferox Dippel.

SYN.—*Mespilus monogyna horrida* Koch, Dendrol., erst. Th., 160 (1869).
Cratægus horrida Regel, in Act. Hort. Petrop., II, 119 (1871), not Medic. (1829).
Cratægus monogyna f *horrida* Dippel, Handb. Laubh., dritt. T., 459 (1893).
Cratægus oxyacantha horrida hort. ex Dippel, l. c. (1893).
Cratægus oxycantha spinosissima hort. ex Dippel, l. c. (1893), not *C. spinosissima* Lodd. (1833).
Cratægus oxycantha ferox hort. ex Dippel, l. c. (1893).
Cratægus Oxycantha subsp. *monogyna* var. *horrida* Hort., ex Hand-list Trees and Shr. Arb. Kew, Pt. I, 205 (1894).
Cratægus spinosissima Hort., ex Hand-list, l. c. (1894), not Lodd. (1833).

Cratægus oxycantha curtispina nom. nov.

SYN.—*Cratægus monogyna* g *brevispina* Dippel, Handb. Laubh., dritt. T., 459 (1893), not *C. brevispina* Steud. (1841), nor Kunze (1846).
Cratægus brevispina hort. ex Dippel, l. c. (1893).

Cratægus oxycantha flexuosa Loud.

SYN.—*Cratægus oxyacantha 30 flexuosa* Loudon, Arb. Frut., II, 835 (1838).
Mespilus flexuosa Koch, Dendrol., erst. Th., 160 (1869).
Cratægus monogyna h *flexuosa* Dippel, Handb. Laubh., dritt. T., 459 (1893).
Cratægus Oxyacantha subsp. *monogyna* var. *flexuosa* Loud., ex Hand-list Trees and Shr. Arb. Kew, Pt. I, 203 (1894).

Cratægus oxyacantha stricta Loud.

SYN.—*Cratægus Oxyacantha 21 stricta* Lodd. Cat. ex Loudon, Arb. Frut., II, 832 (1838).

18158—No. 14——15

Cratægus oxyacantha stricta Loud.—Continued.

SYN.—*Cratægus Oxyacantha rigida* Ronalds ex Loud., 1. c. (1838).
Cratægus Oxyacantha 22 Celsiana Hort. ex Loud., 1. c. (1838).
Cratægus Oxyacantha 29 capitata Loud.. 1. c.. 834 (1838).
Mespilus stricta Koch. Dendrol., erst. Th., 160 (1869).
Mespilus fastigiata Koch. 1. c. (1869).
Cratægus monogyna **k** *fastigiata* Dippel. Handb. Laubn.,
dritt. T., 459 (1893).
Cratægus oxyacantha stricta hort. ex Dippel, 1. c. (1893).
Cratægus Oxyacantha subsp. *monogyna* var. *stricta* Lodd.. ex
Hand-list Trees and Shr. Arb. Kew, pt. I. 205 (1894).
Cratægus fastigiata Hort., ex Hand-list. 1. c. (1894).

Cratægus oxyacantha pendula Loud.

SYN.—*Cratægus Oxyacantha 23 pendula* Lodd. Cat. ex London, Arb.
Frut., II, 832 (1838).
Cratægus Oxyacantha 24 reginæ Hort. ex Loud.. 1. c. (1838).
Cratægus monogyna **i** *pendula* Dippel. Handb. Laubh., dritt.
T. 459 (1893).
Cratægus Reginæ hort. Angl. ex Dippel. 1. c. (1893).
Cratægus Oxyacantha subsp. *monogyna* var. *Reginæ* Hort., ex
Hand-list Trees and Shr. Arb. Kew. Pt. 1, 205 (1894).—?

Cratægus oxyacantha aurea Loud.

SYN.—*Cratægus Oxyacantha 12 aurea* Hort. ex Loudon. Arb. Frut.,
II, 831, f. 610 (1838).
Cratægus flava Hort. ex Loud.. 1. c. (1838). not Solander in
Ait. (1789).
Cratægus Oxyacantha subsp. *monogyna* var. *aurea* Loud.. ex
Hand-list Trees and Shr. Arb. Kew. Pt. I, 203 (1894).

Cratægus oxyacantha variegata (Dippel) nom. nov.

SYN.—*Cratægus Oxyacantha foliis aureis* Lodd. Cat. ex London,
Arb. Frut.. II. 832 (1838).
Cratægus Oxyacantha foliis argenteis Hort. ex Loud.. 1. c.
(1838).
Cratægus monogyna variegata Dippel. Handb. Laubh., dritt.
T., 460 (1893).
Cratægus Oxyacantha variegata Hort.. ex Hand-list Trees
and Shr. Arb. Kew. Pt. 1. 205 (1894).
Cratægus Oxyacantha subsp. *monogyna* var. *foliis argenteis*
Hort., ex Hand-list.. 1. c. (1894).
Cratægus Oxyacantha subsp. *monogyna* var. *fol. aureis* Lodd.
ex Hand-list. 1. c. (1894).
Cratægus Oxyacantha lutescens Hort., ex Hand-list, l.c. (1894).

Cratægus oxyacantha splendens (Koch) nom. nov.

SYN.—*Mespilus monogyna splendens* Koch, Dendrol., erst. Th., 159 (1869).
Cratægus monogyna l *splendens* Dippel, Handb. Laubh., dritt. T., 459 (1893).

Cratægus oxyacantha eriocarpa Loud.

SYN.—*Cratægus Oxyacantha 8 eriocarpa* Lindl. ex Loudon, Arb. Frut., II, 831, f. 607 (1838).
Cratægus eriocarpa Lodd. Cat. ex Loud., l. c. (1838).
Cratægus Oxyacantha subsp. *monogyna* var. *eriocarpa* Hort., ex Hand-list Trees and Shr. Arb. Kew, Pt. I, 203 (1894).

Cratægus oxyacantha oliveriana (Poir.) Loud.

SYN.—*Mespilus Oliveriana* Poiret, in Lamarck, Enc. Méth. Bot. Suppl., IV, 72 (1816).
Cratægus melanocarpa de Candolle, Prodr., II, 629 (1825).
Cratægus Oliveriana Bosc, in de C., l. c., 630 (1825).
Cratægus Oxyacantha 10 Oliveriana Loudon, Arb. Frut., II, 831 (1838).
Cratægus Oliveria Lodd. Cat. ex Loud., l. c. (1838).
Cratægus orientalis Lodd. Cat. ex Loud., l. c. (1838).
Cratægus Oxyacantha 11 melanocarpa Loud., l. c., (1838).
Cratægus fissa Lee ex Loud., l. c. (1838), not Poir. (1816).
Cratægus platyphylla Lodd. Cat. ex Loud., l. c. (1838), not Lindl. (1836).
Mespilus microphylla Koch, in Wochenschr., V, 405 (1862).

Cratægus oxyacantha aurantiaca Loud.

SYN.—*Cratægus Oxyacantha 13 aurantiaca* Booth ex Loudon, Arb. Frut., II, 831 (1838).

Cratægus oxyacantha leucocarpa Loud.

SYN.—*Cratægus Oxyacantha 14 leucocarpa* Loudon, Arb. Frut., II, 831 (1838).

Cratægus oxyacantha apetala Loud.

SYN.—*Cratægus Oxyacantha 27 apetala* Lodd. Cat. ex Loudon, Arb. Frut., II, 834 (1838).
Cratægus Oxyacantha petala Nicholson, Dic. Gard., I, 393 (1885—?).

Cratægus oxyacantha præcox Loud.

SYN.—*Cratægus Oxyacantha 25 præcox* Hort. ex Loudon, Arb. Frut., II, 833 (1838).
Cratægus monogyna d *præcox* Dippel, Handb. Laubh., dritt. T., 459 (1893).

Cratægus oxyacantha præcox Loud.—Continued.

SYN.—*Cratægus Oxyacantha* subsp. *monogyna* var. *præcox* Hort., ex Hand-list Trees and Shr. Arb. Kew, Pt. I, 205 (1894).

Cratægus oxyacantha multiplex Loud.

SYN.—*Cratægus Oxyacantha 15 multiplex* Hort. ex Loudon, Arb. Frut., II, 832, f. 609 (1838).
Cratægus Oxyacanthus flore pleno Hort. ex Loud., l. c. (1838).
Cratægus Oxyacantha subsp. *oxyacanthoides* var. *flore pleno albo*, ex Fl. Serres, XV, t. 1509 (1865).

Cratægus oxyacantha rosea Loud.

SYN.—*Cratægus rosea* ex de Candolle, Prodr., II, 628 (1825).—?
Cratægus Oxyacanthus 16 rosea Hort. ex Loudon, Arb. Frut., II, 832, f. 612 (1838).

Cratægus oxyacantha punicea Loud.

SYN.—*Cratægus Oxyacantha 17 punicea* Lodd. Cat. ex Loudon, Arb. Frut., II, 832 (1838).
Cratægus Oxyacantha rosea superba Hort. ex Loud., l. c. (1838).
Cratægus Oxyacantha subsp. *oxyacanthoides* var. *flore puniceo*, ex Fl. Serres, XV, t. 1509 (1865).
Cratægus punicea Fisch. ex Steudel, Nom. Bot., ed. sec., I, 433 (1841).
Cratægus atropurpurea Stev. ex Nyman, Consp., 244 (1878).
Cratægus Oxyacantha rosea-superba Nicholson, Dic. Gard., I, 394 (1885 ?).[1]

Cratægus oxyacantha punicea plena nom. nov.

SYN.—*Cratægus Oxyacantha punicea flore pleno* Hort. ex Loudon, Arb. Frut., II, 832 (1838).
Cratægus Oxyacantha subsp. *oxyacanthoides* var. *pleno puniceo*, ex Hand-list Trees and Shr. Arb. Kew, Pt. I, 207 (1894).

Cratægus oxyacantha semperflorens Dippel.

SYN.—*Cratægus monogyna* m *semperflorens* Dippel, Handb. Laubh., dritt. T., 460 (1893).
Cratægus oxyacantha semperflorens Hort. ex Dippel, l. c. (1893).
Cratægus Oxyacantha subsp. *monogyna* var. *semperflorens* Carr., ex Hand-list Trees and Shr. Arb. Kew, Pt. I., 205 (1894).
Cratægus Bruantii Hort., ex Hand-list, l. c. (1894).

---

[1] See footnote, p. 146.

Cratægus oxyacantha gratanensis (Boiss.) nom nov.

SYN.—*Cratægus gratanensis* Boissier, Elench. Pl., 41 (1838).
Cratægus monogyna **b** *gratanensis* Dippel, Handb. Laubh.,
dritt. T., 458 (1893).
Cratægus Oxyacantha subsp. *monogyna* var. *gratanensis*, ex
Hand-list Trees and Shr. Arb. Kew, Pt. I, 205 (1894).

Cratægus oxyacantha macrocarpa (Hegetsch.) nom. nov.

SYN.—*Cratægus macrocarpa* Hegetschweiler, Fl. Schweiz., 464
(1840).
Cratægus Oxyacantha subsp. *monogyna* var. *macrocarpa*, ex
Hand-list Trees and Shr. Arb. Kew, Pt. I, 205 (1894).

Cratægus tomentosa Linn.   **Black Haw.**

SYN.—*Cratægus tomentosa* Linnæus, Spec. Pl., ed. 1, I, 476 (1753).
Cratægus leucophlœos Moench, Verzeich. Bäume, 31, t. 2
(1785).
Mespilus Calpodendron Ehrhart, Beitr., II, 67 (1788).
Cratægus pyrifolia Solander, in Aiton, Hort. Kew., ed. 1, II,
168 (1789).
Mespilus tomentosa Castiglioni, Viag. Stati Uniti, II, 293
(1790).
Mespilus latifolia Hort. ex Poiret, in Lamarck, Enc. Méth.
Bot., IV, 444 (1797).
Cratagus latifolia Persoon, Syn. Bot., II, 37 (1807).
Mespilus pyrifolia Willdenow, Enum. Pl., 523 (1809).
Mespilus lobata Poiret, in Lamarck, Enc. Méth. Bot., Suppl.,
IV, 71 (1816).
Mespilus lutea Poiret, in Lamarck, l. c., 72 (1816).
Mespilus pruinosa Wendland f., in Flora., VI, 701 (1823).
Cratægus lobata Bosc, in de Candolle, Prodr., II, 628 (1825).
Halmia tomentosa Roemer, Fam. Nat. Syn., III, 135 (1847).
Halmia tomentosa β pyrifolia Roem., l. c. (1847).
Halmia tomentosa δ leucophlœa Roem., l. c. (1847).
Halmia lobata Roem., l. c., 136 (1847).
Halmia tomentosa ε Calpodendron Roem., l. c. (1847).
Cratægus tomentosa var. *pyrifolia* Gray, Man. N. States, ed.
5, 160 (1867).
Mespilus leucophlœos Koch, Dendrol., erst. Th., 136 (1869).
Cratægus lutea Poiret ex Jackson, Ind. Kew., I, 636 (1893).
Cratægus Downingii Hort., ex Hand-list Trees and Shr. Arb.
Kew, Pt. I, 209 (1894).

COMMON NAMES.

Black Thorn (R. I., N. J., Pa., Del., Ga., Fla., La., Miss., Ky.,
Ill., Ind., Ohio.).

Cratægus tomentosa Linn.—Continued.

Syn.—Pear Haw (Miss., Ohio).
    Red Haw (Miss., Mo.).
    Pear Thorn (R. I., N. J., Mich.).
    White Thorn.
    Thorn (N. Y., Ky.).
    Common Thorn (Pa.).
    Haw Thorn.
    Thorn Apple (Ill.).
    Thorn Plum.

Cratægus punctata Jacq.   **Dotted Haw.**

Syn.—*Cratægus punctata* Jacquin, Hort. Vindob., I, 10, t. 28
    (1770).
    Mespilus cornifolia Muenchhausen, Hausv., V, 145 (1770).
    Mespilus cuneiformis Marshall, Arb. Am., 88 (1785).
    Cratægus Crus-galli Wangenheim, Nordam. Holz., 52
    (1787).—?
    Mespilus cuneifolia Ehrhart, Beitr., III, 21 (1788), not
    Moench (1785).
    Cratægus punctata var. *rubra* Aiton, Hort. Kew., ed. 1, II,
    170 (1789).
    Cratægus punctata var. *aurea* Ait., l. c. (1789).
    Mespilus pyrifolia Desfontaines, Hist. Arb., II, 156 (1809),
    not Willd. (1809).
    Mespilus punctata Loiseleur, in Nouveau Duhamel, IV, 152
    (1819).
    Cratægus latifolia de Candolle, Prodr., II, 627 (1825).
    Cratægus flexuosa Schweinitz, in Long's Second Exped., II,
    Appx., 112 (1825).—?
    Cratægus flava Darlington, Fl. Cestrica, ed. 2, 292 (1837),
    not Ait. (1789).
    Mespilus Treiciana Tausch, Reg. Fl., XXI, 716 (1838).
    Cratægus cuneifolia Borkhausen, in Roemer, Fam. Nat. Syn.,
    III, 118 (1847).
    Cratægus oboratifolia Roem., l. c., 120 (1847).
    Halmia punctata Roem., l. c., 134 (1847).
    Halmia cornifolia Roem., l. c., 135 (1847).
    Phœnopyrum Treicianum Roem., l. c., 154 (1847).
    Cratægus tomentosa var. *plicata* Wood, Cl. Book, 330 (1855).
    Cratægus tomentosa var. punctata Gray, Man. N.
    States, ed. 2, 124 (1856).

COMMON NAME.

Dotted-fruited Thorn (lit.).

Cratægus punctata canescens Britton.   **White Dotted Haw.**

SYN.—*Cratægus punctata canescens* Britton, in Bull. Torr. Bot. Club, XXI, 231 (1894).

VARIETY DISTINGUISHED IN CULTIVATION.

Cratægus punctata xanthocarpa (Medic.) Lav.

SYN.—*Cratægus xanthocarpos* Medicus, Gesch., 85 (1829).
Cratægus punctata var. *xanthocarpa* Lavallée, Arb. Segrez., I, 53, t. 16 (1880).
Cratægus aurea Hort., ex Hand-list Trees and Shr. Arb. Kew, Pt. I, 209 (1894).

Cratægus spathulata Michx.   **Spatulate Haw.**

SYN.—*Cratægus spathulata* Michaux, Fl. Bor.-Am., I, 288 (1803).
Mespilus spathulata Poiret, in Lamarck, Enc. Méth. Bot., Suppl., IV, 68 (1816).
Cratægus microcarpa Lindley, in Bot. Reg., XXII, t. 1846 (1836).
Phænopyrum spathulatum Roemer, Fam. Nat. Syn., III, 155 (1847).
Cotoneaster spathulata Wenzig, in Linnæa, XXXVIII, 201 (1874).

Cratægus cordata (Mill.) Ait.   **Washington Haw.**

SYN.—*Mespilus cordata* Miller, Fig. Pl. Gard. Dict., II, 119, t. 179 (1760).
Mespilus cordifolia Mill., Gard. Dict., ed. 8, No. 4 (1768).
Mespilus Phænopyrum Linnæus f., Syst., Suppl., ed. 13, 254 (1781).
Cratægus acerifolia Moench, Verzeich. Bäume, 31 (1785).
Mespilus acerifolia Burgsdorf, Anleit. Holz., II, 147 (1787).
Cratægus populifolia Walter, Fl. Caroliniana, 147 (1788).
Cratægus cordata Solander in Aiton, Hort. Kew., II, 168 (1789).
Cratægus Phænopyrum Borkhausen, in Roemer, Arch., I, III, 86 (1796).
Mespilus corallina Desfontaines, Tab. Bot. Mus., 174 (1804).
Phænopyrum cordatum Roemer, Fam. Nat. Syn., III, 157 (1847).
Phænopyrum acerifolium Roem., l. c. (1847).
Phalacros cordatus Wenzig, in Linnæa, XXXVIII, 164 (1874).
Cratægus flexispina Hort., ex Hand-list Trees and Shr. Arb. Kew, Pt. I, 197 (1896), not Moench (1785), nor Sarg. (1889).

Cratægus cordata (Mill.) Ait.—Continued.

SYN.—Washington Thorn (N. J., Pa., Del., N. C., S. C., Ill.).
Virginia Thorn (Del.).
Heart-leaved Thorn (Tenn.).
Thorn (Ky.).
Red Haw.

Cratægus viridis Linn.   **Tree Haw.**

SYN.—*Cratægus viridis* Linnæus, Spec. Pl., ed. 1, I, 476 (1753).
Cratægus arborescens Elliott, Sk. Bot. S. C. and Ga., I, 550
(1821).
Phænopyrum arborescens Roemer, Fam. Nat. Syn., III, 153
(1847).
Cratægus Crus-galli var. *pyracanthifolia* Regel, in Act. Hort.
Petrop., I, 109 (1871), in part.
Mespilus arborescens Koch, in Wochenschr., V, 380 (1862).
Cratægus trilobata Hort., ex Hand-list Trees and Shr. Arb.
Kew, Pt. I, 211 (1894), not Labill (1812).

Tree Haw (Ala., Miss., La., S. C.).
Red Haw (Ala., Miss., La.).
Haw (Ala.).
Senellier (La.).
Tree Thorn (Fla.).

Cratægus apiifolia (Marsh.) Michx.   **Parsley Haw.**

SYN.—*Mespilus apiifolia* Marshall, Arb. Am., 89 (1785).
Cratægus Oxyacantha Walter, Fl. Caroliniana, 147 (1788),
not Linn. (1753).
Cratægus Oxyacantha var. *Americana* Castiglioni, Viag.
Stati Uniti, II, 292 (1790).
Cratægus apiifolia Michaux, Fl. Bor.-Am., I, 287 (1803).
Cratægus apiifolia minor Loudon, Arb. Frut., II, 825 (1838).
Cratægus Oxyacantha var. *apiifolia* Regel, in Act. Hort.
Petrop., I, 119 (1871), in part.

Parsley Haw (N. C., Ala., Fla., Miss., La.).
Red Haw (Miss.).
Parsley-leaved Haw (S. C.).

Cratægus flava Ait.   **Yellow Haw.**

SYN.—*Mespilus virginiana* Miller, Gard. Dict., ed. 8, No. 11
(1768).—?

Cratægus flava Ait.—Continued.

SYN.—*Cratægus glandulosa* Solander in Aiton, Hort. Kew., ed. 1, II,
168 (1789), not Moench (1785), nor Georgi (1802), nor Michx.
(1803).

Cratægus flava Solander in Aiton, l. c., 169 (1789).

Mespilus Caroliniana Poiret, in Lamarck, Enc. Méth. Bot.,
IV, 442 (1804).

Cratægus Caroliniana Persoon, Syn. Pl., II, 36 (1807).

Mespilus flava Willdenow, Enum. Pl., 523 (1809).

Cratægus turbinata Pursh, Fl. Am. Sept., 735 (1814).

Mespilus turbinata Sprengel, Syst. Veg., II, 506 (1825).

Cratægus flava var. *lobata* Lindley, in Bot. Reg., XXIII, t.
1932 (1837).

Anthomeles glandulosa Roemer, Fam. Nat. Syn., III, 141
(1847).

Anthomeles flava Roem., l. c., 142 (1847).

Anthomeles turbinata Roem., l. c., (1847).

Phænopyrum Carolinum Roem., l. c., 152 (1847.)

Mespilus flexispina Koch, Dendrol., erst. Th., 139 (1369), not
Moench (1785).

Cratægus flexispina Sargent, in Gard. and For., II, 424 (1889).

COMMON NAMES.

Haw (Fla.).
Yellow Haw (Fla.).
Summer Haw.
Red Haw.

Cratægus flava elliptica (Ait.) Sargent.   **Downy Yellow Haw.**

SYN.—*Cratægus viridis* Walter, Fl. Caroliniana, 147 (1788), not Linn.
(1753).

Mespilus hyemalis[1] Walt., l. c., 148 (1788).—?

Cratægus elliptica Solander in Aiton, Hort. Kew., ed. 1, II,
168 (1789).

Mespilus elliptica Poiret, in Lamarck, Enc. Méth. Bot., IV,
447 (1797).

Cratægus glandulosa Michaux, Fl. Bor.-Am., I, 288 (1803),
not Ait. (1789), nor Willd. (1797), nor Georgi (1802).

Cratægus Michauxii Persoon, Syn. Pl., II, 38 (1814.)

[1] Through the kindness of Dr. D. Thistleton Dyer a specimen of this variety
(*elliptica*), sent for examination, was compared with the specimens of Walter's
herbarium, now preserved at the Kew herbarium. Our plant does not agree with
any of his types, and, moreover, Walter's herbarium does not contain any specimen
labeled *Mespilus hyemalis*. Although the oldest name suspected to belong to our
plant, the vagueness of Walter's description (l. c.) is such as not to warrant taking
up *hyemalis*.

Cratægus flava elliptica (Ait.) Sargent—Continued.

SYN.—*Cratægus spathulata* Pursh, Fl. Am. Sept., I, 336 (1814), not Michx. (1803).

Mespilus Michauxii Hornemann, Hort. Hafn.. 455 (1815).

Cratægus flava Elliott, Sk. Bot. S. C. and Ga., I, 551 (1821), not Ait. (1789).

Cratægus Virginica Loddiges ex Loudon, Arb. Frut., II, 842, f. 560, 615 (1838).

Cratægus virginiana Hort. ex Loud.. l. c. (1838).

Phænopyrum Virginicum Roemer, Fam. Nat. Syn., III, 155 (1847).

Phænopyrum ellipticum Roem., l. c. (1847).

CRATÆGUS FLAVA Ait. var. PUBESCENS Gray, Man. N. States. ed. 5, 160 (1867).

Cratægus flexispina var. *pubescens* (Gray) Millspaugh, Prelim. Cat. Fl. W. Va., 360 (1892).

Cratægus flava var. *elliptica* Sargent, Silva, IV, 114, t. CXC (1892).

Cratægus uniflora Muenchh.   **Small-leaf Haw.**

SYN.—*Cratægus uniflora* Muenchhausen, Hausv.. V. 147 (1770).

Mespilus xanthocarpa Linnæus f., Syst. ed. 13, Suppl.. 254 (1781).

Mespilus flexispina Moench, Verzeich. Bäume, 62, t. 4 (1785).

Mespilus Oxyacantha aurea Marshall. Arb. Am., 89 (1785).

Mespilus laciniata Walter, Fl. Caroliniana, 147 (1788).

CRATÆGUS PARVIFOLIA Solander in Aiton, Hort. Kew., ed. 1, II, 169 (1789).

Cratægus tomentosa Michaux, Fl. Bor.-Am.. I, 289 (1803), not Linn. (1753.)

Cratægus unilateralis Persoon, Syn. Pl., II, 37 (1807).

Mespilus axillaris Pers., l. c., 39 (1807).

Mespilus parvifolia Willdenow, Enum. Pl., 523 (1809).

Mespilus tomentosa Poiret, in Nouveau Duhamel, IV, 153 (1810), not Ait. (1789). nor Castigl. (1790).

Mespilus unilateralis Poiret, in Lamarck, Enc. Méth. Bot., Suppl., IV, 73 (1816).

Mespilus flexuosa Poir. in Lam., l. c. (1816).

Cratægus flexuosa de Candolle, Prodr., II. 627 (1825).

Phænopyrum parvifolium Roemer, Fam. Nat. Syn.. III, 152 (1847).

Phænopyrum uniflorum Roem.. l. c., 153 (1847).

Mespilus uniflora Wensig. in Linnæa, XXXVIII. 123 (1874).

Cratægus æmula Hort. Par., ex Hand-list Trees and Shr. Arb. Kew. Pt. I. 211 (1894.)

Cratægus betulæfolia Lodd., ex Hand-list, l. c. (1894).

Cratægus uniflora Muenchh.—Continued.

SYN.—*Cratægus florida* Lodd., l. c. (1894).
 Cratægus grossulariæfolia Lodd., l. c. (1894).
 Cratægus Pinshow Hort., ex Hand-list, l. c. (1894).

Cratægus æstivalis (Walter) Torr. & Gr.   **Summer Haw.**

SYN.—*Mespilus æstivalis* Walter, Fl. Caroliniana, 148 (1788).
 Cratægus Lucida Elliott, Sk. Bot. S. C. and Ga , I, 548 (1821),
  not Mill. (1768).
 Cratægus Elliptica Elliott, l. c., 549 (1821), not Ait. (1789).
 Cratægus opaca Hooker & Arnott, in Companion Bot. Mag.
  I, 25 (1835).
 Cratægus æstivalis Torrey & Gray, Fl. N. A., I, 468 (1840).
 Cratægus nudiflora Nutt. ex Torr. & Gr., l. c. (1840).
 Anthomeles æstivalis Roemer, Fam. Nat. Syn., III, 141 (1847).

COMMON NAMES.

 May Haw (Tex.).
 Apple Haw (Fla.).
 Summer Haw (Fla.).

HETEROMELES Roem., Fam. Nat. Syn., III, 100 (1847)

Heteromeles arbutifolia (Poir.) Roem.   **Tollon.**

SYN.—*Cratægus arbutifolia* Aiton, Hort. Kew., ed. 2, III, 202 (1811),
  not Poir. (1810).
 Aronia arbutifolia Nuttall, Genera, I, 306 (1818).
 Photinia arbutifolia Lindley, in Trans. Linn. Soc., XIII, 103
  (1822).
 Mespilus arbutifolia Link, Enum. Hort. Berol., II, 36 (1822).
 Heteromeles arbutifolia Roemer, Fam. Nat. Syn., III, 105
  (1847).
 Photinia foliolosa Nutt. ex Roem., l. c. (1847).
 Photinia nudiflora Nutt. ex Roem., l. c. (1847).
 Photinia salicifolia Presl, Epimel. Bot., 204 (1849).
 Heteromeles Fremontiana Decaisne, in Nouv. Arch. Mus.,
  III, 144 (1875).

COMMON NAMES.

 California Holly (Cal., Nev.).
 Christmas Berry (Cal.).
 Chamiso (Cal.).
 Toyon (Cal.).
 Tollon (Cal., Nev.).

CHRYSOBALANUS Linn., Spec. Pl., 513 (1753).

Chrysobalanus icaco Linn.   **Cocoa Plum.**

SYN.—*Chrysobalanus Icaco* Linnæus, Spec. Pl., ed. 1, I, 513 (1753).
Chrysobalanus purpureus Miller, Gard. Dict., ed. 8, No. 2 (1768).
Chrysobalanus Icaco β purpureus Persoon, Syn. Pl., II, 36 (1807).
Chrysobalanus pellocarpus Meyer, Prim. Fl. Esseg., 193 (1818).
Chrysobalanus luteus Sabine, in Trans. Hort. Soc., V, 453 (1824).
Chrysobalanus orbicularis Schumacher, in Schum. Thonnig. Beskr. Guin. Pl., 232 (1827).

COMMON NAMES.

Cocoa Plum (Fla.).
Gopher Plum (Fla.).

PRUNUS Linn., Spec. Pl., 473 (1753).

Prunus nigra Ait.   **Canada Plum.**

SYN.—*Prunus nigra* Aiton, Hort. Kew., ed. 1, II, 165 (1789).
Cerasus nigra Loiseleur, in Nouveau Duhamel, V, 32 (1812), not Mill. (1768).
Prunus mollis Torrey, Fl. U. S., I, 470 (1824).
Prunus americana Torrey & Gray, Fl. N. A., I, 407 (1840), in part.

COMMON NAMES.

Canada Plum (Mass., N. Y., Mich., Ont.).
Red Plum (Me.. Vt., Ont., Mich.).
Horse Plum (Me., Vt.)
Wild Plum (Me., Mass., Vt., N. Y.).

Prunus americana Marsh.   **Wild Plum.**

SYN.—*Prunus Americana* Marshall, Arb. Am., 111 (1785).
Prunus mississippi Marsh., l. c.. 112 (1785).
Prunus spinosa Walter, Fl. Caroliniana, 146 (1788), not Linn. (1753).
Prunus hiemalis Michaux. Fl. Bor.-Am., I, 284 (1803).
Prunus acinaria Desfontaines, Tabl., ed. 1, 179 (1804), not Lam. (1804).
Cerasus canadensis Loiseleur, in Duhamel, Trait. Arb., sec. 6d., V (1812), not Mill. (1768).
Prunus nigra (Americana) Muehlenberg, Cat. Pl. Am. Sept., ed. 2, 48 (1813)—*nomen nudum;* not Ait. (1789).

Prunus americana Marsh.—Continued.

SYN.—*Prunus coccinea* Rafinesque, Fl. Ludovic., 135 (1817).
  Cerasus hiemalis de Candolle, Prodr., II, 538 (1825), in part.
  Cerasus Americana Hooker & Arnott, in Companion Bot.
    Mag., I, 24 (1835).
  Cerasus nigra Hook., l. c. (1835), not Loisel. (1812)

COMMON NAMES.

Wild Plum (R. I., N. J., Del., Pa., Va., W. Va., N. C., S. C.,
  Ga., Fla., Ala., Miss., La., Tex., Ky., Mo., Ark., Ill., Ind.,
  Ohio, Ont., Kans., Nebr., Iowa, Mich., Colo.).
Yellow Plum (N. Y., Del., Pa., Miss., La., Nebr.).
Red Plum (Del., Pa., N. C., Miss., La., Nebr.).
Horse Plum (Miss., Ark., Colo.).
Hog Plum (Colo., Mo.).
August Plum (S. C.).
Native Plum (Iowa).
Plum (Ill.).
Plum Granite.
Goose Plum (Ind.).
Sloe (Fla.).

Prunus americana lanata nom. nov.   **Woollyleaf Plum.**

SYN.—*Prunus Americana β mollis* Torrey & Gray, Fl. N. A., I, 407
  (1840), not *P. mollis* Torr. (1824).

Prunus hortulana Bailey.   **Garden Wild Plum.**

SYN.—*Prunus hortulana* Bailey, in Gard. and For., V, 90 (1892).
  Prunus Americana var.? Patterson, List Pl. Oquawka, Ill., 5
    (1874).
  Prunus Chicasa Watson & Coulter, in Gray, Man. N. States,
    ed. 6, 152 (1889), in part.

Prunus angustifolia Marsh.   **Chickasaw Plum.**

SYN.—*Prunus angustifolia* Marshall, Arb. Am., 111 (1785).
  Prunus insititia Walter, Fl. Caroliniana, 146 (1788), not
    Linn. (1759).
  PRUNUS CHICASA Michaux, Fl. Bor.-Am., I, 284 (1803).
  Cerasus Chicasa Seringe, in de Candolle, Prodr., II, 538
    (1825).

COMMON NAMES.

Chickasaw Plum (Del., W. Va., N. C., Ga., Ala., Fla., Miss.,
  La., Tex., Ill., Kans.).
Hog Plum (Miss., Tex.).
Wild Red Cherry (La.).
Yellow Plum (Fla.).

Prunus alleghaniensis Porter.   **Alleghany Sloe.**

SYN.—*Prunus Alleghaniensis* Porter. in Bot. Gaz., II, 85 (1877).

Prunus[1] subcordata[2] Benth.   **Pacific Plum.**

SYN.—*Prunus subcordata* Bentham. Pl. Hartweg., 308 (1848).

COMMON NAME.

Wild Plum.

Prunus umbellata Elliott.   **Black Sloe.**

SYN.—*Prunus Umbellata* Elliott, Sk. Bot. S. C. and Ga., I, 541 (1821).
Prunus pumila Walter, Fl. Caroliniana. 146 (1788), not Linn.
(1767).
Cerasus umbellata Torrey & Gray, Fl. N. A., I, 409 (1840).

COMMON NAMES.

Black Sloe (S. C., Ga., Ala., Miss.).
Southern Bullace Plum (N. C., S. C., Ala., Miss.).
Hog Plum (Fla.).
Wild Plum (Fla.).

Prunus persica[3] (Linn.) Stokes.   **Peach.**

SYN.—*Amygdalus Persica* Linnæus, Spec. Pl., ed. 1, 472 (1753).
Persica Vulgaris Miller. Gard. Dict., ed. 8, No. 1 (1768).
Persica Amygdalus Miller, l. c., No. 3 (1768).
Amygdalus grata Salisbury. Prodr., 356 (1796).
Persica Paria Delarbre, Fl. Auv., ed. sec., 328 (1800).
Persica lævis de Candolle, Fl. Fran., IV, 487 (1805).
Persica ispahanensis Thouin. in Ann. Mus. Par., VIII, 433
(1806).
Prunus Persica Stokes, Bot. Mat. Med.. III, 100 (1812).
Amygdalus ispahanensis Thouin, in Ann. Mus. Par.. ser. 1,
VIII. 425 (1826).

---

[1] I have not seen specimens of the following species which Professor Greene states
was found by Mrs. Austin in 1893 on the Yanex Indian Reservation, in southeastern
Oregon. It is not yet known whether it is a tree or shrub, but appears to be nearly
related to *Prunus subcordata*.

Prunus Oregona Greene.

SYN.—*Prunus Oregona* Greene, in Erythea, III, pt. 13, 21 (1896).

[2] The following is a shrubby variety:

Prunus subcordata kelloggii Lemmon.   **Graybranch Plum.**

SYN.—*Prunus subcordata* var. *Kelloggii* Lemmon, in Pittonia, II, 67 (1890).

[3] In many parts of the Eastern States, and especially in the South Atlantic region,
this species has escaped from cultivation and become established among other forest
growths; the fruit borne is of a very degenerate form, with thin, bitter flesh.

Prunus persica (Linn.) Stokes—Continued.

SYN.—*Persica domestica* Risso, Hist. Nat., II, 104 (1828).
Persica violacea Risso, l. c., 119 (1828).
Amygdalus collinus Wallich, Cat. No. 723 (1828).
Amygdalus Nuci-persica Reichenbach, Fl. Germ. Exc., 647 (1832).
Amygdalus communis Bunge, Enum. Pl. Chin., 21 (1831), not Linn. (1753)
Prunus salix Royle, Ill. Bot. Him., 204 (1839)—*nomen nudum*.
Amygdalus lævis Dietrich, Syn. Pl., III, 42 (1852).
Persica communis Duhamel ex Steudel, Nom. Bot., ed. sec., I, 81 (1840).
Persica necturina Steudel, l. c. (1840).
Persica nucicarpa Steudel, l. c. (1840).
Amygdalus camellæflora Morr., in Belg. Hort., VIII, 97, t. 26 (1858).
Amygdalus dasylepis Miquel, in Journ. Bot. Néerl., I, 122 (1861).
Cerasus diffusa Boiss. & Haussk. ex Boissier, Fl. Orient., II, 647 (1867).
Amygdalus grandiflora in Hort. Gall. ex Koch. Dendrol., erst. Th., 81 (1869).
Amygdalus persicoides ex Koch, l. c., 84 (1869).
Amygdalus caryophyllacea ex Koch, l. c. (1869).
Amygdalus rosæflora ex Koch, l. c., 85 (1869).
Amygdalus dianthiflora Hort. ex Koch, l. c. (1869).
Amygdalus Persica versicolor ex Koch, l. c. (1869).
Amygdalus chinensis ex Koch, l. c. (1869), not Koch, l. c., (1869), ante!
Persica Davidiana Carrière, in Rev. Hort., 74 (1872).

Prunus emarginata (Dougl.) Walp.   **Bitter Cherry.**

SYN.—*Cerasus emarginata* Douglas, Mss. in Hooker, Fl. Bor.-Am., I, 169 (1840).
Prunus emarginata Walpers, Rep. Bot. Syst., II, 9 (1843).
Cerasus erecta Presl, Epimel. Bot., 194 (1849).
Prunus erecta Walpers, Ann. Bot. Syst., III, 854 (1853).
Cerasus glandulosa Kellogg, in Proc. Cal. Acad., I, 59 (1855).
Cerasus Pattoniana Carrière, in Rev. Hort., 135, f. 17 (1872).
Cerasus Californica Greene, Fl. Franciscana, I, 50 (1891).
Prunus Pattoniana Hort., ex Hand-list Trees and Shr. Arb. Kew, Pt. I, 143 (1894).

COMMON NAMES.

Wild Plum (Cal.).
Bitter Cherry (Idaho, Cal.).
Wild Cherry (Utah).

Prunus emarginata villosa nom. nov. **Woollyleaf Cherry.**

SYN.—*Cerasus mollis*[1] Douglas, Mss. in Hooker, Fl. Bor.-Am., I, 169 (1840).[2]
Prunus mollis Walpers, Rep. Bot. Syst., II, 9 (1843), not *Pr. mollis* Torr. (1824).
Prunus emarginata var. *mollis* Brewer, in Brewer & Watson, Bot. Cal., I, 167 (1880).

Prunus pennsylvanica Linn. f. **Wild Red Cherry.**

SYN.—*Prunus Pennsylvanica* Linnæus f., Syst. Nat., ed. 13, Suppl., 252 (1781).
Prunus-Cerasus montana Marshall, Arb. Am., 113 (1785).
Prunus lanceolata Willdenow, Berl. Baumz., ed. 1, 240, t. 3, f. 3 (1796).
Cerasus borealis Michaux, Fl. Bor.-Am., I, 286 (1803).
Prunus borealis Poiret, in Lamarck, Enc. Méth. Bot., V, 674 (1804), not Salisb. (1796).
Prunus persicifolia Desfontaines, Hist. Arb., II, 205 (1809).
Cerasus persicifolia Loiseleur, in Nouveaux Duhamel, V, 9 (1812).
Cerasus Pennsylvanica Loisel., l. c. (1812).
Prunus rupestris Rafinesque, in Am. Month. Mag., II, 206 (1818).
Prunus cerasifolia Desf. ex Watson, Bib. Ind. N. A. Bot., Pt. 1, 306 (1878).

COMMON NAMES.

Wild Red Cherry (Me., Vt., N. H., Mass., R. I., Conn., N. Y., N. J., Pa., Va., N. C., Mo., Mich., Ohio, Ont., Ill., Wis., Iowa, Kans., Minn., N. Dak.).
Pin Cherry (N. H., Vt., N. Y., Mich., Ohio, Iowa, N. Dak.).
Pigeon Cherry (Vt., N. H., R. I., N. Y., Ont., N. Dak.).
Wild Cherry (N. Y., Ga., S. C.).
Bird Cherry (Me., N. H., N. Y., Pa., Minn., Iowa).
Red Cherry (Me., R. I.).
Fire Cherry (N. Y.).

Prunus cerasus[3] Linn. **Sour Cherry.**

SYN.—*Prunus Cerasus* Linnæus Spec. Pl., ed. 1, I, 474 (1753).
Cerasus vulgaris Miller, Gard. Dic., ed. 8, No. 1 (1768).

[1] In taking up the earliest specific name for this plant a combination would be produced identical with an earlier name applied to a different species in the same genus, leaving the Woollyleaf Cherry without a name.
[2] Pritzel (Thes. Lit. Bot.) gives the date of publication for Vol. I as 1833; the date of the copy here examined is 1840.
[3] Naturalized.

nus cerasus Linn.—Continued.

SYN.—*Cerasus hortenses* Mill., l. c., No. 3 (1768).
Cerasus rubra Gilibert, Fl. Lithuan., II, 229 (1781).
Prunus austera Ehrhart, Beitr., V, 160 (1790).
Prunus acida Ehr., l. c. (1790).
Cerasus acida Borkhausen, in Roemer, Arch. Bot., I, 11, 38 (1796).
Cerasus austera Borkh., in Roem., l. c. (1796).
Prunus æstiva Salisbury, Prodr., 356 (1796).
Prunus serotina Roth, Catalec. Bot., I, 58 (1797), not Ehr. (1788).
Prunus plena Poiret, in Lamarck, Enc. Méth. Bot., V, 671 (1804).
Prunus rosea Poir., in Lam., l. c. (1804).
Cerasus Caproniana de Candolle, Fl. Fran.,trois. éd., IV, 482 (1805).
Prunus hortensis Persoon, Syn. Pl., II, 34 (1807).
Cerasus nicotianæfolia Hort. ex de Candolle, Prodr., II, 536 (1825).
Cerasus caproniana δ polygna Seringe, in de C., l. c., 537 (1825).
Cerasus polygna de C.! Herb. ex Ser. in de C., l. c. (1825).
Cerasus bigarella Dumortier, Fl. Belg., 91 (1827).
Cerasus effusa Host, Fl. Austr., II, 6 (1831).
Cerasus Marasca Host., l. c. (1831).
Cerasus collina Lejeune et Courtois, Comp. Fl. Belg., II, 130 (1831).
Prunus Juliana Reichenbach, Fl. Germ. Exc., 643 (1832), not Poir. in Lam. (1805).
Prunus Marasca Reichenbach, l. c., 644 (1832).
Cerasus rosea Hort. ex Steudel, Nom. Bot., ed. sec., I, 331 (1840).
Prunus oxycarpa Bechstein, Forst. Bot., fünft. Ausg., 424 (1843).
Cerasus Bungei Walpers, Rep., II, 9 (1843).
Cerasus Heaumiana Roemer, Syn. Rosifl., 69 (1847).
Cerasus tridentina Roem., l. c., 76 (1847).
Prunus vulgaris Schur, Enum. Pl. Transsilv., 954 (1866).
Cerasus Rhexii (Hort.) Gall. ex Van Houtte, Fl. Serres, sér. 2, VII, 159 (1868).
Cerasus cucullata Hort. ex Koch, Dendrol., erst. Th., 6 (1869).
Cerasus domestica Cat. Perck ex Wesmael, in Bull. Cong. Bot. Brux., 255 (1864).
Cerasus multicarpa Hort. (Auct. ?) ex Rev. Hort., 409 (1875).
Cerasus beroliensis Nyman, Consp. Eu., 213 (1878).
Cerasus ebroliensis Lamotte, Prodr., I, 238 (1877).

Prunus avium[1] Linn.   **Sweet Cherry.**

Syn.—*Prunus Avium* Linnæus, Fl. Suec., ed. 2, 165 (1755).
Cerasus nigra Miller, Gard. Dic., ed. 8, No. 2 (1768).
Prunus nigricans Ehrhart, Beitr., VII, 126 (1792).
Prunus varia Ehrh., l. c., 127 (1792).
Cerasus Avium Moench, Méth., 672 (1794).
Cerasus varia Borkhausen, in Roemer. Arch. i., II, 38 (1796).
Cerasus Juliana de Candolle, Fl. Fran., IV, 483 (1805).
Cerasus duracina de C., l. c. (1805).
Prunus sylvestris Persoon, Syn. Pl., II, 35 (1807).
Cerasus rubicunda Bechstein. Forstb., 160, 335 (1810).
Cerasus intermedia Host. Fl. Austr., II, 7 (1831), not Loisel.
    in Duham. (1812).
Prunus dulcis Miller ex Reichenbach, Fl. Germ. Exc., 644
    (1832).
Cerasus decumana Delaunay ex Seringe, in de Candolle,
    Prodr., II, 536 (1825).
Cerasus macrophylla Sweet, Hort. Brit., ed. 1, 485 (1827).
Cerasus dulcis Borkhausen ex Steudel, Nom. Bot., ed. sec.,
    I, 331 (1840).
Cerasus pallida Roemer, Syn. Rosifl., 69 (1847).
Cerasus avicularis Dulac, Fl. Haut.-Pyr., 301 (1867).
Cerasus heterophylla in Hort. ex Koch., Dendrol., erst. Th.,
    106 (1869).
Cerasus asplenifolia in Hort. ex Koch, l. c. (1869).
Cerasus salicifolia in Hort. ex Koch, l. c. (1869), not Ser. in
    de C. (1825).

Prunus virginiana Linn.   **Choke Cherry.**

Syn.—*Prunus Virginiana* Linnæus, Spec. Pl., ed. 1, I, 473 (1753).
Padus rubra Miller, Gard. Dict., ed. 8, No. 2 (1768).
Prunus-Cerasus canadensis Marshall, Arb. Am., 113 (1785).
Prunus rubra Aiton, Hort. Kew., ed. 1, II, 162 (1789).
Padus oblonga Moench, Méth., 671 (1794).
Padus virginica Borkhausen, in Roemer, Arch. i., II, 38
    (1798).
Prunus dumosa Salisbury, Prodr., 356 (1796).
Prunus serotina Poiret, in Lamarck, Enc. Méth. Bot., V, 665
    (1804), not Ehrh. (1788).
Cerasus Virginiana Loiseleur, in Nouveau Duhamel, V, 3
    (1812), not Michx. (1803).
Prunus Hirsuta Elliott, Sk. Bot. S. C. and Ga., I, 541 (1821).
Prunus obovata Bigelow, Fl. Bost., ed. 2, 192 (1824).

---

[1]Naturalized.

Prunus virginiana Linn.—Continued.

SYN.—*Prunus virginalis* Wenderoth, in Schrift. Marb., II, 253 (1830).

Cerasus obovata Beck, Bot., 37 (1833).

Cerasus micrantha Spach, Hist. Vég., I, 414 (1834).

Cerasus densiflora Spach, l. c., 415 (1834).

Cerasus fimbriata Spach, l. c., 416 (1834).

Cerasus hirsuta Spach, l. c., 417 (1834).

Cerasus serotina Hooker, Fl. Bor.-Am, I, 169 (1840), not Loisel., in Nouv. Duham. (1812).

Cerasus Virginiana var. β Torrey & Gray, Fl. N. A., I, 410 (1840).

Cerasus Duerinckii Martens, Sel. Sem. Hort. Lovan. (1840); in Bull. Acad. Belg., VIII, 68 (1841).

Cerasus virginica Michaux ex Steudel, Nom. Bot., ed. sec., I, 331 (1840).

Prunus virginica Steudel, l. c., II, 246 (1841).

Prunus fimbriata Steudel, l. c., II, 403 (1841).

Prunus micrantha Steudel, l. c. (1841).

Prunus densiflora Steudel, l. c. (1841).

Prunus Duerinckii Walpers, Rep. Bot. Syst., II, 10 (1843).

Padus fimbriata Roemer, Fam. Nat. Syn., III, 84 (1847).

Padus densiflora Roem., l. c. (1847).

Padus micrantha Roem., l. c. (1847).

Padus Duerinckii Roem., l. c. (1847).

Padus obovata Roem., l. c., 86 (1847).

Padus virginalis Roem., l. c. (1847).

Padus hirsuta Roem., l. c., 87 (1847).

Prunus montana Hort. ex Koch, Dendrol., erst. Th., I, 122 (1869), not *Pr.-Cerasus montana* Marsh. (1785).

Prunus virginiana **a** *rubra* Dippel, Handb. Laubh., dritt. T., 643 (1893).

Prunus virginiana **b** *salicifolia* Dippel, l. c. (1893), not H., B. & K. (1823).

Prunus heterophylla variegata in Hort. ex Dippel, l. c. (1893).

Prunus virginiana var. *asplenifolia*,[1] ex Hand-list Trees and Shr. Arb. Kew, Pt. I, 145 (1894), not *P. serot. asplenifolia* Dippel (1893).

COMMON NAMES.

Choke Cherry (Mich.).
Wild Cherry.

---

[1] A form cultivated as a distinct variety in the Royal Kew Gardens. The distinctive features may be surmised from the significance of the name, but the latter is unaccompanied by a description, and appears to be unknown in American gardens.

VARIETIES DISTINGUISHED IN CULTIVATION,

Prunus virginiana leucocarpa Wats.   **White-fruit Choke Cherry.**

Syn.—*Prunus Virginiana* var. *leucocarpa* Watson. in Bot. Gaz.,
XIII. 233 (1888).

Prunus virginiana pendens nom. nov.   **Weeping Choke Cherry.**

Syn.—*Prunus virginiana* d *pendula* Dippel. Handb. Laubh.. dritt.
T., 643 (1893), not *P. pendula* Maxim. (1878).
Prunus caroliniana hort. ex Dippel. l. c. (1893). not Ait.
(1789).

Prunus virginiana nana (Du Roi) Dippel.   **Dwarf Choke Cherry.**

Syn.—*Prunus nana* Du Roi. Obs. Bot.. 12 (1771).
Prunus virginiana c *nana* Dippel. Handb. Laubh.. dritt. T..
643 (1893).
Prunus Padus nana monstrosa hort. ex Dippel, l. c. (1893).
Prunus Padus racemosa nana monstrosa hort. ex Dippel. l. c.
(1893).
Prunus virginiana var. *nana monstrosa* Hort., ex Hand-list
Trees and Shr. Arb. Kew. Pt. I. 145 (1894).

Prunus demissa (Nutt.) Walp.   **Western Choke Cherry.**

Syn.—*Cerasus serotina* Hooker. Fl. Bor.-Am.. I. 169 (1833). in part;
not Loisel. (1812?).
Cerasus demissa Nuttall. mss. in Torrey & Gray. Fl. N. A..
I. 411 (1840).
Prunus demissa Walpers. Rep. Bot. Syst.. II. 10 (1843).
Padus demissa Roemer. Fam. Nat. Syn.. III, 87 (1847).
Prunus Virginiana var. *demissa* Torrey. Bot. Wilkes' Exped..
284 (1854).

COMMON NAMES.

Wild Cherry (Cal.. N. Mex., Utah. Idaho, Mont.. Oreg.)
Choke Cherry (Cal.. Nev.. Idaho. Utah).
California Cherry (Cal.).
Western Choke Cherry.

Prunus serotina Ehrh.   **Black Cherry.**

Syn.—*Prunus Virginiana* Miller. Gard. Dict.. ed. 8. No. 3 (1768).
not Linn. (1753).
Prunus Cerasus virginiana Marshall. Arb. Am.. 113 (1785).
Prunus serotina Ehrhart. Beitr.. III. 20 (1788).
Cerasus Virginiana Michaux. Fl. Bor.-Am.. I. 285 (1803), not
Loisel. (1812).

Prunus serotina Ehrh.—Continued.

SYN.—*Cerasus serotina* Loiseleur, in Nouveau Duhamel, V, 3 (1812).
Cerasus serotina β retusa Ser. mss. ex de Candolle, Prodr.,
II, 540 (1825).
Prunus cartilaginea Lehmann, Ind. Sem. Hamburg (1833)—
Cf. Linnæa X, 76 (1836).
Padus serotina Agardh., Theor. Syst. Pl., t. 14, f. 8 (1858).
Padus Virginiana Roemer, Fam. Nat. Syn., III, 86 (1847).
Padus cartilaginea Roem., l. c. (1847).
Prunus serotina a *cartilaginea* Dippel, Handb. Laubh., dritt.
T., 645 (1893).

COMMON NAMES.

Wild Black Cherry (Vt., Mass., R. I., N. Y., N. J., Del., Pa.,
N. C., Ala., Miss., La., Ky., Mo., Ill., Iowa, Wis., Kans.,
Nebr., Minn., Ohio, S. Dak. (cult.), Ont.).
Wild Cherry (Conn., N. J., Pa., Del., W. Va., N. C., S. C.,
Ala., Fla., Tex., Ark., Ky., Ind., Ill., Iowa, Wis.).
Black Cherry (Me., N. H., Vt., R. I., N. Y., Miss., Ky., Mich.,
Wis., Ind., Nebr.).
Rum Cherry (N. H., Mass., R. I., Miss., Nebr.).
Whisky Cherry (Minn.).
Choke Cherry (Mo., Wis., Iowa, Mont., Colo.).

Prunus serotina neomontana nom. nov.　**Mountain Black Cherry.**

SYN.—*Cerasus serotina* var. *montana*[1] Small, in Small & Vail, in
Mem. Torr. Bot. Club, IV, 114 (1893).
Prunus serotina montana (Small) Britton, Mem. Torr. Bot.
Club, V, 357 (1894), not *P. cerasus montana* Marsh. (1785),
not *P. montana* Koch (1854).

VARIETIES DISTINGUISHED IN CULTIVATION.

Prunus serotina penduliformis nom. nov.　**Weeping Black Cherry.**

SYN.—*Prunus serotina* c *pendula* Dippel, Handb. Laubh., dritt. T.,
645 (1893), not *P. pendula* Maxim. (1878).
Prunus serotina var. *pendula*, ex Hand-list Trees and Shr.
Arb. Kew, Pt. I, 143 (1894).
Padus serotina pendula Hort., ex Hand-list, l. c. (1894).

Prunus serotina asplenifolia Dippel.　　**Fernleaf Black Cherry.**

SYN.—*Prunus serotina* b *asplenifolia* Dippel, Handb. Laubh., dritt.
T., 645 (1893).

[1] A tree 25 feet high found on the "balds" near the summit of White Top Mountain, southwestern Virginia, at 5,500 feet elevation. Also in Alabama. (Dr. Mohr.)

Prunus salicifolia H., B. K.　　　　　　　　**Willowleaf Cherry.**

SYN.—*Prunus salicifolia* Humboldt, Bonpland & Kunth, Nov. Gen.
Sp., VI, 241, t. 563 (1823).

PRUNUS CAPULI Cavanilles, in Sprengel, Syst. Veg., II, 477
(1825).

Cerasus Capollin (D. C. mss.) Seringe, in de Candolle, Prodr.,
II, 539 (1825).

Cerasus salicifolia de C., l. c., 540 (1825).

Cerasus Capuli Seringe, mss. in de C., l. c., 541 (1825).

Prunus Capulin Zuccarini, in Abhandl. Akad. Muench., II,
345, t. 8 (1836).

Prunus Canadensis Mocino & Sesse ex de Candolle, Prodr.,
II, 539 (1825), not Linn. (1762).

Padus Capulin Roemer, Fam. Nat. Syn., III, 89 (1847).

Laurocerasus salicifolia Roem., l. c. (1847).

Padus Capulinos Hamel, in Rev. Hort., 111 (1884).

Prunus salicifolia var. *acutifolia* Watson, in Proc. Am. Acad.,
XXII, 411 (1887).

Prunus caroliniana (Mill.) Ait.　　　　　　　**Laurel Cherry.**

SYN.—*Padus Caroliniana* Miller, Gard. Dict., ed. 8, No. 6 (1768).

Padus Carolina Du Roi, Harbk. Baumz., II, 198 (1772).

Prunus-Lauro-Cerasus serratifolia Marshall, Arb. Am., 114
(1785).

Prunus Lusitanica Walter, Fl. Caroliniana, 146 (1788), not
Linn. (1753).

Prunus Caroliniana Aiton, Hort. Kew., ed. 1, II, 163 (1789).

Prunus Lusitanica var. *serratifolia* Castiglioni, Viag. Stati
Uniti, II, 340 (1794).

Prunus nitida Salisbury, Prodr., 356 (1796).

Cerasus Caroliniana Michaux, Fl. Bor.-Am., I, 285 (1803).

Prunus sempervirens Willdenow, Enum. Pl. Suppl., 33 (1813).

Bumelia serrata Pursh, Fl. Am. Sept., 155 (1814).

Achras serrata Poiret, in Lamarck, Enc. Méth. Bot., Suppl.,
V, 36 (1817).

Chimanthus amygdalina Rafinesque, Fl. Ludovic., 26 (1817).

Laurocerasus Caroliniana Roemer, Fam. Nat. Syn., III, 90
(1847).

COMMON NAMES.

Wild Peach (Miss., La., Tex.).
Wild Orange (N. C., S. C., Miss., Tex.).
Mock Orange (N. C., S. C., Ala., Miss., La., Tex.).
Laury Mundy (La.).
Laurii amande (La.).
Cherry Laurel (Fla.).
Evergreen Cherry (Tex.).

Prunus caroliniana (Mill.) Ait.—Continued.

SYN.—Mock Olive (Fla.).
Carolinian Cherry (lit.).
Laurel Cherry.

Prunus sphærocarpa Swartz. **West-India Cherry.**

SYN.—*Prunus sphærocarpa* Swartz, Prodr., 81 (1788).
Cerasus sphærocarpa Loiseleur, in Nouveau Duhamel, V, 4 (1812).
Cerasus Braziliensis Chamisso & Schlechtendal, in Linnæa, II, 540 (1827).
Prunus Braziliensis Schott ex Sprengel, Syst. Veg., IV, 406 (1827).
Cerasus reflexa Gardner, in Hooker, London Journ. Bot., II, 342 (1843).
Laurocerasus sphærocarpa Roemer, Fam. Nat. Syn., III, 89 (1847).
Laurocerasus sphærocarpa β Brazilieneis Roem., l. c. (1847).
Prunus pleuradenia Grisebach, Fl. Brit. West Ind., 231 (1864).

COMMON NAME.

West India Cherry (Fla.).

Prunus ilicifolia (Nutt.) Walp. **Hollyleaf Cherry.**

SYN.—*Cerasus ilicifolia* Nuttall, in Hooker & Arnott, Bot. Beechey's Voyage, 340, t. 83 (1841).
Prunus ilicifolia Walpers, Rep. Bot. Syst., II, 10 (1843).
Laurocerasus ilicifolia Roemer, Fam. Nat. Syn., III, 92 (1847).

COMMON NAMES.

Spanish Wild Cherry (Cal.).
Islay (Cal.).
Evergreen Cherry (Cal.).
Holly-leaved Cherry (Cal.).
Oak-leaf Cherry (Cal.).
Holly Cherry (Cal.).
Wild Cherry (Cal.).
Holly (Cal.).
Mountain Evergreen Cherry (Cal.).

Prunus ilicifolia integrifolia Sudworth. **Entire-leaf Cherry.**

SYN.—*Prunus occidentalis* Lyon, in Bot. Gaz., XI, 202, 333 (1886), not Swartz (1800).
Prunus ilicifolia var. *occidentalis* Brandegee, in Proc. Cal. Acad. Sci., ser. 2, I, 209 (1889).
Prunus ilicifolia var. *integrifolia* Sudworth, in Gard. and For., IV, 51 (1891).

Family LEGUMINOSÆ.

ZYGIA[1] Browne, Nat. Hist. Jam., 279 (1756).

Zygia unguis-cati (Linn.) nom. nov.   **Florida Catsclaw.**

SYN.—*Mimosa Unguis-cati* Linnæus. Spec. Pl.. ed. 1. I, 517 (1753).
Inga microphylla Willdenow. Sp. Pl.. IV. Par. II, 1004 (1805).
Inga Unguis-cati Willd., l. c., 1006 (1805).
Mimosa rosea Vahl. Eclogæ. III. 33. t. 25 (1807).
Inga forfex Kunth. Mim.. 52. t. 16 (1819).
Inga rosea de Candolle. Prodr.. II. 437 (1825).
Pithecolobium forfex Bentham, in Hooker. London Journ. Bot., III, 199 (1844).
PITHECOLOBIUM UNGIUS-CATI Benth.. l. c.. 200 (1844).
Pithecolobium microphyllum Benth.. l. c. (1844).

COMMON NAMES.

Cat's Claw (Fla.).
Long Pod (Fla.).

Zygia brevifolia (Benth.) nom. nov.   **Huajillo.**

SYN.—*Pithecolobium brevifolium* Bentham. in Gray. in Smithsonian Contr., III. 67 (1852).

Zygia flexicaulis (Benth.) nom. nov.   **Texan Ebony.**

SYN.—ACACIA FLEXICAULUS Bentham. in Hooker. London Journ. Bot., I. 505 (1842).
Pithecolobium Texense Coulter. in Contr. U. S. Nat. Herb., No. II. 37 (1890).
Pithecolobium flexicaule Coulter, l. c.. 101 (1891).

LYSILOMA Benth., in Hook.. Journ. Bot.. III, 82 (1844).

Lysiloma latisiliqua (Linn.) Benth.   **Wild Tamarind.**

SYN.—*Mimosa latisiliqua* Linnæus, Spec. Pl.. ed. 1, 519 (1753).
Acacia latisiliqua Willdenow. Sp. Pl., IV. Par. II, 1067 (1805).
Lysiloma Bahamensis Bentham, in Hooker. London Journ. Bot., III, 82 (1844).
Acacia Bahamensis Grisebach, Fl. Brit. West Indies, 221 (1864).

---

[1] = *Pithecolobium* Martius, in Flora XX, II. Beibl., 114 (1837).

Lysiloma latisiliqua (Linn.) Benth.—Continued.

SYN.—*Lysiloma latisiliqua* Benthem, in Trans. Linn. Soc., XXX, 534 (1875).

COMMON NAME.

Wild Tamarind (Fla.).

ACACIA Adanson, Fam. Pl., II, 319 (1763).

Acacia farnesiana (Linn.) Willd. **Huisache.**

SYN.—*Mimosa farnesiana* Linnæus, Spec. Pl., ed. 1, I, 521 (1753).
Mimosa scorpioides Forskål, Fl. Ægypt.-Arab., LXXVII, (1775).
Acacia Farnesiana Willdenow, Sp. Pl., IV, Par. II, 1083 (1805).
Acacia pedunculata Willd., l. c. 1084 (1805).
Mimosa pedunculata Poiret, in Lamarck, Enc. Méth. Bot., Suppl., I, 81 (1810).
Acacia (?) leptophylla de Candolle, Cat. Hort. Monsp., 74 (1813).
Acacia armata Heyne ex Wallich, Cat. No. 2564 (1828).
Acacia Farnesiana var. *pedunculata* Don, Gen. Syst., II, 414 (1832).
Acacia adenopa Hooker & Arnott, in Hooker, Bot. Misc., III, 206 (1823).
Farnesia odora Gasparini, Descr. Nuov. Gen. Leg., t. 95 (1838).
Vachellia Farnesiana Wight, Icones Pl. Ind. Orient., t. 300 (1840).
Acacia lenticellata Mueller, in Journ. Linn. Soc., III, 147 (1859).
Acacia acidularis Humboldt et Bonpland ex Willdenow, Enum. Hort. Berol., 1056 (1809).
Acacia edulis Humb. et Bonpl. ex Willd., l. c. (1809).
Acacia aromatica Poell. ex Bentham, in Trans. Linn. Soc., XXX, 502 (1875).
Acacia Farnesiana var. *brachycarpa* Kuntze, Rev. Gen. Pl., Par. I, 156 (1891).

COMMON NAMES.

Huisache (Tex.).
Cassie (Tex.).

Acacia wrightii Benth. **Texas Catsclaw.**

SYN.—*Acacia Wrightii* Bentham, in Smithsonian Contr., III, 64 (1852).

COMMON NAME.

Cat's Claw (Tex.).

Acacia greggii Gray.   **Devils Claws.**

SYN.—*Acacia Greggii* Gray, in Smithsonian Contr., III, 65 (1852).
    Acacia Durandiana Buckley, in Proc. Acad. Sci. Phila. 1861,
    453 (1862).

COMMON NAMES.

Cat's Claw (Tex., Cal., Ariz.).
Paradise Flower (N. Mex.).
Devil's Claws (Nev.).
Ramshorn.
Uña de Gato.

LEUCÆNA[1] Hooker, in Journ. Bot.. IV. 416 (1842).

Leucæna glauca (Linn.) Benth.   **Leucæna.**

SYN.—*Mimosa glauca* Linnæus. Spec. Pl., ed. 1, I, 520 (1753).
    Mimosa leucocephala Lamarck, Enc. Méth. Bot., I. 12 (1783).
    Acacia glauca Moench, Meth. 466 (1794).
    Acacia biceps Willdenow, Sp. Pl., IV, Par. I1, 1075 (1805).
    Acacia frondosa Willd., l. c., 1076 (1805).
    Mimosa biceps Poiret, in Lamarck. Enc. Méth. Bot.. Suppl.,
    I, 75 (1810).
    Mimosa frondosa Klein, in Poir., l. c., 76 (1810).
    Acacia leucocephala Link, Enum. Hort. Berl., II, 444 (1822).
    Leucæna glauca Bentham, in Hooker, London Journ. Bot.,
    IV, 416 (1842).

Leucæna[2] **pulverulenta** (Schlecht.) Benth.   **Chalky Leucæna.**

SYN.—*Acacia pulverulenta* Schlechtendal, in Linnæa, XII, 517
    (1838).
    Acacia esculenta Martins & Galeotti, in Bull. Acad. Brux.,
    X, Pt. 2. 312 (1843).
    Leucaena pulverulenta Bentham, in Hooker, London Journ.
    Bot., IV, 417 (1845).

COMMON NAME.
Mimosa.

---

[1] The only other species known to occur in the United States is the following slender shrub, found in Texas from the Colorado River to New Mexico:

Leucæna retusa Benth.
    SYN.—*Leucæna retusa* Bentham, in Gray, in Contr. U. S. Nat. Mus., III, 64 (1852).

[2] *Leucæna macrocarpa* Rose (in Contr. U. S. Nat. Herb., I, No. 29, 327, 1895), is a new arborescent species first detected in the State of Jalisco, Mexico, in 1886, and referred by Watson (Proc. Am. Acad. Sci., XXII, 409, 1887) to Bentham's *L. macrophylla* (Bot. Voy. Sulph., 90 (1844), another Mexican species, from which the former has since proved to be distinct. It occurs as a shrub or tree 25 feet or less in height.

PROSOPIS Linn., Mant., 10 (1767).

Prosopis odorata Torr. & Frem.   **Screw Bean.**

SYN.—*Prosopis odorata* Torrey & Fremont, in Fremont's Rep., 313,
    t. 1, f. 3 (1845), in part.
    PROSOPIS PUBESCENS Bentham, in Hooker, London Journ.
    Bot., V, 82 (1846).
    Prosopis Emoryi Torrey, in Emory's Rep., 139 (1848).
    Strombocarpa pubescens Gray, in Smithsonian Contr., III, 60
    (1852).
    Strombocarpa odorata Torrey, in Sitgreaves' Rep., 158 (1853).

COMMON NAMES.

Screw Bean (Tex., Cal., N. Mex., Ariz., Nev., Utah).
Screw-Pod Mesquit (Tex., Ariz., N. Mex., Cal., Nev., Utah).
Tornillo (Tex., N. Mex., Ariz., Nev., Utah).
Mescrew (Nev.).
Screw Bean Mesquit (Ariz.).

Prosopis juliflora (Swartz) de C.   **Mesquit.**

SYN.—*Mimosa juliflora* Swartz, Prodr., 85 (1788).
    Acacia Cumanensis Humboldt et Bonpland ex Wildenow,
    Sp. Pl., IV, Par. II, 1058 (1805).
    Acacia pallida Humb. et Bonpl. ex Willd., l. c., 1059 (1805).
    Acacia lævigata Humb. et Bonpl. ex Willd., l. c. (1805).
    Acacia juliflora Willd., l. c., 1076 (1805).
    Acacia falcata Desfontaines, Tabl., ed. 2, 207 (1815), not
    Willd. (1805).
    Mimosa salinarum Vahl, Eclog. Am., III, 35 (1807).
    Acacia diptera Willdenow, Enum. Hort. Berol., 1051 (1809).
    Mimosa pallida Poiret, in Lamarck, Enc. Méth. Bot., Suppl.,
    I, 65 (1810).
    Mimosa Cumana Poiret, in Lam., l. c. (1810).
    Mimosa lævigata Poiret, in Lam., l. c. (1810).
    Mimosa furcata Desfontaines, Cat. Hort. Paris., ed. 2, 207
    (1812).
    Acacia flexuosa Lagasca, Elench. Hort. Matrit., 16 (1816).
    Acacia Siliquastrum Lagasca, l. c. (1816).
    Prosopis horrida Kunth, Mim., 106, t. 33 (1819).
    Prosopis pallida Kunth, l. c. (1819).
    Prosopis Cumanensis Kunth, l. c. (1819).
    Prosopis dulcis Kunth, l. c., 110, t. 34 (1819).
    Prosopis inermis Humboldt, Bonpland & Kunth, Nov. Gen.
    Sp., VI, 307 (1823).

Prosopis juliflora (Swartz) de C.—Continued.

Syn.—*Prosopis Siliquastrum* de Candolle. Prodr., II, 447 (1825).
    Prosopis flexuosa de C.. l. c. (1825).
    Prosopis bracteolata de C.. l. c. (1825).
    Prosopis Domingensis de C.. l. c. (1825).
    Prosopis juliflora de C.. l. c. (1825).
    Acacia ? salinarum de C.. l. c. 456 (1825).
    Prosopis affinis Sprengel. Syst. Veg.. II, 326 (1825).
    Prosopis glandulosa Torrey. in Ann. Lyc. N. Y.. II, 192. t. 2.
       (1828).
    Algarobia dulcis Bentham, Pl. Hartweg., 13 (1839).
    Algarobia glandulosa Torrey & Gray, Fl. N. A.. I. 339
       (1840).
    Prosopis fruticosa Meyen, Reise, I, 376 (1843).
    Prosopis odorata Torrey, in Fremont's Rep., 313, t. 1 (excl.
       fruit) (1845).

COMMON NAMES.

Mesquit (Cal., Tex.. N. Mex.. Ariz.).
Algaroba (Cal.. Tex.. N. Mex., Ariz.).
Honey Locust (Tex., N. Mex.).
Honey Pod (Tex.).
Ironwood (Tex.).

CERCIS Linn.. Spec. Pl.. 374 (1753).

Cercis canadensis Linn.   **Redbud.**

Syn.—*Cercis Canadensis* Linnæus. Spec. Pl.. ed. 1. I, 374 (1753).
    Siliquastrum canadense Medicus. in Vorl. Chur. Phys. Ges.,
       II, 339 (1787).
    Siliquastrum cordatum Moench. Meth., 54 (1794).

COMMON NAMES.

Red Bud (Mass., N. Y. (cult.), N. J., Pa., Del.. D. C., Va.,
    W. Va., N. C.. S. C.. Ala., Fla.. Ark.. Miss., La., Tex., Mo.;
    Ill.. Ind.. Mich.—cult.).
Judas Tree (Mass., R. I., N. Y. (cult.), N. J., Del., Pa., D. C.,
    Va., N. C., S. C., Miss.. La., Tex., Ky.; Ill., Ind., Ohio.
    Mich., Minn.—cult.).
Red Judas Tree.
Salad Tree (Del.).
Canadian Judas Tree (lit.).

Cercis canadensis pubescens Pursh.   **Downy Redbud.**

Syn.—*Cercis Canadensis* var. *pubescens* Pursh, Fl. Am. Sept., I, 308
    (1814).

VARIETY DISTINGUISHED IN CULTIVATION.

Cercis canadensis plena[1] nom. nov.

SYN.—*Cercis canadensis* var. *flore pleno,* ex Hand-list Trees and Shr.
Arb. Kew, Pt. I, 129 (1894).

Cercis reniformis Engelm.   Texas Redbud.

SYN.—*Cercis occidentalis* var. (Torrey ined.) Gray, in Bost. Journ.
Nat. Hist., VI (Pl. Lindh.), No. 2, 177 (1850).
Cercis reniformis Engelmann, Mss. ex Gray, l. c. (1850).
Cercis Californica Torrey ex Bentham, Pl. Hartweg., 361
(1857)—*nomen nudum.*
Cercis occidentalis var. *Texensis* Watson, Bib. Ind. N. Am.
Bot., Pt. I, 209 (1878).
Cercis Texensis Sargent, in Gard. and For., IV, 448 (1891).

COMMON NAMES.

Redbud (Tex.).
Texas Redbud.

GLEDITSIA[2] Linn., Spec. Pl., 1056 (1753).

Gleditsia triacanthos Linn.   Honey Locust.

SYN.—*Gleditsia triacanthos* Linnæus, Spec. Pl., ed. 1, II, 1056 (1753).
Gleditsia spinosa Marshall, Arb. Am., 54 (1785).
Gleditsia Meliloba Walter, Fl. Caroliniana, 254 (1788).
Gleditsia elegans Salisbury, Prodr., 323 (1796).
Gleditsia ferox Desfontaines, Hist. Arb., II, 247 (1809).
Gleditschia polysperma Stokes, Bot. Mat. Med., I, 228 (1812).
Gleditsia heterophylla Rafinesque, Fl. Ludovic., 99 (1817).
Gleditschia latisiliqua Loddiges Cat. ex Don, in Loud., l. c.,
414 (1830).
Meliolobus heterophylla Rafinesque, Syl. Tell., 121 (1838).
Acacia lævis Hort. ex Steudel, Nom. Bot., ed. sec., I, 6 (1840).
Gleditschia flava Hort. ex Koch, Dendrol., erst. Th., 9 (1869).
Cæsalpiniodes triacanthum Kuntze, Rev. Gen. Pl. Par., I,
167 (1891).

COMMON NAMES.

SYN.—Honey Locust (Vt., N. H., Mass., R. I., N. Y., N. J., Pa., Del.,
D. C., Va., W. Va., N. C., S. C., Ga., Fla., Ala., Miss., La.,
Tex., Ark., Ky., Mo., Ohio, Ill., Ind., Kans., Nebr., Mich.,
Iowa.)

---

[1] A variety with double flowers.
[2] Linnæus's original spelling is preserved.

Gleditsia triacanthos Linn.—Continued.

SYN.—Black Locust (Miss., Tex., Ark., Kans., Nebr.).
Sweet Locust (S. C., La., Kans., Nebr.).
Three-Thorned Acacia (Mass., R. I., La., Tex., Mich., Ont., Nebr.).
Thorn Locust (N. Y., Ind., La.).
Thorntree (N. Y., Ind., La.).
Thorny Locust (N. J.).
Locust (Nebr.).
Honey (R. I., N. J., Iowa).
Honey Shucks (R. I., N. J., Va., Iowa, Fla.).
Thorny Acacia (Tenn.).
Honey-Shucks Locust (Ky.).
Piquant Amourette (La.).
Confederate Pintree (Fla.).

Gleditsia triacanthos lævis (Loud.) nom. nov.

SYN.—*Gleditsia inermis* Moench, Meth., 69 (1794) not Mill. (1768).
GLEDITSIA TRIACANTHUS var. INERMIS Willdenow, Berl. Baumz., ed. 1, 163 (1796).
Gleditschia lævis Hort. ex Loud. Arb., Frut., II, 650 (1838).

Gleditsia triacanthos brachycarpos Michx.

SYN.—*Gleditsia triacanthos* var. *brachycarpos* Michaux, Fl. Bor.-Am., II, 257 (1803).
Gleditsia brachycarpa Pursh, Fl. Am. Sept., I, 221 (1814).

Gleditsia aquatica Marsh. **Water Locust.**

SYN.—*Gleditsia triacanthos* β Linnæus, Spec. Pl., ed. 1, II, 1057 (1753).
GLEDITSIA INERMIS Miller, Gard. Dic., ed. 8, No. 2 (1768). not Linn. (1759).
Gleditsia aquatica Marshall, Arb. Am., 54 (1885).
Gleditsia Carolinensis Lamarck, Enc. Méth. Bot., II, 465 (1786).
Gleditsia monosperma Walter, Fl. Caroliniana, 254 (1788).
Gleditsia triacanthos β *aquatica* Castiglioni, Viag. Stati Uniti, II, 249 (1790).
Gleditsia triacantha Gaertner, Fruct., II, 311, t, 146, f. 3 (1791), not *G. triacanthos* Linn. (1753).
Asacara aquatica Rafinesque, Syl. Tell., 121 (1838).
Cæsalpiniodes monospermum Kuntze, Rev. Gen. Pl., Par. I, 167 (1891).

COMMON NAME.

Water Locust (Fla., La., Tex., Mo., Ind., Ill.).

GYMNOCLADUS Lam., Enc. Méth. Bot., I, 733 (1783).

Gymnocladus dioicus (Linn.) Koch. **Coffeetree.**

SYN.—*Guilandina dioica* Linnæus, Spec. Pl., ed. 1, I, 381 (1753).
GYMNOCLADUS CANADENSIS Lamarck, Enc. Méth. Bot., I, 733 (1783).
Hyperanthera dioica Vahl, Symbol. Bot., I, 31 (1790).
Gymnocladus dioica Koch, Dendrol., erst. Th., 5 (1869).

COMMON NAMES.

Kentucky Coffee Tree (Mass., R. I. (cult.), N. Y., Pa. (cult.), Del., Va., W. Va., N. C., Miss., Ark., Mo., Ill., Kans., Ont., Mich.).
Coffeenut (Ky., Mo., Ill, Ind., Nebr.).
Coffeetree (W. Va., Ark., Ky., Nebr.).
Coffee Bean (Ill., Kans., Nebr.).
Coffee Bean Tree (Ky., Ark.).
Mahogany (N. Y.).
Virgilia (Tenn.).
Nickertree (Tenn.).
Stumptree (Tenn.).

PARKINSONIA Linn., Spec. Pl., 375 (1753).

Parkinsonia aculeata Linn. **Horse Bean.**

SYN.—*Parkinsonia aculeata* Linnæus, Spec. Pl., ed. 1, I, 375 (1753).

COMMON NAME.

Horse Bean.
Retama (Tex.).

Parkinsonia microphylla Torr. **Small-leaf Horse Bean.**

SYN.—*Parkinsonia microphylla* Torrey, in Pacif. R. R. Rep., IV, 82 (1857).

COMMON NAMES.

Desert Bush (Ariz.).
Jerusalem Thorn.
Palo Verde (Cal.).

CERCIDIUM Tulasne, in Arch. Mus. Par., IV, 133 (1844).

Cercidium floridum Benth. **Greenbark Acacia.**

SYN.—*Cercidium floridum* Bentham, in Gray, Pl. Wright., I, 58 (1852).

Cercidium floridum Benth.—Continued.

SYN.—*Parkinsonia florida* Watson, in Proc. Am. Acad. Sci., XI, 135 (1876).

COMMON NAMES.

Green Barked Acacia (Ariz.).
Palo Verde (Tex., Ariz., N. Mex.).
Acacia.

Cercidium[1] torreyanum (Wats.) Sargent.   **Palo Verde.**

SYN.—*Cercidium floridum* Torrey, in Pacif. R. R. Rep., IV, 82 (1857), not Benth. (1852).
PARKINSONIA TORREYANA Watson, in Proc. Am. Acad. Sci., XI, 135 (1876).
Cercidium Torreyanum Sargent, in Gard. and For., II, 388 (1889).

COMMON NAMES.

Green-barked Acacia.
Palo Verde.

SOPHORA Linn., Spec. Pl., 373 (1753).

Sophora secundiflora (Cav.) de C.   **Frigolito.**

SYN.—*Virgilia secundiflora* Cavanilles, Icones, V, t. 401 (1799).
Broussonetia secundiflora Ortega, Decades, V, 61, t. 7 (1800).
Sophora secundiflora de Candolle, Cat. Hort. Monsp., 148 (1818).
Cladrastis secundiflora Rafinesque, Neogen., 1 (1825).
Agastianis secundiflora Rafinesque, New Fl. and Bot., 3d pt., 86 (1836).
Dermatophyllum speciosum Scheele, in Linnæa, XXI, 459 (1848).
Sophora speciosa Bentham, Mss. in Gray, in Bost. Journ. Soc. Nat. Hist., VI (Pl. Lindheim.), 178 (1857).
Sophora sempervirens Engelmann, in Gray, l. c. (1857).

COMMON NAME.

Coral Bean.

Sophora affinis Torr. & Gr.   **Sophora.**

SYN.—*Sophora affinis* Torrey & Gray, Fl. N. A., I, 390 (1840).
Styphnolobium affine Walpers, Rep., I, 807 (1842).

---

[1] The following low shrub occurs in western Texas:

Cercidium texanum Gray.
SYN.—*Cercidium Texanum* Gray, in Smithsonian Contr., III, 58 (1852).
Parkinsonia Texana Watson, in Proc. Am. Acad. Sci., XI, 136 (1876).

Sophora affinis Torr. & Gr.—Continued.

COMMON NAMES.

SYN.—Pink Locust (Tex.).
Beaded Locust (Tex.).

CLADRASTIS Raf., Neogen., 1 (1825).

Cladrastis lutea (Michx. f.) Koch.   **Yellow-wood.**

SYN.—*Virgilia lutea* Michaux f., Hist. Arb. Am., III, 266, t. 3
(1813).
Cladrastis fragrans Rafinesque, Cat. Bot. Gard. Trans., 12
(1824)—*nomen nudum.*
CLADRASTIS TINCTORIA Raf., Neogen., 1 (1825).
Cladrastis kentuckensis Raf., l. c. (1825).
Cladrastis lutea Koch, Dendrol., erst. Th., 6 (1869).

COMMON NAMES.

Yellow-Wood (Tenn., N. C.).
Yellow Locust (Ky., Tenn.).
Yellow Ash.
Gopherwood.

EYSENHARDTIA Humboldt, Bonpland & Kunth, Nov. Gen. Sp., VI, 489 (1823).

Eysenhardtia orthocarpa (Gray) Wats.   **Eysenhardtia.**

SYN.—*Eysenhardtia amorphoides* var. *orthocarpa* Gray, in Smith-
sonian Contr., III, 46 (1852).
Eysenhardtia amorphoides Torrey, in Bot. Mex. Bound. Sur-
vey, 51 (1858), in part.
Eysenhardtia orthocarpa Watson, in Proc. Am. Acad. Sci.,
XVII, 339 (1882).

DALEA Willd., Sp. Pl., III, 1336 (1801).

Dalea spinosa Gray.   **Indigo Bush.**

SYN.—*Dalea spinosa* Gray, in Mem. Am. Acad., new. ser., V, 315
(1854).
Asagræa spinosa Baillon, Adansonia, IX, 233 (1870).

COMMON NAMES.

Dalea (Cal.).
Indigo Bush (Cal.).

ROBINIA Linn., Spec. Pl., 722 (1753).

Robinia pseudacacia Linn.   **Locust.**

SYN.—*Robinia Pseudacacia* Linnæus, Spec. Pl., ed. 1. II, 722 (1753).
Pseudacacia odorata Moench, Meth., 145 (1794).
Robina fragilis Salisbury, Prodr., 336 (1796).
Robinia pseudacacia Audubon, Bird, III, Pl. 165 (1856).
Robinia bullata ex Koch, Dendrol., erst. Th.. 56 (1869).

COMMON NAMES.

Locust (Me., N. H., Vt., Mass., R. I., Conn., N. Y., N. J., Pa.,
Del., W. Va., N. C., S. C., Ga., Ala., Miss., Tex., Ky., Ark.,
Ariz., Ill., Wis., Ohio., Ind., Kans., Nebr., Mich., Iowa,
Minn.).
Black Locust (Pa., Va., W. Va.. N. C., S. C., Ala., Miss., La.,
Tex., Ark., Ky., Mo., Ill., Ohio, Ind., Iowa, Kans., Nebr.,
Mich., Minn.).
Yellow Locust (Vt., Mass., N. Y., Pa., Del., Va., W. Va., Miss.,
La., Ill., Ind., Kans., Nebr., Minn.).
White Locust (R. I., N. Y., Tenn.).
Red Locust (Tenn.).
Green Locust (Tenn.).
Acacia (La.).
False Acacia (S. C., Ala., Tex., Minn.).
Honey Locust (Minn.).
Bastard Acacia (lit.).
Peaflower Locust.
Post Locust (Md.).

PRINCIPAL VARIETIES DISTINGUISHED IN CULTIVATION.

Robinia pseudacacia decaisneana Carr.   **Pink Locust.**

SYN.—*Robinia Pseudacacia* var. *Decaisneana* Carrière, in Rev.
Hort. 1863, 151, t. (1863).
Robinia Decaisneana Verlot, in Rev. Hort., 155 (1873).
Robinia Pseudacacia var. *bella-rosea*.[1] ex Hand-list Trees
and Shr. Arb. Kew, Pt. I, 119 (1894), not *R. bella-rosea*
Nich. (1887).

Robinia pseudacacia crispa de C.   **Crinkleleaf Locust.**

SYN.—*Robinia Pseudacacia* γ *crispa* de Candolle, Prodr., II, 261
(1825).
Robinia pseudacacia revoluta ex Koch, Dendrol., erst. Th., 56
(1869).

---

[1] Cultivated as a distinct variety in the Royal Kew Gardens. Name published
without description.

Robinia pseudacacia crispa de C.—Continued.

SYN.—*Robinia undulata* ex Koch, l. c. (1869).
Robinia monstrosa Loddiges Cat. Pl. (ed. 1830) ex Loudon,
Arb. Frut., II, 610 (1838).
Robinia Pseudacacia var. *revoluta*, ex Hand-list, Trees and
Shr. Arb. Kew, Pt. I, 121 (1894).

Robinia pseudacacia amplifolia nom. nov. **Broadleaf Locust.**

SYN.—*Robinia Pseudacacia* var. *macrophylla* Loddiges Cat. Pl. (ed.
1830) ex Loudon, Arb. Frut., II, 610 (1838), not *R. macrophylla* Roxb. (1832).

Robinia pseudacacia angustifolia (Loud.) Lav.
Small-leaf Locust.

SYN.—*Robinia Pseudacacia* var. *microphylla* Loddiges Cat. Pl. (ed.
1830) ex Loudon, Arb. Frut., II, 610 (1838), not Pall.
(1800).
Robinia angustifolia Hort. ex Loud., l. c. (1838).
Robinia pseudacacia sophoræfolia Lk. ex Loud., l. c. (1838).
Robinia pseudacacia amorphæfolia Lk. ex Loud., l. c. (1838).
Robinia elegans ex Koch, Dendrol., erst. Th., 56 (1869).
Robinia coluteoides ex Koch, l. c. (1869).
Robinia tragacanthoides ex Koch, l. c. (1869).
Robinia myrtifolia ex Koch, l. c. (1869).
Robinia insignis ex Koch, l. c. (1869).
Robinia linearis ex Koch, l. c. (1869).
Robinia pseudacacia angustifolia Lavallée, Arb. Seg., 136
(1880).
Robinia pseudacacia cotuloides in Hort. ex Dippel, Handb.
Laubh., dritt. T., 702 (1893).
Robinia pseudacacia tragacanthoides in Hort. ex Dippel, l. c.
(1893).
Robinia pseudacacia myrtifolia in Hort. ex Dippel, l. c. (1893).
Robinia pseudacacia elegans in Hort. ex Dippel, l. c. (1893).
Robinia pseudacacia insignis in Hort. ex Dippel, l. c. (1893).
Robinia angustifolia elegans Hort., ex Hand-list Trees and
Shr. Arb. Kew, Pt. I, 119 (1894).
Robinia microphylla Hort., ex Hand-list, l. c. (1894), not
Pall. (1800).
Robinia umbraculifera angustifolia Hort., l. c. (1894).

Robinia pseudacacia monophylla Petz. & Kirchn.
Singleleaf Locust.

SYN.—*Robinia pseudacacia monophylla* Petzold & Kirchner, Arb.
Musc., 377 (1864).
Robinia Pseudacacia var. *unifoliata*, ex Hand-list Trees and
Shr. Arb. Kew, Pt. I., 121 (1894).

Robinia pseudacacia tortuosa (Hoffm.) de C.

Twistbranch Locust.

Syn.—*Robinia tortuosa* Hoffmannsegg, Verz. Pfl., 193 (1824).
Robinia pseudacacia ε tortuosa de Candolle, Prodr., II, 261 (1825).
Robinia tortuosa ex Koch, Dendrol., erst. Th., 57 (1869).
Robinia tortuosa elegans ex Koch, l. c. (1869), not *R. elegans* ex Koch, l. c., 56 (1869).
Robinia tortuosa microphylla ex Koch, l. c. (1869), not *R. pseud. microphylla* Pall. (1800), nor Loud. (1838).
Robinia volubilis ex Koch, l. c. (1869).
Robinia pseudacacia volubilis Hort. ex Dippel, Handb. Laubh., dritt. T., 702 (1893).

Robinia pseudacacia pyramidalis Petz. & Kirchn.

Pyramid Locust.

Syn.—*Robinia pseudacacia pyramidalis* Petzold & Kirchner, Arb. Musc., 378 (1864).
Robinia pyramidalis ex Koch, Dendrol., erst. Th., 57 (1869).
Robinia stricta ex Koch, l. c. (1869).
Robinia Gondouini ex Koch, l. c. (1869).
Robinia pseudacacia stricta Lk. ex Loudon, Arb. Frut., II, 610 (1838).
Robinia fastigiata Hort. ex Sargent, Silva, III, 42 (1892).
Robinia pseudacacia Gondouini ex Dippel, Handb. Laubh., dritt. T., 702 (1893).

Robinia pseudacacia pendula (Ortega) Loud. **Weeping Locust.**

Syn.—*Robinia pendula* Ortega, Decades 26, (1800).
Robinia pseudacacia pendula Loudon, Arb. Frut., II, 610 (1838).
Robinia pendulifolia ex Koch, Dendrol., erst. Th., 57 (1869).
Robinia pseudacacia Ulriciana ex Dippel, Handb. Laubh., dritt. T., 702 (1893).

Robinia pseudacacia inermis (Jacq.) nom. nov. **Parasol Locust.**

Syn.—*Robinia inermis* Jacquin, Select. Am., 210 (1763); Du Mont de Courset, Bot. Cult., VI, 140 (1802).
Robinia pseudacacia δ umbraculifera de Candolle, Cat. Monsp., 137 (1813).
Robinia pseudacacia patula Petzold & Kirchner, Arb. Musc., 374 (1864).
Robinia pseudacacia Bessoniana Petz. & Kirchn., l. c. (1864).
Robinia pseudacacia inermis rubra Petz. & Kirchn., l. c. (1864).

Robinia pseudacacia inermis (Jacq.) nom. nov.—Continued.

SYN.—*Robinia pseudacacia inermis nigra* Petz. & Kirchn., l. c. (1864).
Robinia pseudacacia nigricans Baumann, ex Petz. & Kirchn., l. c. (1864).
Robinia Pseudacacia Rhederii Petz. & Kirchn., l. c. (1864).
Acacia Parasol ex Koch, Dendrol., erst. Th., 57 (1869).
Robinia Bessoniana ex Koch, l. c., 58 (1869).
Robinia Pseudacacia bessoniana latifolia,[1] ex Hand-list Trees and Shr. Arb. Kew, Pt. I, 119 (1894), not *R. latifolia* Mill. (1768), nor Poir. in Lam. (1804).

Robinia pseudacacia spectabilis (Du Mont Cour.) Koch.   **Thornless Locust.**

SYN.—*Robinia spectabilis* Du Mont de Courset, Bot. Cult., ed. 2, VI, 140 (1802).
Robinia pseudacacia inermis de Candolle, Cat. Monsp., 136 (1813), not Jacq. (1763), nor Du Mont (1802).
Robinia Utterharti Hort. ex Verlot, in Rev. Hort. 1873, 155 (1873).
Robinia pseudacacia spectabilis ex Koch, Dendrol., erst. Th., 55 (1869).
Robinia pseudacacia mitis ex Koch, l. c. (1869).
Robinia formosa ex Koch, l. c. (1869).
Robinia procera ex Koch, l. c., 56 (1869).
Robinia pseudacacia formosa Hort. ex Dippel, Handb. Laubh., dritt. T., 702 (1893).
Robinia pseudacacia formosissima Hort. ex Dippel, l. c. (1893).
Robinia pseudacacia speciosa Hort. ex Dippel, l. c. (1893).
Robinia pseudacacia procera Loddiges Cat. Pl. (ed. 1830) ex Loudon, Arb. Frut., II, 610 (1838).

Robinia pseudacacia latisiliqua Loud.   **Broadpod Locust.**

SYN.—*Robinia pseudacacia latisiliqua* Prince in Cat. 1829 ex Loudon, Arb. Frut., II, 610 (1838).
Robinia latisiliqua Verlot, in Rev. Hort. 1873, 155 (1873).

Robinia pseudacacia dissecta (Koch) Sargent.   **Cutleaf Locust.**

SYN.—*Robinia linearis dissecta* ex Koch, Dendrol., erst. Th., 56 (1869).
Robinia dissecta Verlot, in Rev. Hort. 1873, 155 (1873).
Robinia pseudacacia dissecta Hort. ex Sargent, Silva, III, 42 (1892).

---

[1] Cultivated as a distinct variety in the Royal Kew Gardens. Name published without description.

Robinia pseudacacia glaucescens Koch. **Blue Locust.**

SYN.—*Robinia pseudacacia glaucescens* ex Koch, Dendrol., erst. Th., 56 (1869).

Robinia pseudacacia aurea (Koch) Dippel. **Goldenleaf Locust.**

SYN.—*Robinia Pseudacacia 2 flore luteo* Dumont ex London, Arb. Frut. II, 609 (1838).
Robinia aurea ex Koch, Dendrol., erst. Th., 56 (1869).
Robinia pseudacacia aurea ex Dippel, Handb. Laubh., dritt. T., 703 (1893).

Robinia pseudacacia purpurea Dippel. **Purpleleaf Locust.**

SYN.—*Robinia pseudacacia purpurea* ex Dippel, Handb. Laubh., dritt. T., 703 (1893).
Robinia pseudacacia atropurpurea Dippel, l. c. (1893).

Robinia pseudacacia argenteo-variegata (Koch) nom. nov.
Spotted Locust.

SYN.—*Robinia pseudacacia fol. argenteo-variegatis* ex Koch, Dendrol., erst. Th., 57 (1869).
Robinia pseudacacia fol. aureo-variegatis ex Koch, l. c. (1869).

Robinia neo-mexicana[1] Gray. **New-Mexican Locust.**

SYN.—*Robinia Neo-Mexicana* Gray, in Mem. Am. Acad. Sci., new ser., V, 314 (1854).

COMMON NAME.

Locust (Ariz., N. Mex.).

Robinia viscosa Vent. **Clammy Locust.**

SYN.—*Robinia viscosa* Ventenat, Descript. Pl. Jard. Cels., 4, t. 4 (1800).
Robinia glutinosa Sims, in Bot. Mag., XVI, No. 560 (1801).
Robinia amœna Hort. ex Koch, Dendrol., erst. Th., 60 (1869).

COMMON NAMES.

Clammy Locust[2] (N. H., Mass., R. I., N. Y., N. J., Pa., N. C., S. C., La., Miss., Ill., Nebr.).
Honey Locust (N. Y., N. J.).
Red-flowering Locust (Ala.).

---

[1] *Robinia Neo-Mexicana* var. *luxurians* Dieck ex Goeze, in Gard. Chron., ser. 3, XII, 669 (1892), is a dwarf form occurring in the Colorado plateau and southern Rocky Mountain region.

[2] The Clammy Locust is known in its natural state only in the high mountains of North Carolina, but is elsewhere widely naturalized by cultivation.

Robinia viscosa Vent.—Continued.

SYN.—Rose-Flowering Locust (Tenn.).
Rose Acacia (Vt., R. I., Pa.).

VARIETIES DISTINGUISHED IN CULTIVATION.

Robinia viscosa albiflora Dippel. **White-flower Clammy Locust.**

SYN.—*Robinia viscosa* a *albiflora* Dippel, Handb. Laubh., dritt. T.,
703 (1893).

Robinia dubia[1] Fouc.

SYN.—*Robinia dubia* Fouc., in Desvaux, Journ. Bot., II, 204 (1813).
Robinia ambigua Poiret, in Lamarck, Enc. Méth. Bot. Suppl.,
690 (1816).
Robinia hybrida Audib. ex de Candolle, Prodr., II, 262 (1825).

Robinia bella-rosea[2] Nichol.

SYN.—*Robinia bella-rosea* Nicholson, Dic. Gard., III, 310 (1887).

OLNEYA Gray, in Mem. Am. Acad. Sci., new ser., V, 328 (1855).

Olneya tesota Gray. **Sonora Ironwood.**

SYN.—*Olneya Tesota* Gray, in Mem. Am. Acad. Sci, new ser., V,
313 (1854).

Tesota Mueller, in Walp., Ann. IV, 479 (1860).

COMMON NAMES.

Iron Wood (Cal.).
Arbol De Hierro (Cal.).
Palo de Hierro (Ariz., etc.).

ICHTHYOMETHIA Browne, Nat. Hist. Jam., 296 (1756).

Ichthyomethia piscipula (Linn.) Kuntze. **Jamaica Dogwood.**

SYN.—*Erythrina Piscipula* Linnæus, Spec. Pl., ed. 1, I, 707 (1753).
PISCIDIA ERYTHRINA Linn., Syst. Nat., ed. 10, 1155 (1759).
Piscidia Carthagenensis Jacquin, Enum. Pl. Carib., 27 (1760).
Piscidia inebrians Medicus, in Vorl. Chur. Rhys. Ges., 11,
394 (1787).
Piscidia taxicaria Salisbury, Prodr., 336 (1796).
Piscidia Piscipula Sargent, Gard. and For., IV, 436 (1891).

---

[1] A form occasionally found in cultivation. Its origin is uncertain, but is commonly thought to be a hybrid from *Robinia pseudacacia* and *R. viscosa*.

[2] Possibly a hybrid from *Robinia pseudacacia* and *R. viscosa*. As cultivated this form is entirely without the viscid glands of *R. viscosa*.

Ichthyomethia piscipula (Linn.) Kuntze—Continued.

SYN.—*Ichthyomethia Piscipula* Kuntze. Revis. Gen. Pl., Par. I, 191, (Sept., 1891); Hitchcock, in Gard. and For., IV, 272 (Nov., 1891).

COMMON NAME.

Jamaica Dogwood (Fla.).

Family ZYGOPHYLLACEÆ.

GUAJACUM Linn., Spec. Pl., 381 (1753).

Guajacum sanctum Linn.   **Lignumvitæ.**

SYN.—*Guajacum sanctum* Linnæus, Spec. Pl., ed. 1, I, 382 (1753).
Guajacum multijugum Stokes, Bot. Mat. Med., II, 488 (1812).
Guaiacum sanctum var. *parvifolium* Nuttall, Sylva, III, 17 (1849).
Guaiacum verticale Richard, Fl. Cub., 321 (1853).

COMMON NAMES.

Lignum Vitæ (Fla.).
Ironwood (Fla.).

Family RUTACEÆ.

XANTHOXYLUM [1] Linn., Spec. Pl., 270 (1753).

Xanthoxylum clava-herculis Linn.   **Prickly Ash.**

SYN.—*Xanthoxylum Clava-Herculis* Linnæus, Spec. Pl., ed. 1, I, 270 (1753)—exclusive "Habitat in Jamaica."

---

[1] The following shrubby species complete the list of North American species:

Xanthoxylum americanum Mill.

SYN.—*Xanthoxylum Americanum* Miller, Gard. Dic., ed. 8, No. 2 (1768).
Xanthoxylum fraxinifolium Marshall, Arb. Am., 167 (1785).
Xanthoxylum clava-Herculis Lamarck, Enc. Méth. Bot., II, 38 (1786), not Linn. (1753).
Xanthoxylum caribæum Gaertner, Frut., I, 333 (1788), not Lam. (1786).
Xanthoxylum ramiflorum Michaux, Fl. Bor.-Am., II. 235 (1803).
Zanthoxylum fraxineum Willdenow, Sp. Pl., IV, Par. II, 757 (1805).
Zanthoxylon cauliflorum Steudel, Nom. Bot., ed. 1, 897 (1821).
Xanthoxylum tricarpum Hooker, Fl. Bor.-Am., I, 118 (1830).

Xanthoxylum emarginatum Swartz.

SYN.—*Zanthoxylum emarginatum* Swartz, Fl. Ind. Occ., I, 572 (1797).
Xanthoxylum sapindoides de Candolle, Prodr., I, 728 (1824).

The latter species, described as a shrub or small tree, is indigenous in the West Indies, and was once admitted into the arborescent flora of North America on the authority of Dr. A. P. Garber, who discovered it as a shrub on an island in Biscayne Bay, southern Florida. It has not, however, been found in the United States since reported by Dr. Garber.

Xanthoxylum clava-herculis Linn.—Continued.

SYN.—*Xanthoxylum Carolinianum* Lamarck, Enc. Méth. Bot., II, 39 (1786).
Xanthoxylum fraxinifolium Walter, Fl. Caroliniana, 243 (1788), not Marsh. (1785).
Fagara fraxinifolia Lamarck, Ill., t. 334 (1791).
Xanthoxylum tricarpum Michaux, Fl. Bor.-Am., II, 235 1803.
Xanthoxylum aromaticum Willdenow, Sp. Pl., IV, Par. I, 755 (1805), excl. syn.
Kampmania fraxinifolia Rafinesque, Med. Rep., V, 352 (1808).
Xanthoxylum lanceolatum Poiret, in Lamarck, Enc. Méth. Bot., Suppl. II, 293 (1811).
Pseudopetalon glandulosum Rafinesque., Fl. Ludovic., 108 (1817).
Pseudopetalon tricarpum Raf., Med. Bot. II, 114 (1830).
Xanthoxylum Catesbyanum Raf., l. c. (1830).

COMMON NAMES.

Prickly Ash (N. C., S. C., Ga., Fla., Miss., La., Tex, Ark.).
Toothache-tree (N. C., S. C., Fla., Miss., La., Ark.).
Pepper-wood (Miss.).
Sea Ash (Miss., Fla.).
Southern Prickly Ash (Ala.).
Ash (Va.).
Frêne-piquant (La.).
Sting-tongue (Fla. negroes, Ark.).
Wait-a-bit, Tear-blanket (Ark.).
Wild Orange.

Xanthoxylum clava-herculis fruticosum Gray.

SYN.—*Xanthoxylum Clava-Herculis* var. *fruticosum* Gray, Pl. Wright., Pt. I, 30 (1852).
Xanthoxylum hirsutum Buckley, in Proc. Acad. Sci. Phila. 1861, 450 (1861).

Xanthoxylum cribrosum Spreng. **Satinwood.**

SYN.—*Xanthoxylum cribrosum* Sprengel, Syst. Veg., I, 946 (1825).
Xanthoxylum Floridanum Nuttall, Sylva, III, 14, t. 85 (1849).
Xanthoxylum Caribæum Watson, Bib. Index N. A. Bot., Pt. I, 155 (1878), not Lam. (1786).
Xanthoxylum Caribæum var. *Floridanum* Gray, in Proc. Am. Acad. Sci., new ser., XXIII, 225 (1888).

Xanthoxylum cribrosum Spreng.—Continued.

COMMON NAMES.

Syn.—Yellow-wood (Fla.).
Satinwood (Fla.).

Xanthoxylum fagara (Linn.) Sargent.   **Wild Lime.**

Syn.—*Schinus Fagara* Linnæus, Spec. Pl., ed. 1, I, 389 (1753).
Pterota subspinosa Browne, Nat. Hist. Jam.. ed. 1. 146. t. 5,
f. 1 (1756).
Fagara Pterota Linnæus, Amœn.. V, 393 (1760).
Fagara tragodes Jacquin, Enum. Pl. Carib., 12 (1760).
Fagara lentiscifolia Willdenow, Enum. Pl.. I. 165 (1809).
Xanthoxylum Pterota Humboldt, Boupland & Kunth,
Nov. Gen. Sp., VI, 3 (1823).

COMMON NAME.

Wild Lime (Fla.).

PTELEA[1] Linn., Spec. Pl., 118 (1753).

Ptelea trifoliata Linn.   **Hoptree.**

Syn.—*Ptelea trifoliata* Linnæns, Spec. Pl., ed. 1, I, 118 (1753).
Ptelea pentaphylla Fabricius, Enum. Pl. Helm., 416 (1759).
Ptelea viticifolia Salisbury, Prodr.. 68 (1796).
Ptelea trifoliata β pubescens Pursh. Fl. Am. Sept., I, 107
(1814).
Ptelea trifoliata β pentaphylla Moench ex de Candolle,
Prodr., II, sect. 2, 83 (1825).
Ptelea glauca (in Nurs.) ex Koch, Dendrol.. erst. Th., 566
(1869).
Ptelea trifoliata a *heterophylla* Dippel, Handb. Laubh., zw.
T., 354 (1892).
Ptelea triafoliata glauca Hort. ex Dippel, l. c. (1892).

---

[1] The following are shrubs:

Ptelea aptera Parry.

Syn.—*Ptelea aptera* Parry, in Proc. Davenport Acad. Sci., IV. 39 (1884).
A low shrub detected on Todos-Santos Bay, Lower California.

Ptelea baldwinii Torr. & Gr.

Syn.—*Ptelea Baldwinii* Torrey & Gray, Fl. N. A., I, 215 (1838).
Ptelea angustifolia Bentham, Pl. Hartweg., 9 (1839).
South Atlantic coast region, Texas to California, southern Rocky Mountains, and northern Mexico.

Ptelea trifoliata mollis Torr. & Gr.

Syn.—*Ptelea trifoliata* var. *mollis* Torrey & Gray, Fl. N. A.. I, 680 (1840).
Ptelea mollis Curtis, in Am. Journ. Sci., ser. 2, VII, 406 (1849).
Atlantic coast region, Florida, western Texas, and New Mexico.

Ptelea trifoliata Linn.—Continued.

COMMON NAMES.

SYN.—Hoptree.
Wafer Ash.
Whahoo.
Quinine-tree (Mich.).

VARIETY DISTINGUISHED IN CULTIVATION.

Ptelea trifoliata aurea Nich.   **Golden Hoptree.**

SYN.—*Ptelea trifoliata aurea* Nicholson, Dic. Gard., III, 240 (1887).
Ptelea trifoliata fol. variegatis Dippel, Handb. Laubh., zw.
T., 354 (1892).

HELIETTA Tulasne, in Ann. Sc. Nat., sér. 3, VII, 280 (1847).

Helietta parvifolia (Hemsl.) Benth.   **Baretta.**

SYN.—*Ptelea parvifolia* Hemsley (ex Mss. Gray in Herb. Kew.),
Biol. Cent. Am., I, 170 (1879).
Helietta parvifolia Bentham, in Hooker, Icon. Pl., XIV, 66
(1882).

AMYRIS Browne, Hist. Jam., 208 (1756).

Amyris maritima Jacq.   **Torchwood.**

SYN.—*Amyris maritima* Jacquin, Enum. Pl. Carib., 23 (1760).
Amyris Elemifera Linnæus, Spec. Pl., ed. 2, I, 495—exclusive
habitat "in Carolina"—(1762).
Amyris dyatripa Sprengel, Neue Entdeck., III, 48 (1822).
Amyris Floridana Nuttall, in Am. Journ. Sci., V, 294 (1822).
AMYRIS SYLVATICA de Candolle, Prodr., II, 81 (1825).
Amyris Lunani Sprengel, Syst. Veg., II, 217 (1825).
Amyris maritima var. *angustifolia* Gray, in Proc. Am. Acad.
Sci., XXIII, 226 (1888).

COMMON NAME.

Torchwood (Fla.).

CANOTIA Torr., in Pac. R. R. Rep., IV, 68 (1857).

Canotia holacantha Torr.   **Canotia.**

SYN.—*Koeberlinia* (?) Engelmann, in Emory's Rep., 158, f. 14 (1848).
[C]*anotia holacantha* Torrey, in Pac. R. R. Rep., IV, 68 (1857).

COMMON NAME.

Canotia (N. Mex.).

Family SIMAROUBACEÆ.

SIMAROUBA[1] Aublet, Pl. Guian., II, 859 (1775).

Simarouba glauca de C.   **Paradise-tree.**

SYN.—*Simaruba glauca* de Candolle, in Diss. Ann. Mus., XVII, 323 (1811).
Simaruba officinalis Macfadyen, Flora Jam., 198 (1837), not de C. (1824).
Simaruba medicinalis Endlicher, Mediz. Pharm., 525 (1842).

COMMON NAMES.

Paradise-tree (Fla.).
Gumbo Limbo (Fla.).
Bitterwood (Fla.).

KŒBERLINIA Zucc., in Abh. Akad. Muench., I, 358 (1832).

Kœberlinia spinosa Zucc.   **Kœberlinia.**

SYN.—*Kœberlinia spinosa* Zuccarini, in Abh. Akad. Muench., I, 359 (1832).

AILANTHUS Desf., in Mém. Acad. Sc. Par., 1786, 265 (1789).

Ailanthus glandulosa[2] Desf.   **Ailanthus.**

SYN.—*Ailanthus glandulosa* Desfontaines, in Mém. Acad. Sc. Par., 1786, 265, t. 8 (1789).
Ailanthus procera Salisbury, Prodr., 171 (1796).
Ailanthus rhodoptera Mueller, Fragm., III, 43 (1863).
Ailanthus Pongelion Gmelin, Syst., II, 726 (1867).
Ailanthus Japonica Hort. ex Koch, Dendrol., erst. Th., 569 (1869).
Ailanthus macrophylla Hort., ex Hand-list Trees and Shr. Arb. Kew, Pt. I, 53 (1894).

---

[1] Aublet's original spelling is preserved. Linnæus previously used *Simaruba* in the *Materia Medica* (188, 1749) for a *Bursera*, and later, as the specific name of *Pistacia simaruba* (Spec. Pl., 1026, 1753). In taking up Aublet's genus subsequent authors have followed Linnæus's spelling—*Simaruba*.

[2] Native of China, but thoroughly naturalized by cultivation in this country; in many localities, especially in Central and Southern States, escaped and forming dense thickets.

VARIETIES DISTINGUISHED IN CULTIVATION.

Ailanthus glandulosa rubra Dippel. **Redfruit Ailanthus.**

SYN.—*Ailanthus glandulosa* a. *rubra* Dippel, Handb. Laubh., zw. T., 365 (1892).
Ailanthus purpurascens Hort., ex Hand-list Trees and Shr. Arb. Kew, Pt. I, 53 (1894).
Ailanthus rubra Hort., l. c. (1894).

Ailanthus glandulosa pendulifolia Dippel. **Drooping Ailanthus.**

SYN.—*Ailanthus glandulosa* b. *pendulifolia* Dippel, Handb. Laubh., zw. T., 365 (1892).
Ailanthus mascula pendula Hort., ex Hand-list Trees and Shr. Arb. Kew, Pt. I, 53 (1894).

Ailanthus glandulosa ancubæfolia Dippel.

SYN.—*Ailanthus glandulosa* c. *ancubæfolia*, Dippel, Handb. Laubh., zw. T., 365 (1892).

Family BURSERACEÆ.

BURSERA Jacquin, Stirp. Am. 94 (1763).

Bursera simaruba (Linn.) Sargent. **Gumbo Limbo.**

SYN.—*Pistacia Simaruba* Linnæus, Spec. Pl., ed. 1, II, 1026 (1753).
BURSERA GUMMIFERA Linn., l. c., ed. 2, I, 471 (1762).
Elaphrium integerrimum Tulasne, in Ann. Sc. Nat. sér. 3, VI, 369 (1846).
Bursera Simaruba Sargent, in Gard. and For., III, 260 (1890).

COMMON NAMES.

Gum Elemi (Fla.).
Gumbo Limbo (Fla.).
West Indian Birch (Fla.).

Family MELIACEÆ.

MELIA Linn., Spec. Pl., 384 (1753).

Melia azedarach[1] Linn. **China-tree.**

SYN.—*Melia Azedarach* Linnæus, Spec. Pl., ed. 1, I, 384 (1753).
Melia sempervirens Swartz, Prodr., 67 (1788).

[1]This species, probably a native of Persia, but long cultivated in the warmer parts of the Old and New World, has become thoroughly naturalized in the south Atlantic and southwestern coast regions of the United States.

Melia azedarach Linn.—Continued.

SYN.—*Azederach delecteria* Moench, Meth., 171 (1794).
Melia Florida Salisbury, Prodr., 317 (1796).
Melia sambucina Blume, Bijdrag. Fl., 162 (1826).
Melia guineensis Don in Loudon, Hort. Brit., 168 (1830).
Melia Japonica Don, Gen. Hist. Dichl. Pl., I, 680 (1831).
Melia Bukayun Royle, Ill. Bot. Him., 144 (1839).
Melia Commelini Medicus ex Steudel, Nom. Bot., ed. sec.,
  II, 118 (1841).
Melia cochinchinensis Rœmer, Syn. Hesp., 95 (1846).
Melia orientalis Rœm., l. c., (1846).
Melia composita Bentham, Fl., Austral., 1, 380 (1863), not
  de C., (1824).

Melia azedarach umbraculifera Sargent.   **Umbrella China-tree.**

SYN.—*Melia Azedarach* var. *umbraculifera* Sargent (?), in Gard. &
  For., VII, 92, f. 20 (1894).

SWIETENIA Jacq., Enum. Pl. Carib., 4 (1760).

Swietenia mahagoni Jacq.   **Mahogany.**

SYN.—*Swietenia Mahagoni* Jacquin, Enum. Pl. Carib., 20 (1760).
Cedrus Mahogoni Miller, Gard. Dict., ed. 8, No. 2 (1768).
Swietenia macrophylla King, in Hooker, Icon. Pl. XVI, t. 1550
  (1886).
COMMON NAMES.
Mahogany (Fla.).
Madeira (Fla.).
Redwood (Fla.).

Family EUPHORBIACEÆ.

DRYPETES Vahl, Ecl., III, 49 (1805).

Drypetes lateriflora (Swartz) Urban.   **Guiana Plum.**

SYN.—*Schæfferia lateriflora* Swartz, Prodr., 30 (1788).
Kælera laurifolia Willdenow, Sp. Pl., IV, Par. 1, 750 (1805),
  in part.
Bessera spinosa Sprengel, Pugill., II, 91 (1815).
DRYPETES CROCEA Poiteau, Mém. Mus., I, 159, t. 8 (1815).
Limacia laurifolia Dietrich, Lex. Gart. Bot. Nachtr., IV, 334
  (1818).
Roumea coriacea Steudel, Nom. Bot., ed. sec., II, 475 (1841),
  not Poiteau (1815).

Drypetes lateriflora (Swartz) Urban—Continued.

SYN.—*Drypetes Glauca* Richard, Fl. Cub., III, 218 (1855), not Vahl
(1798).
Drypetes sessiliflora Baillion, Étud. Gén. Euphorb. Atlas, 45,
t. f. 34–36, 38, 40 (1858).
Drypetes alba var. *latifolia* Grisebach, in Nachr. Kgl. Gesell.
Göt., t. 165 (1865).
Drypetes crocea β longipes Mueller, in de Candolle, Prodr.,
XV, sect. 2, 456 (1866).
Drypetes crocea γ latifolia Mueller, in de C., l. c. (1866).
Drypetes latifolia Suavalle, Fl. Cub., 127 (1873).
Drypetes laterifolia Urban, in Bot. Jahrb., fünftz. B., 357
(1893).
Xylosma nitidum Hooker & Jackson, Ind. Kew., II, 802
(1893), not Griseb. (1865).

COMMON NAMES.

Guiana Plum (Fla.).
Whitewood (Fla.).

Drypetes keyensis Urban.   **Guiana Plum.**

SYN.—*Drypetes glauca* Nuttall, Sylva, II, 68 (1846), not Vahl(1798).
Drypetes crocea var. *latifolia* Sargent, in Tenth Cen. U. S.,
IX (Cat. For. Trees), 121 (1884), not Mueller (1866).

COMMON NAMES.

Guiana Plum (Fla.).
Whitewood (Fla.).

GYMNANTHES Swartz, Prodr., 95 (1788).

Gymnanthes lucida Swartz.   **Crabwood.**

SYN.—*Gymnanthes lucida* Swartz, Prodr., 96 (1788).
Excœcaria lucida Swartz, Fl. Ind. Occ. II, 1122 (1800).
SEBASTIANA LUCIDA Mueller, in de Candolle, Prodr., XV,
sect. 2, 1181 (1866).

COMMON NAMES.

Crabwood (Fla.).
Poisonwood (Fla.).

HIPPOMANE Linn., Spec. Pl., 1191 (1753).

Hippomane mancinella Linn.   **Manchineel.**

SYN.—*Hippomane mancinella* Linnæus, Spec. Pl., ed. 1, II, 1191
(1753).

272

Hippomane mancinella Linn.—Continued.

Syn.—*Hippomane dioica* Rottbœll, in Act. Lit. Univ. Hafn., I, 301 (1778).
Mancinella venenata Tussac. Fl. Antilles, III, 21, t. 5 (1824)

COMMON NAME.

Manchineel (Fla.).

Family CHEIRANTHODENREÆ.[1]

FREMONTODENDRON[2] Coville, in Contr. U. S. Nat. Herb., IV, 74 (1893).

Fremontodendron californicum (Torr.) Coville.   **Fremontia.**

Syn.—FREMONTIA CALIFORNICA Torrey, Pl. Frem., 6 (1853).
Cheiranthodendron Californicum Baillon, Hist. Pl., IV, 70 (1873).
Fremontodendron Californicum (Torr.) Coville. in Contr. U. S. Nat. Herb.. IV (Bot. Death Val. Exp.), 74 (1893).

COMMON NAMES.

Slippery Elm (Cal., Idaho).
Silver Oak (Cal.).
Leatherwood (Cal.).
Fremontia (Cal.).

Family THEACEÆ.[3]

GORDONIA Ellis, in Phil. Trans.. LX., 518 (1770).

Gordonia lasianthus (Linn.) Ellis.   **Loblolly Bay.**

Syn.—*Hypericum Lasianthus* Linnæus, Spec. Pl., ed. 1, 783 (1753).
Gordonia Lasianthus Ellis. in Phil. Trans., LX, 518, t. 11 (1770).
Gordonia pyramidalis Salisbury. Prodr., 386 (1796).

COMMON NAMES.

Loblolly Bay (N. C., S. C.. Ga.. Ala.. Fla.. Miss.. La.).

---

[1] Usually included in *Sterculiaceæ*.
[2] Mr. Coville (l. c.) has pointed out that the genus *Fremontia* is untenable for this plant. having been first applied by Torrey to a plant already named *Sarcobatus*, of which *Fremontia* becomes a synonym, and therefore can not be maintained for this small tree, so long known as "Fremontia," and for which he proposes *Fremontodendron*.
[3] Usually included in *Ternstræmiaceæ*.

Gordonia lasianthus (Linn.) Ellis—Continued.

SYN.—Tan Bay (Miss., Fla., La.).
　　Black Laurel (N. C.).

Gordonia altamaha (Marsh.) Sargent. . **Franklinia.**

SYN.—*Franklinia altamaha* Marshall, Arb. Am., 49 (1785).
　　GORDONIA PUBESCENS L'Héritier, Stirp. Nov, 156 (1785).
　　Gordonia Franklini L'Héritier, l. c. (1785).
　　Michauxia sessilis Salisbury, Prodr., 386 (1796).
　　Franklinia Americana Marshall ex Persoon, Syn. Pl., II, 259
　　　(1807).
　　Gordonia Altamaha Sargent, in Gard. & For., 11, 616 (1889).

Family CANELLACEÆ.

CANELLA Browne, Nat. Hist. Jam., 275 (1756).

Canella winterana (Linn.) Gærtn. **Cinnamon-bark.**

SYN.—*Laurus Winterana* Linnæus, Spec. Pl., ed. 1, 371 (1753).
　　Winterania Canella Linn., Spec. Pl., ed. 2, 636 (1763).
　　CANELLA ALBA Murray in Linn., Syst. Veg., ed. 14, IV, 443
　　　(1784).
　　Canella Winterana Gærtner, Fruct., I, 377, t. 77 (1788).
　　Canella laurifolia Lodd. ex Sweet, Hort. Brit., 65 (1827).
　　Canella canella (Linn.) Sudworth, in Bull. Torr. Bot. Club,
　　　XX, 46 (1893).

COMMON NAMES.

Cinnamon-bark (Fla.).
Whitewood.
Wild Cinnamon.

Family ANACARDIACEÆ.

COTINUS Adanson, Fam. Pl., II., 345 (1763).

Cotinus cotinoides (Nutt.) Britton. **American Smoke-tree.**

SYN.—*Rhus Cotinus* Torrey & Gray, Fl. N. Am., I, 216 (1838), not
　　Linn. (1753).
　　RHUS COTINOIDES (Nutt.! in Herb. Acad. Philad.) Torrey
　　　& Gray, Fl. N. A., I, 217 (1838).
　　Cotinus Americana Nuttall, Sylva, III, 1, t. 81 (1849).
　　Cotinus Coggygria Engler, in de Candolle, Monog. Phan., IV,
　　　351 (1878), in part.

Cotinus cotinoides (Nutt.) Britton—Continued.

Syn.—*Rhus Americanus* (Nutt.) Sudworth, in Bull. Torr. Bot. Club,
XIX, 80 (1892).
Cotinus cotinoides (Nutt.) Britton, in Mem. Torr. Bot. Club,
V, Sig. 14, 216 (1894).

COMMON NAMES.

Chittamwood (Ala.).
Yellowwood (Ala.).
Smoke-tree (Ark.: R. I., cult.).

RHUS Linn.. Spec. Pl., 265 (1753).

Rhus metopium Linn.   **Poisonwood.**

Syn.—*Rhus Metopium* Linnæus. Amœn.. V, 395 (1760).
Rhus oxymetopium Grisebach. Cat. Pl. Cub., 67 (1866).
Metopium Linnæi Engler, in de Candolle. Monog. Phan.,
IV. 367 (1878).
Metopium Linnæi var. *oxymetopium* Engl.. in de C.. l. c. (1878).

COMMON NAMES.

Poisonwood (Fla.).
Coral Sumach.
Mountain Manchineel.
Bumwood.
Hog Plum.
Doctor Gum.

Rhus hirta (Linn.) Sudworth.   **Staghorn Sumach.**

Syn.—*Datisca hirta* Linnæus, Spec. Pl., ed. 1, II. 1037 (1753).
RHUS TYPHINA Linn., Amœn., IV, 311 (1760).
Rhus Canadensis Miller, Gard. Dict., ed. 8, No. 5 (1768).
Rhus viridiflora Duhamel, Trait. Arb., nouv. éd.. 163 (1808).
Rhus typhina var. *arborescens* Willdenow, Enum.. 323 (1809).
Rhus typhina var. *frutescens* Willd.. l. c. (1809.)
Rhus gracilis Hort. ex Engler, in de Candolle. Monog. Phan.,
IV, 377 (1883).
Toxicodendron typhinum Kuntze. Rev. Gen. Pl.. Par. I, 154
(1891).
Rhus hirta (Linn.) Sudworth, in Bull. Torr. Bot. Club, XIX,
81 (1892), not Harv. MSS. in Herb. Kew., Engler (1883).
Rhus Americana Dippel. Handb. Laubh.. zw. T., 367 (1892).
Rhus coriaria Hort., ex Hand-list Trees and Shr., Arb.
Kew. Pt. I. 103 (1894). not Linn. (1753).
Rhus frutescens Hort.. l. c. (1894).
Rhus glabra Hort.. l. c. (1894). not Linn. (1753).

Rhus hirta (Linn.) Sudworth—Continued.

SYN.—Staghorn Sumach (Vt., N. H., Mass., R. I., Conn., N. Y., N. J., Del., Pa., N. C., S. C., Miss., Mo., Mich., Wis., Ohio, -Ont.).
Sumach (Me., Vt., N. Y., Pa., W. Va., Ark., Ky., Ind., Wis.).
Virginia Sumach (Tenn.).
Hairy Sumach.

Rhus copallina Linn.   **Dwarf Sumach.**

SYN.—*Rhus copallina* Linnæus, Spec. Pl., ed. 1, I, 266 (1753).
Rhus copallina α *latifolia* Engler, in de Candolle, Monog.
Phan., IV, 384 (1883).
Rhus copallina α *1 angustialata* Engl., l. c. (1883).
Rhus copallina α *2 latialata* Engl., l. c. (1883).
Rhus copallina β, *angustifolia* Engl., l. c. (1883).
Rhus copallina β *2 serrata* Engl., l. c. (1883).
Toxicodendrum copallinum Kuntze, Rev. Gen. Pl., Par. I,
154 (1891).
Toxicodendrum copallinum O K. (L.) var. *latifolium* Engl.
Kuntze, l. c. (1891).

Dwarf Sumach (Vt., N. H., R. I., Mass., N. Y., Del., Pa., Ala.,
Fla., Miss., La., Kans.).
Sumach (Vt., Pa., W. Va., S. C., Fla., Ga., Miss., La., Tex.,
Ky., Mo., Kans.).
Smooth Sumach.
Mountain Sumach (Vt., Tenn.).
Black Sumach (Ark., Tex.).
Wing-rib Mountain Sumach (S. C.).
Common Sumach (S. C.).

Rhus copallina leucantha (Jacq.) de C.   **White-flower Dwarf Sumach.**

SYN.—*Rhus leucantha* Jacquin, Plant Hort. Schönb., III, 50, t.
342 (1798).
Rhus copallina L. β *leucantha* de Candolle, Prodr., II, 68
(1825).

Rhus copallina lanceolata Gray.   **Lanceleaf Dwarf Sumach.**

SYN.—*Rhus copallina L.*, var. *lanceolata* Gray, in Journ. Boston
Soc. Nat. Hist., VI, 158 (1850).
Rhus copallina β *1 integrifolia* Engler, in de Candolle,
Monog. Phan., IV, 384 (1883), not *R. integrifolia* T. &
Gr. (1838).
Rhus lanceolata Gray Mss. ex Engl., l. c. (1883).

Rhus vernix Linn.   **Poison Sumach.**

Syn.—*Rhus Vernix* Linnæus. Spec. Pl.. ed. 1. I. 265. excl. hab.
 "Japonia" (1753).
Toxicodendron pinnatum Miller. Gard. Dict.. ed. 8. No. 4
 (1768).
Rhus venenata de Candolle. Prodr.. II. 68 (1825).
Toxicodendron vernix Kuntze. Rev. Gen. Pl.. Par. I. 153
 (1891).

COMMON NAMES.

Poison Sumach (Vt.. N. H.. Mass., R. I., Conn.. N. Y., N. J.,
 Del.. N. C.. S. C.. Ala.. Miss.. La.. Mo., Iowa. Wis.. Mich.,
 Minn., Ohio. Ont.. Nebr.).
Poison Elder (Vt., Mass., R. I.. N. Y.. Del.. S. C.. Ga.. Ala..
 Miss.. La., Mo., Nebr.. Minn. .
Dogwood (Vt.. Mass.. R. I., Wis.. Mich., Iowa. Nebr., Minn..
 La.).
Poison Dogwood (N. H.. Vt.. N. J.. Pa.. D. C.. Mo.. Mich.,
 Minn.).
Swamp Sumach (R. I.. N. Y.. N. J.. Tenn., Minn.).
Sumach (R. I.).
Poison Oak (La.).
Poison Ash (Pa.).
Poisonwood (Tenn.).
Poisontree.

Rhus integrifolia (Nutt. Benth. & Hook. **Western Sumach.**

Syn.—*Styphonia integrifolia* (Nutt.! mss.). in Torrey & Gray. Fl. N.
 A.. I, 220 (1838).
Styphonia serrata (Nutt.! mss.). in Torr. & Gr., l. c. (1838).
Rhus integrifolia Bentham & Hooker. Genera Pl.. I, 419
 (1862).
Rhus integrifolia β serrata Engler. in de Candolle. Monog.
 Phan.. IV. 388 (1883).

Family CYRILLACEÆ.

CYRILLA Linn.. Mantiss.. I. 5 (1767).

Cyrilla racemiflora Linn.   **Ironwood.**

Syn.—*Cyrilla racemiflora* Linnæus, Mantiss.. I. 50 1767 .
 Andromeda plumata Bartram Cat. ex Marshall. Arb. Am.. 9
 1785.

Cyrilla racemiflora Linn.—Continued.

SYN.—*Cyrilla polystachia* Rafinesque, Autikon Bot.. 8 (1840).
Cyrilla parrifolia Raf., l. c. (1840).
Cyrilla fuscata Raf., l. c. (1840).

COMMON NAMES.

Ironwood (N. C., S. C., Ga., Fla., Miss., La.).
Leatherwood (Ala., Fla.).
He Huckleberry (N. C., S. C.).
Burnwood (N. C.).
Burnwood Bark (S. C.).
Red Titi (Fla.).
White Titi.

CLIFTONIA Gærtn. f., Frut., 246 (1805).

Cliftonia monophylla (Lam.) Sargent. **Titi.**

SYN.—*Ptelea monophylla* Lamarck, Ill., I, 336 (1791).
Cliftonia nitida Gærtner f., Frut.. Suppl., III, 247, t. 225, f. 5
(1805).
Mylocaryum ligustrinum Willdenow, Enum. Hort. Berol., 454
(1809).
CLIFTONIA LIGUSTRINA Sims ex Sprengel, Syst. Veg., II,
316 (1825).
Cliftonia monophylla Sargent,[1] Silva. II, 7 (1891).

COMMON NAMES.

Titi (S. C., Ga., Ala., Fla., Miss.).
Buckwheat-tree (Fla.. La.).
Black Titi (Fla.).
Ironwood.

Family AQUIFOLIACEÆ.

ILEX Linn., Spec. Pl., 125 (1753).

Ilex opaca Ait. **American Holly.**

SYN.—*Ilex Aquifolium* Linnæus, Spec. Pl. ed. 1, I, 125 (1753), in part.
Ilex Canadensis Marshall. Arb. Am.. 64 (1785).—?

---

[1]Professor Sargent (Silva, II, 7) assigns this combination to Britton (Bull. Torr.
Bot. Club.. XVI, 310, 1889). Dr. Britton does not (l. c.) make this change, but calls
attention to the fact that *Ptelea monophylla*, as adduced by Watson (Bull. Torr. Bot.
Club, XIV., 167), is the oldest name for this plant. Professor Sargent (l. c.) publishes *C. monophylla* for the first time.

Ilex opaca Ait.—Continued.

SYN.—*Ilex opaca* Solander in Aiton, Hort. Kew., I, 169 (1789).
Ilex laxiflora Lamarck, Enc. Méth. Bot., III, 147 (1789).
Ilex Americana Lamarck, in Journ. Nat. Hist., Par. I. 416
(1792).—?
Ilex quercifolia Meerburgh, Pl. Select. Ic., t. 5 (1798).
Ageria (*mac ?*) *opaca* Rafinesque, Sylva Tellur., 47 (1838).
Ilex opaca 2 macrodon Loudon,[1] Arb. Frut., II, 517 (1838).
Ilex opaca 3 latifolia Loud., l. c. (1838).
Ilex opaca 4 acuminata Loud., l. c. (1838).
Ilex opaca 5 globosa Loud., l. c. (1838).

COMMON NAMES.

Holly (R. I., Del., W. Va., Pa., N. C., S. C., Ga., Fla., Miss.,
La., Ark.).
American Holly (Mass., R. I., Conn., N. Y., N. J., Pa., Del.,
N. C., Ala., Miss., La.).
White Holly (Va.).

Ilex cassine Linn.   **Dahoon.**

SYN.—*Ilex Cassine* Linnæus. Spec. Pl., ed. 1, I, 125, exclusive var.
β (1753).
Ilex caroliniana Miller, Gard. Dict., ed. 8, No. 3 (1768).
ILEX DAHOON Walter, Fl., Caroliniana, 241 (1788).
Ilex Cassine var. *latifolia* Aiton. Hort. Kew., I. 170 (1789).
Ilex cassinoides Link, Enum. Pl. Berol., I. 148 (1821), not
Du Mont de Cour. (1811).
Ilex laurifolia Nuttall, in Am. Journ. Sci., V, 289 (1822)
Ageria palustris Rafinesque, Sylva Tellur., 47 (1838).
Ageria obovata Raf., l. c., 48 (1838).
Ageria (*Dah.*) *heterophylla* Raf., l. c. (1838).

COMMON NAMES.

Dahoon (N. C., S. C., Ga., Ala., Fla., Miss., La.).
Dahoon Holly (N. C., S. C., Fla.).
Yaupon (Fla.).

Ilex cassine angustifolia Willd.   **Narrow-leaf Dahoon.**

SYN.—*Ilex Cassine* var. *angustifolia* Willdenow. Sp. Pl., I, 709
(1797).
Ilex angustifolia Willd., Enum. Hort. Berol., 172 (1809).
Ilex Ligustirna Elliott. Sk. Bot. S. C. and Ga., II, 680 (1824),
not Jacq. (1790).

---

[1] Loudon (l. c.) states that Rafinesque is the author of the following varieties, but
does not cite the publication, nor have I been able to find it.

279

Ilex cassine angustifolia Willd.—Continued.

SYN.—*Ilex Watsoniana* Spach, Hist. Vég., II, 429 (1834).
Ilex Dahoon var. *angustifolia* Watson, Bib. Ind. N. A. Bot., 158 (1878).

Ilex cassine myrtifolia (Walter) Sargent. **Myrtle-leaf Dahoon.**

SYN.—*Ilex myrtifolia* Walter, Fl. Caroliniana, 241 (1788).
Ilex rosmarifolia Lamarck, Ill., I, 256 (1791).
Ilex ligustrifolia Don, Hist. Dichl. Pl., II, 19 (1832).
ILEX DAHOON var. MYRTIFOLIA Chapman, Fl. S. States, ed. 1, 269 (1860).
Ilex Cassine var. *myrtifolia* Sargent, in Gard. & For., II, 616 (1889).

Ilex vomitoria Ait. **Yaupon.**

SYN.—*Ilex Cassine* β Linnæus, Spec. Pl., ed. 1, I, 125 (1753).
Cassine Peragua Linn., Mantiss., 220 (1767), in part.
Cassine Caroliniana[1] Lamarck, Enc. Méth. Bot., I, 652 (1783).
ILEX CASSINE Walter, Fl. Caroliniana, 241 (1788), not Linn. (1753).
Ilex vomitoria Solander in Aiton, Hort. Kew., I, 170 (1789).
Ilex ligustrina Jacquin, Coll. Bot., IV, 105 (1790).
Ilex Floridana Lamarck, Ill., I, 356 (1791).
Ilex Cassena Michaux, Fl. Bor.-Am., II, 229 (1803).
Ilex cassinoides Du Mont de Courset, Bot. Cult., ed. 2, VI, 251 (1811).
Ilex religiosa Barton, Fl. Virg., 66 (1812).
Cassine ramulosa Rafinesque, Fl. Ludovic., 110 (1817).
Hierophyllus Cassine Raf., Med. Bot., II, 8 (1830).
Ilex (Emetila) ramulosa Raf., Sylva Tellur., Cent. III, 45 (1838).
Ageria (Dah.) Cassena Raf., l. c., 47 (1838).
Ilex Caroliniana (Lam.) Lœsner, in Bot. Centralb., XLVII, 163 (1891), not Mill. (1768).

COMMON NAMES.

Yopon (N. C., Ga., Ala., Miss., Tex.).
Yaupon (N. C., S. C., Fla., Miss., La.).
Cassena (N. C., S. C., Fla., La.).
Cassine (La.).
True Cassena.
Evergreen Cassena.

[1] This would seem to be the oldest specific name, but in taking it up under *Ilex* a combination is produced identical with Miller's *I. Caroliniana* (1768), a synonym of *I. Cassine* Linn. (1753).

Ilex vomitoria Ait.—Continued.

Syn.—Cassio-berry Bush.
Emetic Holly[1] (S. C.).

Ilex decidua Walter.   **Deciduous Holly.**

Syn.—*Ilex decidua* Walter, Fl. Caroliniana, 241 (1788).
Ilex prinoides Solander in Aiton, Hort. Kew., I, 169 (1789).
Ilex æstivalis Lamarck, Enc. Méth. Bot., III, 147 (1789).
Ilex Prionitis Willdenow, Enum. Hort. Berol., Suppl., 8 (1813).
Prinos ambiguus Elliott, Sk. Bot. S. C. and Ga., II, 705 (1824).
Prinos deciduous de Candolle, Prodr., II, 16 (1825).

COMMON NAMES.

Holly (Tex., Ark., Mo.).
Bearberry (Miss.).
Possum Haw (Fal.).

Ilex[2] monticola Gray.   **Mountain Holly.**

Syn.—*Ilex ambigua* Torrey, Fl. N. Y., II, 2, excl. syn. (1843), not Elliott (1824).
Ilex montana Gray, Man. Bot. N. U. S., ed. 1, 276 (1848), not *Prinos montanus*[3] Swartz (1788).
Ilex monticola Gray, l. c., ed. 2, 264 (1856).
Ilex mollis Gray, l. c., ed. 5, 306 (1867).
Prinos pubescens Michaux ex Gray, l. c. (1867).
Ilex dubia B. S. P., in Prelim. Cat. Anth. Pterid., 11 (1888).
Ilex montana T. & G. var. *mollis* (A. Gray) Britton, in Bull. Torr. Bot. Club. XVII, 313 (1890).

---

[1] A very fitting local name, indicating (as in *vomitoria*) the medicinal character of the foliage.

[2] Up to the present time the genus *Ilex* has had no representatives in the West, all the known species being confined to the region east of the Rocky Mountains, and although beyond the confines of the United States, it is of interest to here include the following recently discovered species.

Ilex triflora Brand.

Syn.—*Ilex triflora* Brandegee, in Gard. & For., VII, 414, f. 65 (1894).

A species detected in the vicinity of La Chuparosa, Baja, Lower California. It attains "a height of 50 feet or less," with a diameter of "a foot or more." Reported to be rare.

Ilex californica Brand.

Syn.—*Ilex californica* Brandegee, in Gard. & For., VII, 414, f. 66 (1894).

This species was detected in the same region as the above, but is said to be more common than *Ilex triflora*, and oftener a large bush, occasionally becoming a small tree.

[3] This plant is the same as the *Ilex montana* (Swartz) Grisebach (1861), a West Indian species, and therefore has an older specific name than the *Ilex montana* Gray (1848).

Family CELASTRACEÆ.

EVONYMUS[1] Linn., Spec. Pl., 197 (1753).

Evonymus atropurpureus Jacq.   **Waahoo.**

SYN.—*Euonymus atropurpureus* Jacquin, Hort. Vind., II. 55, t. 120 (1772).
Euonymus caroliniensis Marshall, Arb. Am., 43 (1785).
Euonymus latifolius Marsh., l. c., 44 (1785), not Mill. (1768).
Euonymus tristis Salisbury, Prodr., 142 (1796).

COMMON NAMES.

Burning Bush (R. I., N. Y., N. J., Pa., Del., Md., N. C., S. C., Miss., Ark., Ky., Ohio, Ill., Ind., Iowa, Kans., Nebr., Mich.).
Waahoo (N. Y., N. J., Pa., W. Va., N. C., S. C., Miss., Ky., Ark., Mo., Nebr., Ill., Iowa, Kans., Ohio, Ind.
Spindle-tree (R. I., Del., Pa., N. C., Ill.).
Arrow-wood (Miss., La., Ill., Mo.).
Strawberry-tree (N. Y.).
Strawberry Bush (Tenn.).
"Moses in the Burning Bush" (N. J.).
Bleeding Heart (N. C.).
Indian Arrow (Ind.).

GYMINDA Sarg., in Gard. & For., IV, 4 (1891).

Gyminda grisebachii Sargent.   **Gyminda.**

SYN.—*Myginda integrifolia* Humboldt, Bonpland & Kunth, Nov. Gen. Sp., VII, 66 (1825), not Lam. (1797).
Myginda latifolia Chapman, Fl. S. States, 76 (1865), not Swartz (1806).
MYGINDA PALLENS Smith ex Sargent, in Tenth Cen. U. S., IX, (Cat. For. Trees N. A.) 38 (1884), not Smith. (1820).
Gyminda Grisebachii Sargent, in Gard. & For., IV, 4 (1891).

COMMON NAMES.

False Boxwood (Fla.).
Gyminda.

Gyminda grisebachii glaucifolia (Griseb.) nom. nov.   **Pale Gyminda.**

SYN.—*Myginda latifolia* var. Grisebach, in Mem. Am. Acad. Sci. VIII, Pt. I, 171 (1860).

---

[1] Linnæus' original spelling is *Euonymus* (Hort. Cliff., 39, 1737).

Gyminda grisebachii glaucifolia (Griseb.) nom. nov.—Continued.

Syn.—*Myginda latifolia, forma glaucifolia* Grisebach, Cat. Pl. Cub., 55 (1866).
Gyminda Grisebachii var. *glaucescens* Sargent, in Gard. & For., IV, 4 (1891).

SCHÆFFERIA Jacq., Enum. Pl. Carib., 10 (1760).

Schæfferia frutescens Jacq.   **Boxwood.**

Syn.—*Schæfferia frutescens* Jacquin, Enum. Pl. Carib., 33 (1760).
Schæfferia completa Swartz, Fl. Ind. Occ., I, 327, t. 7, f. A. (1797).
Schæfferia buxifolia Nuttall, Sylva, II, 42, t. 56 (1846).

COMMON NAMES.

Yellow-wood (Fla.).
Boxwood (Fla.).

Family ACERACEÆ.[1]

ACER Linn., Spec. Pl., 1055 (1753).

Acer spicatum Lam.   **Mountain Maple.**

Syn.—*Acer Pennsylvanicum* Du Roi, Diss. Obs. Bot., 61 (1771), not Linn. (1753).
Acer spicatum Lamarck, Enc. Méth. Bot., II, 381 (1786).
Acer parriflorum Ehrhart, Beitr. Nat., IV, 25 (1789).
Acer montanum Aiton, Hort. Kew., ed. 2, III, 435 (1811), not Lam. (1778).
Acer ukurunduense Trautvetter & Meyer, Fl. Ochot., 24 (1856).
Acer rugosum Hort., ex Hand-list Trees and Shr. Arb. Kew, Pt. I, 101 (1894).

COMMON NAMES.

Mountain Maple (Vt., N. H., R. I., Conn., N. Y., N. J., Pa., N. C., S. C., Mich., Minn.).
Moose Maple (Vt.).
Low Maple (Tenn.).
Water Maple (Ky.).

Acer pennsylvanicum Linn.   **Striped Maple.**

Syn.—*Acer Pennsylvanicum* Linnæus, Spec. Pl., ed. 1, II, 1055 (1753).
Acer striatum Du Roi, Diss. Obs. Bot., 58 (1771).

---
[1] Usually included in *Sapindaceæ*.

Acer pennsylvanicum Linn.—Continued.

SYN.—*Acer canadense* Marshall, Arb. Am., 3 (1785).
Acer tricuspifolium Stokes, Bot. Mat. Med., II, 370 (1812).
Acer hybridum Bosc, Dic. Agr., V, 251 (1821).

COMMON NAMES.

Striped Maple (Vt., N. H., R. I., Mass., N. Y., Pa., N. J., S. C.,
Ga., Ky., Mich., Minn., Ont.).
Moosewood (Me., Vt., N. H., R. I., Mass., N. Y., Pa., N. C.,
Mich., Minn.).
Northern Maple (Minn.).
Striped Dogwood (N. Y., N. C.).
Mountain Alder (N. C.).
Whistlewood (Mich.).
Goosefoot Maple.

Acer macrophyllum Pursh.   **Oregon Maple.**

SYN.—*Acer macrophyllum* Pursh, Fl. Am. Sept., I., 267 (1814).
Acer palmatum Rafinesque, New Fl. and Bot., 1st Pt., 48
(1836), not Thunb. (1784).
Acer Murrayanum Hort. (auct. ?), ex Gard. Chron., t. 1632
(1873).
Acer macrophyllum α normalis Kuntze, Rev. Gen. Pl., Par. I,
146 (1891).
Acer macrophyllum β brevialatum Kuntze, l. c. (1891).
Acer macrophyllum γ imbricatum Kuntze, l. c. (1891).
Acer speciosum Hort., ex Hand-list Trees and Shr., Arb.
Kew, Pt. I, 91 (1894).

COMMON NAMES.

Bigleaf Maple (Oreg.).
Broad-leaved Maple (central Cal., Oregon - Willamette
Valley).
Oregon Maple (Oreg., Wash.).
White Maple (Oreg., Wash.).
Maple (Cal.).

Acer circinatum Pursh.   **Vine Maple.**

SYN.—*Acer circinatum* Pursh, Fl. Am. Sept., I, 266 (1814).
Acer virgatum Rafinesque, New Fl. and Bot., 1st Pt., 48
(1836).

COMMON NAMES.

Vine Maple (central and northern Cal., Oregon-Willamette
Valley).
Mountain Maple.

Acer glabrum Torr.   **Dwarf Maple**.

SYN.—*Acer glabrum* Torrey, in Ann. Lyc. N. Y., II, 172 (1828).
Acer barbatum Douglas, in Hooker, Fl. Bor.-Am., I, 113 (1833), not Michx. (1803).
Acer tripartitum (Nutt.! mss.). in Torrey & Gray. Fl. N. A., I, 247 (1838).
Acer Douglasii Hooker, in Lond. Journ. Bot., VI, 77, t. 6 (1847).
Acer glabrum var. *tripartitum* (Nutt.) Pax, in Bot. Jahrb., sieb. B., 218 (1886).
Acer glabrum b. *Douglasii* Dippel, Handb. Laubh., zw. T., 438 (1892).

COMMON NAMES.

Dwarf Maple (Oreg., Utah, Cal., Colo.).
Mountain Maple (Colo., Mont.).
Soft Maple (Utah).
Shrubby Maple (Utah).
Bark Maple (Idaho).
Maple (Mont.).

Acer saccharum Marsh.   **Sugar Maple**.

SYN.—*Acer Saccharum* Marshall, Arb. Am., 4 (1785).
ACER SACCHARINUM Wangenheim, Beitr Nordam. Holz., 36, t. 11, f. 26 (1787), not Linn. (1753).
Acer barbatum Michaux, Fl. Bor.-Am., II., 252 (1803).
Acer saccharophorum Koch, Hort. Dendrol., 80 (1853).

COMMON NAMES.

Sugar Maple (Me., N. H., Vt., Mass., R. I., Conn., N. Y., N. J., Pa., Del., Va., W. Va., N. C., S. C., Ala., La., Ky., Mo., Ohio, Ill., Ind., Iowa, Kans., Nebr., Mich., Minn., Wis., Ont.).
Hard Maple (Vt., R. I., N. Y., N. J., Pa., Va., Ala., Ky., Mo., Kans., Nebr., Ill.. Ind., Iowa. Mich., Ohio, Minn., Ont.).
Sugar-tree (Me., Vt., R. I., Pa., Va., W. Va., Ala., Miss., La., Ark., Ky., Mo., Ill., Ind., Ohio, Kans.).
Rock Maple (Me., Vt., N. H., Conn., Mass., R. I., N. Y., Tenn., Ill., Mich., Iowa, Kans., Wis., Minn., Ont.).
Black Maple (Fla., Ky., N. C.).
Maple (S. C.).

Acer saccharum nigrum (Michx. f.) Britton.   **Black Maple**.

SYN.—*Acer nigrum* Michaux f., Hist. Arb. Am., II., 238. t. 16 (1810).
ACER SACCHARINUM Wang. var. NIGRUM Torrey & Gray, Fl. N. A., I., 248 (1838).

Acer saccharum nigrum (Michx. f.) Britton—Continued.

Syn.—*Acer saccharinum* var. *pseudoplatanoides* Pax, in Bot. Jahrb.
  sieb. B., 242 (1886).
  Acer saccharinum var. *glaucum* Pax, l. c. (1886), not *A. glau-*
  cum Marsh. (1785).
  Acer Rugelii Pax, l. c. 243 (1886).
  Acer saccharinum var. *Rugelii* Wesmæl,. Gen. Acer, 45
  (1890).
  Acer saccharum var. *nigrum* (Michx. f.) Britton, in Trans.
  N. Y. Acad. Sci., IX., 9 (1889).
  Acer barbatum Mx. var. *nigrum* Sargent, in Gard. & For.,
  IV., 148 (1891).
  Acer saccharum var. *barbatum* (Michx.) Trelease, in 5th Ann.
  Rep. Mo. Bot. Gard., 94 (1894).

COMMON NAMES.

Black Sugar Maple (Mich., Mo.).
Black Maple (Mich., Iowa).
Hard Maple (Minn.).

Acer saccharum leucoderme [1] (Small) nom. nov.
Whitebark Maple.

Syn.—*Acer Floridanum* var. *acuminatum* Trelease, Sug. Map. and
  Map. in Wint., 12 (1894), not *A. acuminatum* Don (1825).
  Acer leucoderme Small, in Bull. Torr. Bot. Club, vol. 22,
  p. 367 (1895).

Acer saccharum floridanum (Chapm.) Small & Heller.
Florida Maple.

Syn.—Acer saccharinum Wang. var. Floridanum Chapman,
  Fl. S. States, ed. 1, 81 (1860).
  Acer Mexicanum Gray, in Proc. Am. Acad. Sci., V., 176
  (1862).
  Acer floridanum (Chapm.) Pax, in Bot. Jahrb., sieb. B., 243
  (1886).
  Acer barbatum Chapm. *Floridanum* Sargent, in Gard. & For.,
  IV., 148 (1891).
  Acer Saccharum var. *Floridanum* Small & Heller, in Mem.
  Torr. Bot. Club, III, 24 (1892).

---

[1] A shrub or low-branched tree reported by Dr. J. K. Small in the rocky canyons of
Yadkin River, Stanley County, N. C., and in similar locations on the Yellow River,
Guinett County, Ga. It has also been found sparingly on the sides of a deep rocky
ravine near Pullman, Ala., by Dr. Charles Mohr and the writer. The characters of
this maple are chiefly the habit of branching very near the collar and the white bark.

Acer saccharum grandidentatum (Nutt.) nom. nov.

Large-tooth Maple.

SYN.—ACER GRANDIDENTATUM (Nutt.! mss.), in Torrey & Gray,
Fl. N. A., 247 (1838).
Acer barbatum var. *grandidentatum* Sargent, in Gard. & For.,
IV, 148 (1891).
Acer barbatum var. *floridanum* Sarg., ex Hand-list Trees
and Shrubs Arb. Kew, Pt. I, 89 (1894), not Sargent (1891).
Acer saccharinum var. *floridanum* Chapman, ex Arb. Kew,
l. c. (1894), not Chapm. (1865).

COMMON NAMES.

Western Sugar Maple.
Hard Maple (Utah).
Large-toothed Maple.

FORM DISTINGUISHED IN CULTIVATION.

Acer saccharum nigrum monumentale (Temple) nom. nov.

SYN.—*Acer Saccharum Columnare*[1] Temple, Cat. Trees, Shr. Pl., 5
(1889), not *A. columnaris* Pax (1886).
Acer Saccharinum Monumentale Temple. l. c. (1889).

Acer saccharinum Linn.

Silver Maple.

SYN.—*Acer saccharinum* Linnæus, Spec. Pl., ed. 1, II, 1055 (1753).
Acer rubrum Lauth, De Acere, 11 (1781), not Linn. (1753).
ACER DASYCARPUM Ehrhart, Beitr. Natur., IV, 24 (1789).
Acer rubrum var. *pallidum* Aiton, Hort. Kew., ed. 1, III, 434
(1789).
Acer rubrum mas Schmidt, Oestr. Baumz., I, 11, t. 7 (1792).
Acer eriocarpum Michaux, Fl. Bor.-Am., II, 253 (1803).
Acer virginianum Mill. ex Steudel, Nom. Bot., ed. 1, I, 5
(1821).
Acer tomentosum Desfontaines, Cat. Hort. Par., ed. 3, 136
(1829).
Acer Saira ex Koch, l. c. (1869).
Acer Pavia ex Koch, l. c. (1869).
Acer Pallidum Aiton ex Steudel, Nom. Bot., ed. sec., I, 11
(1840).
Acer palmatum in Hort. ex Nicholson, in Gard. Chron., new
ser., XV, 137 (1881), not Thunb. (1784).

---

[1] This form is very distinct in the narrow columnar shape of its crown. Mr.
Temple describes his original tree as 30 feet high, with its crown only 3 feet in diameter. Trees similar in form of crown are, however, not uncommon in parks and
gardens where the Black Sugar Maple has been extensively planted. A good example
of this narrow-crowned form can be seen in the Smithsonian Grounds, Washington,
D. C.

Acer saccharinum Linn.—Continued.

SYN.—*Acer coccineum* in Hort. ex Nich., l. c. (1881), not Michx.
  (1810).
Acer dasycarpum monstrosum in Hort. ex Nich., l. c. (1881).
Acer floridanum in Hort. ex Nich., l. c. (1881), not *A. sacch.*
  Floridanum Chapm. (1860).
Acer floridum in Hort. ex Nich., l. c. (1881).
Acer macrocarpum in Hort. ex Nich., l. c. (1881), not Opiz
  (1824).
Acer saccharinum floridanum macrophyllum in Hort. ex
  Nich., l. c. (1881), not *A. macrophyllum* Pursh (1814).
Acer saccharinum floridanum palmatum in Hort. ex Nich.,
  l. c. (1881), not *A. palmatum* Thunb. (1784).
Acer virginicum rubrum in Hort. ex Nich., l. c. (1881), not
  A. rubrum Linn. (1753).
Acer dasycarpum macrophyllum Nich., l. c. (1881).
Acer album in Hort. ex Nich., l. c. (1881).
Acer dasycarpum **a.** *normale* Pax, in Bot. Jahrb., sieb. B., 180
  (1885).
Acer macrophyllum Hort ex Pax, l. c. (1885), not Pursh (1814).
Acer spicatum Hort. ex Pax, l. c. (1885), not Lam. (1786).
Acer spec. Kiæchta Hort. ex Dippel, Handb. Laubh., zw. T.,
  436 (1892).
Acer dasycarpum **b.** *longifolium* Dippel, l. c. (1892).
Acer Douglasi Arbor. Zœsch., ex Hand-list Trees and Shr.
  Arb. Kew, Pt. I, 87 (1894), not Hook. (1846).
Acer saccharinum floridanum Hort., ex Arb. Kew, l. c. (1894),
  not Chapm. (1865).
Acer saccharum Marsh., ex Arb. Kew, l. c. (1894), not Marsh.
  (1785).

COMMON NAMES.

Silver Maple (Me., Vt., Mass., R. I., N. Y., N. J., Pa., Del.,
  Va., W. Va., N. C., S. C., Fla., Miss., Ky., Ohio, Ill., Ind.,
  Mo., Kans., Nebr., Iowa, Mich., Minn., S. Dak.).
Soft Maple (Vt., N. H., Mass., R. I., N. Y., N. J., Pa., W. Va.,
  Ala., Miss., La., Tex., Mo., Ohio, Mich., Ont., Ill., Ind.,
  Kans., Nebr., Iowa, Wis., Minn., S. Dak.).
White Maple (Me., Vt., R. I., N. Y., N. J., Pa., W. Va., N. C.,
  S. C., Ga., Fla., Ala., Miss., La., Ky., Mo., Ill., Ind., Kans.,
  Nebr., Minn.).
River Maple (Me., N. H., R. I., W. Va., Minn.).
Silver-leaved Maple (Del., N. J.).
Water Maple (Pa., W. Va.).
Creek Maple (W. Va.).
Swamp Maple (W. Va., Md.).

VARIETIES DISTINGUISHED IN CULTIVATION.

Acer saccharinum pendulum (Nich.) nom. nov.

Weeping Silver Maple.

SYN.—*Acer dasycarpum* var. *pendula* Nicholson, in Gard. Chron., new ser., XV, 137 (1881).
Acer dasycarpum **b.** *cuneatum* Pax, in Bot. Jahrb.. sieb. B., 180 (1885).
Acer pendulum Hort. ex Pax, l. c. (1885).
Acer longifolium Hort. ex Pax. l. c. (1885).

Acer saccharinum aureo-variegatum (Nich.) nom. nov.

Variegated Silver Maple.

SYN.—*Acer dasycarpum* var. *aureo-variegata* Nicholson, in Gard. Chron., new ser., XV, 137 (1881).
Acer pulverulentum Wittmack, in Gartenz., 513 (1883), not Koch (1869).
Acer albo-variegatum Hort. ex Wittmack, l. c. (1883).
Acer dasycarpum **c.** *albo-maculatum* Pax, in Bot. Jahrb., sieb. B., 180 (1885).
Acer dasycarpum **e.** *pulverulentum* Dippel, Handb. Laubh., zw. T., 436 (1892).
Acer dasycarpum fol. variegatum Dippel, l. c. (1892).
Acer dasycarpum fol. albo-variegatis Späth ex Dippel, l. c. (1892).
Acer dasycarpum fol. aureo-variegatis Hort. ex Dippel l. c., (1892).

Acer saccharinum wierii (Pax) nom. nov.   **Cutleaf Silver Maple.**

SYN.—*Acer Wageneri laciniatum* (French Nurs.) ex Koch, Dendrol., erst. Th., 541 (1869), not Lauth (1781), nor *A. laciniatum* Borkh. (1795).
Acer Wagneri laciniatum in Hort. ex Nicholson, in Gard. Chron., new ser., XV, 137 (1881).
Acer Weiri laciniatum in Hort. ex Nich., l. c. (1881), not of auth. cit. supra.
Acer heterophyllum laciniatum (Auct.?) ex Garden, XX, 167, t. (1881), not of auth. cit. supra.
Acer dasycarpum **d.** *laciniatum* Pax, in Bot. Jahrb., sieb. B., 180 (1885), not of auth. cit. supra.
Acer laciniatum Wierii Hort. ex Pax. l. c. (1885).
Acer dasycarpum Wierii laciniatum Hort. ex Dippel, Handb. Laubh., zw. T., 436 (1892), not of auth. cit. supra.
Acer heterophyllum Hort. ex Dippel. l. c. (1892), not Willd. (1796).

Acer saccharinum dissectifolium nom. nov.

SYN.—*Acer Wagneri dissectum* in Hort. ex Nicholson, in Gard
Chron., new. ser., XV, 137 (1881), not *A. dissectum* Thunb.
(1784).
Acer dasycarpum e. *dissectum* Pax, in Bot. Jahrb., sieb. B.,
180 (1885), not *A. dissectum* Thumb. (1784).
Acer dissectum Wagneri Hort. ex Pax, l. c. (1885), not *A.
Wageneri* ex Koch (1869).

Acer saccharinum lutescens (Pax) nom. nov.

SYN.—*Acer lutescens* Pax, Bot. Jahrb., sieb. B., 180 (1885).
Acer dasycarpum d. lutescens Dippel, Handb. Laubh., zw. T.,
436 (1892).

Acer saccharinum novum (Ellw. & Barr.) nom. nov.

Crisp-leaf Silver Maple.

SYN.—*Acer crispum novum* Hort. Ellwanger & Barry, ex Garden,
XX, 167, f. (1881).

Acer rubrum Linn.   **Red Maple.**

SYN.—*Acer rubrum* Linnæus, Spec. Pl., ed. 1, II, 1055 (1753).
Acer glaucum Marshall, Arb. Am., 2 (1785).
Acer Caroliniana Walter, Fl. Caroliniana, 251 (1788).
Acer coccineum Michaux f., Hist. Arb. Am., II, 203 (1810).
Acer dasycarpum β glabrum Ascherson, Fl. Brand.,116(1864),
not *A. glabrum* Torr. (1828).
Acer Wageneri in Hort. ex Koch, Dendrol., erst. Th., 542
(1869).
Acer hybridum ex Koch, l. c. (1869), not Spach (1834), nor
Loud. (1838).
Acer hypoleucum (in Germ. Nurs.) ex Koch, l. c. (1869).
Acer leucophyllum in Hort. ex Koch, l. c. (1869).
Acer microphyllum Pax, in Bot. Jahrb., sieb. B., 180
(1885).
Acer semiorbiculatum Pax, l. c., 181 (1885).
Acer rubrum var. *clausum* Pax, l. c., 182 (1885).
Acer rubrum var. *pallidiflorum* (Koch) Pax, l. c. (1885).
Acer rubrum var. *tomentosum* (Hort.) Pax, l. c. (1885).
Acer tomentosum Hort. ex Pax, l. c. (1885), not Desf. (1829).
Acer rubrum var. *semiorbiculatum* Wesmæl, Gen. Acer., 13
(1890).
Acer rubrum var. *microphyllum* Wesmæl, l. c. (1890).
Acer globosum Hort., ex Hand-list Trees and Shr., Arb. Kew,
Pt. I, 99 (1894).
Acer virginianum Hort., l. c. (1894).

Acer rubrum Linn.—Continued.

COMMON NAMES.

SYN.—Red Maple (Me., N. H., Vt., Mass., R. I., Conn., N. Y., N. J.,
Pa., Del., Va., W. Va., N. C., S. C., Ga., Fla., Ala., Miss.,
La., Tex., Ky., Mo., Ill., Iud., Ohio, Ont., Iowa, Wis.,
Nebr.).
Swamp Maple (Vt., N. H., Mass., Conn., R. I., N. Y., N. J.,
Pa., Del., N. C., S. C., Fla., Ala., Miss., La., Tex., Mo.,
Ind., Ont., Minn.).
Soft Maple (Vt., Mass., N. Y., Va., Miss., Mo., Ill., Ind.,
Ohio, Ont., Mich., Kans., Nebr., Minn.).
Water Maple (Miss., La., Tex., Ky., Mo.).
White Maple (Me., N. H.).
Shoe-peg Maple (W. Va.).
Erable (La.).
Ah-weh-hot-kwah = "Red flower" (Onondaga Indians, N. Y.).
Scarlet Maple (Tex.).

Acer rubrum drummondii (Hook. & Arn.) Sargent.

Drummond Maple.

SYN.—*Acer Drummondii* Hooker & Arnott, in Hooker, Journ. Bot.,
I, 200 (1834).
Acer rubrum L. var. *Drummondii* Sargent, in Tenth Cen.
U. S., IX (Cat. For. Trees N. A.), 50 (1884).

VARIETY DISTINGUISHED IN CULTIVATION.

Acer rubrum sanguineum (Spach) Pax.

SYN.—*Acer sanguineum* Spach, in Ann. Sc. Nat., sér. 2, 176 (1834).
Acer floridanum ex Koch, Dendrol., erst. Th., 542 (1869),
not *A. saccharin.* var. *Floridanum* Chapm. (1860), nor Koch,
l. c., 541 (1869).
Acer splendens ex Koch, l. c. (1869).
Acer glaucum ex Koch, l. c. 543 (1869), not Marsh. (1785).
Acer rubrum var. *eurubrum* Pax, in Bot. Jahrb., sieb. B.,
181 (1885).
Acer fulgens Hort. ex Pax, l. c. 182 (1885).
Acer rubrum var. *sanguineum* (Spach) Pax, l. c. (1885).
Acer coccineum Hort. ex Pax, l. c. (1885), not Michx. (1810).
Acer rubrum a *coccineum* Dippel, Handb. Laubh., zw. T.,
435 (1892).

Acer negundo Linn. **Boxelder.**

SYN.—*Acer Negundo* Linnæus, Spec. Pl., ed. 1, II, 1056 (1753).
NEGUNDO ACEROIDES Moench, Meth., 334 (1794).

Acer negundo Linn.—Continued.

SYN.—*Negundium fraxinifolium* Rafinesque, Med. Rep., V, 354 (1808).
Negundo Fraxinifolium Nuttall, Genera, I, 253 (1818).
Negundo Mexicanum de Candolle, Prodr., I, 596 (1824).
Negundo trifoliatum Rafinesque, New Fl. and Bot., 1st Pt., 48 (1836).
Negundo lobatum Raf., l. c. (1836).
Negundo Californicum Scheele, in Roemer, Texas, 433 (1849), not Torr. & Gr. (1838).
Negundo Negundo Karsten, Pharm. Med. Bot., 596 (1883).
Acer Negundo var. *Texanum* Pax, in Bot. Jahrb., sieb. B., 212 (1886).
Acer Negundo α *normals* Kuntze, Rev. Gen. Pl., Par. I, 146 (1891).
Acer Negundo β *trifoliatum* Kuntze, l. c. (1891).
Rulac Negundo Hitchcock, Spring Fl. Manhattan, 6 (1894).
Acer Negundo var. *Guichardi* [1] Hort., ex Hand-list Trees and Shr., Arb. Kew, Pt. I, 91 (1894).

COMMON NAMES.

Boxelder (Vt., Mass., R. I., Del., N. Y., N. J., Pa., Va., W. Va., N. C., S. C., Ala., Fla., Miss., La., Tex., Ark., Mo., Ariz., N. Mex., Mont., Ill., Ind., Wis., Ohio, Mich., Iowa, Kans., Nebr., N. Dak., S. Dak., Minn.).
Ash-leaved Maple (R. I., Mass., N. J., Pa., Del., Va., S. C., La., Tex., Ill., Wis., Iowa, Ont., Kans., Nebr., Mont., N. Dak., Mich., Minn.).
Cut-leaved Maple (Colo.).
Negundo Maple (Ill.).
Red River Maple (N. Dak.).
Three-leaved Maple (Pa.).
Black Ash (Tenn.).
Stinking Ash (S. C.).
Sugar Ash (Fla.).
Water Ash (Dakotas).

Acer negundo californicum (T. & Gr.) Sargent.

California Boxelder.

SYN.—NEGUNDO CALIFORNICUM Torrey & Gray, Fl. N. A., I, 250 (1838).

---

[1] This form, cultivated as a distinct variety at the Royal Gardens, Kew, England, appears not to be known elsewhere. The catalogue name (l. c.) is unaccompanied characters of the plant.

Acer negundo californicum (T. & Gr.) Sargent—Continued.

SYN.—*Negundo aceroides* Torrey, in Pacif. R. R. Rep., IV, 74 (1857),
not Moench (1794).
Acer Californicum Dietrich, Syn. Pl., II, 1283 (? 1843).
Negundo aceroides var. *Californicum* Sargent, in Gard. & For.,
II, 364 (1889).
Acer Negundo var. *Californicum* Sarg., in Gard & For., IV,
148 (1891).
Acer Negundo γ *mexicanum* Kuntze, Rev. Gen. Pl., Par. I,
146 (1891).
Acer Negundo δ *Parishianum* Kuntze, l. c. (1891).

COMMON NAMES.

Box Elder (Cal., N. Mex.).
Maple (Cal.).
False Maple (Cal.).

VARIETIES DISTINGUISHED IN CULTIVATION.

Acer negundo variegatum Kuntze.

SYN.—*Acer Negundo variegatum* Kuntze, Fl. Leipz., 200 (1867).
Acer Negundo var. *vulgare* a *bicolor* Pax, in Bot. Jahrb.,
sieb. B., 211 (1886).
Acer aureo-variegatum Hort. ex Pax, l. c. (1886).
Acer argenteo-variegatum Hort. ex Pax, l. c. (1886).
Acer Negundo d *variegatum* Dippel, Handb. Laubh., zw. T.,
467 (1892).
Acer foliis albo-marginatis Hort. ex Dippel, l. c. (1892).

Acer negundo angustissimum (Pax) nom. nov.

SYN.—*Acer Negundo* var. *vulgare* b *angustissimum* Pax. in Bot.
Jahrb., sieb. B., 211 (1886).
Acer Negundo heterophyllum Hort. ex Dippel, Handb. Laubh.,
zw. T., 467 (1892), not *A. heterophyllum* Willd. (1796).
Acer Negundo laciniatum Kuntze, Fl. Leipz., 200 (1867), not
A. laciniatum Lauth. (1781).

Acer negundo crispifolium nom. nov. **Curl-leaf Boxelder.**

SYN.—*Negundo fraxinifolium* β *crispa* Don, Hist. Dichl. Pl., I, 651
(1831).
Acer Negundo crispum Kuntze, Fl. Leipz., 200 (1867), not *A.*
crispum Moench (1785).
Acer crispum ex Koch, Dendrol., erst. Th., 544 (1869), not
Moench (1785).

r negundo violaceum (Koch) Dippel.

Syn.—*Negundo violaceum* ex Koch, Dendrol., erst. Th., 545 (1869).
Acer Negundo var. *vulgare* Pax, in Bot. Jahrb., sieb. B., 211 (1886).
Acer californicum Hort. ex Pax, l. c. (1886), not Dietr. (1843).
Acer versicolor Hort. ex Pax, l. c. (1886).
Acer violaceum Hort. ex Pax, l. c. (1886).
Acer Negundo b *violaceum* Dippel, Handb. Laubh., zw. T., 467 (1892).
Negundo californicum Hort. ex Dippel, l. c. (1892), not Torr. & Gr. (1838).

Family HIPPOCASTANACEÆ.

ÆSCULUS Linn., Spec. Pl., ed. 1, I, 344 (1753).

:ulus glabra Willd.   Ohio Buckeye.

Syn.—*Æsculus glabra* Willdenow, Enum. Pl. Hort. Berol., 405 (1809).
Æsculus pallida Willd., l. c., 406 (1809).
Æsculus echinata Muhlenberg, Cat. Pl. Am. Sep., 38 (1813).
Pavia Ohioenses Michaux f., Hist. Arb. Am., III, 242 (1813).
Æsculus Ohioenses de Candolle, Prodr., I, 597 (1824).
Pavia pallida Spach, in Ann. Sc. Nat., sér. 2, II, 54 (1834).
Pavia glabra Spach, l. c. (1834).
Paria Watsoniana Spach, l. c. (1834).—?
Æsculus muricata Rafinesque, Alsograph. Am., 68 (1838).
Æsculus ochroleuca Raf., l. c. (1838).
Æsculus verrucosa Raf., l. c. 69 (1838).
Æsculus alba Raf., l. c. (1838).
Pavia reticulata Raf., l. c. 73 (1838).
Æsculus Hippocastanum var. *Ohioensis* Loudon, Arb. Frut., I, 467 (1838).
Æsculus Hippocastanum glabra Loud., l. c. (1838).
Æsculus rosea Hort. ex Loud., Enc. Trees, 127 (1842).
Æsculus Watsoniana Dietrich, Syn. Pl., II, 1225 (1852).
Æsculus arguta Buckley, in Proc. Acad. Sci. Phila. 1860, 443 (1860).

COMMON NAMES.

Ohio Buckeye (Miss., Ga., Ark., Mo., Ohio).
Buckeye (Pa., Ky., Mo., Ill., Ind., Ohio, Iowa, Kans.).
Fetid Buckeye (W. Va.).
Stinking Buckeye (Ala., Ark.).
American Horse Chestnut (Pa.).

Æsculus octandra Marsh.   **Yellow Buckeye.**

Syn.—*Æsculus octandra* Marshall, Arb. Am., 4 (1785).
Æsculus lutea Wangenheim, in Schr. Ges. Nat. Fr. Berl.,
     VIII, 133, t. 6 (1788).
ÆSCULUS FLAVA Aiton, Hort. Kew., I, 494 (1789).
Pavia flava Moench, Meth., 66 (1794).
Pavia lutea Poiret, in Lamarck, Enc. Méth. Bot., V, 94 (1804).
Paviana flava Rafinesque, Fl. Ludovic., 87 (1817).
Æsculus neglecta Lindley, in Bot. Reg., XII, t. 1009 (1826).
Pavia neglecta Don in Loudon, Hort. Brit., I, 143 (1830).
Pavia fulva Rafinesque, Alsograph. Am., 74 (1838).
Pavia bicolor Raf., l. c. (1838).
Æsculus flava aurantia Browne. Trees Am., 118 (1846).
Pavia octandra Kuntze, Rev. Gen. Pl., Par. I, 146 (1891).
Æsculus lucida Hort. ex Dippel, Handb. Laubh., zw. T., 401
     (1892).
Æsculus Michauxii Hort. ex Dippel, l. c. 403 (1892).
Pavia Michauxii Hort. ex Dippel, l. c. (1892).
Æsculus marylandica Booth ex Dippel, l. c. (1892).
Æsculus lucida Hort., ex Hand-list Trees and Shr., Arb. Kew,
     Pt. I, 83 (1894).
Pavia macrocarpa Hort., ex Arb. Kew, l. c. (1894).
Pavia pallida bicolor Hort., ex Arb. Kew, l. c. (1894).

COMMON NAMES.

Buckeye (N. C., S. C., Ala., Miss., La., Tex., Ky.).
Sweet Buckeye (W. Va., Miss., Tex., Mo., Ind.).
Yellow Buckeye (S. C., Ala.).
Large Buckeye (Tenn.).
Big Buckeye (Tex., Tenn.).
Ohio Buckeye (Pa., cult.).

Æsculus octandra hybrida (de C.) Sargent.   **Purple Buckeye.**

Syn.—*Æsculus hybrida* de Candolle, Cat. Hort. Monsp., 75 (1813).
Æsculus discolor Pursh, Fl. Am. Sept., I, 255 (1814).
Pavia discolor Poiret, in Lamarck. Enc. Méth. Bot., Suppl.,
     V, 769 (1817).
Pavia hybrida de Candolle, Prodr., I, 598 (1824).
Æsculus pavia var. *discolor* Torrey & Gray, Fl. N. A., I,
     252 (1838).
ÆSCULUS FLAVA Ait. var. PURPURASCENS Gray, Man. Bot.
     N. U. S., ed. 5, 118 (1867).
Æsculus octandra var. *hybrida* Sargent, Silva, II, 60 (1891).

Æsculus californica (Spach) Nutt.   **California Buckeye.**

Syn.—*Calothyrsus Californica* Spach, in Ann. Sc. Nat., sér. 2,
     II, 62 (1834).

Æsculus californica (Spach) Nutt.—Continued.

Syn.—*Pavia Californica* Hartweg, in Journ. Hort. Soc. Lond., II, 123 (1847).
Æsculus Californica (Nutt.!), in Torrey & Gray, Fl. N. A., I, 251 (1838).

COMMON NAMES.

California Buckeye (Cal., Nev., Colo.).
Horse Chestnut (Cal.).

Family SAPINDACEÆ.

UNGNADIA Endl., Atacta Bot., t. 36 (1833).

Ungnadia speciosa Endl.   **Texas Buckeye.**

Syn.—*Ungnadia speciosa* Endlicher, Atacta Bot., t. 36 (1833).
Ungnadia heterophylla Scheele, in Linnæa, XXI, 589 (1848).
Ungnadia heptaphylla Scheele, l. c., XXII, 352 (1849).

COMMON NAMES.

Spanish Buckeye (Tex.).
Texas Buckeye (Tex.).

SAPINDUS Linn., Spec. Pl., 367 (1753).

Sapindus saponaria Linn.   **Soapberry.**

Syn.—*Sapindus Saponari* Linnæus, Spec. Pl., ed. 1, I, 367 (1753).

COMMON NAMES.

False Dogwood (Fla.).
Soapberry (Fla.).

Sapindus marginatus Willd.   **Wild China.**

Syn.—*Sapindus Saponaria* Lamarck, Ill., II, 441, t. 307 (1793), not Linn. (1753).
Sapindus marginatus Willdenow, Enum. Pl., I, 432 (1809).
Sapindus falcatus Rafinesque, Med. Bot., II, 261 (1830).
Sapindus acuminatus Raf., New Fl. Bot., 3d Pt., 22 (1836).
Sapindus Drummondi Hooker & Arnott, in Bot. Beechey Voyage, 281, excl. var. (1841).
Sapindus Manatensis Radlkofer, Sitz. Akad. Münch. 1878, 318, 400 (1878).

COMMON NAMES.

Soapberry (N. C., S. C., Fla., Miss., La., Tex., Ark., Kans.).
Wild China (Fla., Miss., La., Tex., Iowa, Kans.).
Chinaberry (N. Mex.).

EXOTHEA Macfadyen. Fl. Jam., 232 (1837).

Exothea paniculata (Jussieu) Radlkofer. Inkwood.

SYN.—*Meliococca paniculata* Jussieu, in Mém. Mus. Hist. Nat., III,
187, t. 5 (1817).
Sapindus lucidus Hamilton. Prodr. Pl. Ind. Occid., 36 (1825).
HYPELATE PANICULATA Cambesedes, in Mém. Mus. Hist.
Nat., XVIII, 32 (1829).
Hypelate oblongifolia Hooker, Lond. Journ. Bot., III, 7 (1844).
Exothea oblongifolia Macfadyen, Fl. Jamaica, 223 (1837).
Exothea paniculata Walpers, Rep., V, 366 (1846).
Exothea paniculata Radlkofer, in Durand, Index Gen. Phan.,
81 (1888).

COMMON NAMES.

Inkwood (Fla.).
Ironwood.

HYPELATE Browne, Nat. Hist. Jam., 208 (1756).

Hypelate trifoliata Swartz. White Ironwood.

SYN.—*Hypelate trifoliata* Swartz, Prodr., 61 (1788).
Amyris Hypelate Robinson, in Lunan. Hort. Jam., I, 149
(1814).
Amyris Lypelata Roxb. ex Stendel, Nom. Bot., ed. sec.. I. 81
(1840).

COMMON NAME.

White Ironwood (Fla.).

Family RHAMNACEÆ.[1]

REYNOSIA Griseb., Cat. Pl. Cub., 33 (1866).

Reynosia latifolia Griseb. Red Ironwood.

SYN.—*Scutia ferrea* Brong. ex Chapman. Fl. S. States, ed. 1, 72
1860), not Brong. (1826).
Reynosia latifolia Grisebach, Cat. Pl. Cub., 34 (1866).

[1] *Zizyphus Mexicana* Rose (in Contr. U. S. Nat. Herb., I, No. 9, 315, 1895) is an interesting arborescent species of this family recently described from the region of Armeria, Mexico. Dr. Edward Palmer collected specimens in 1890-91. It is reported as a tree 30 feet in height, with a diameter of 8 inches. The drupe-like fruit, one-half to two-thirds of an inch long, is much used by the Mexicans as a substitute for soap in washing woolen goods, being sold by the dozen under the name "Amole." All the species of *Zizyphus* thus far known to occur within the borders of the United States are shrubs.

Reynosia latifolia Griseb.—Continued.

SYN.—*Rhamnidium* Wright ex *revolutum* Chapman, Fl. S. States, Suppl., 612 (1887), not Griseb. (1866).

COMMON NAMES.

Red Ironwood (Fla.).
Darling Plum (Fla.).

CONDALIA Cavanilles, in Anal. Cienc. Hist. Nat., I, 39 (1799).

Condalia obovata Hook.   **Bluewood.**

SYN.—*Condalia obovata* Hooker, in Icones, III, t. 287 (1840).

COMMON NAMES.

Bluewood (Tex.).
Logwood (Tex.).
Purple Haw (Tex.).

RHAMNIDIUM Reissek, in Martins, Fl. Braz., XI, I, 94 (1861).

Rhamnidium ferreum (Vahl) Sargent.   **Black Ironwood.**

SYN.—*Rhamnus ferrea* Vahl, Symb. Bot., III, 41, t. 58 (1794).
Myginda integrifolia Lamarck, Enc. Méth. Bot., IV, 396 (1797).
Zizyphus emarginatus Swartz, Fl. Ind. Occ., III, 1954 (1806).
Ceanothus ferreus de Candolle, Prodr., II, 30 (1825).
Scutea ferrea Brongniart, in Mém. Rham., 56 (1826), not Chapm. (1860).
CONDALIA FERREA Grisebach, Fl. Brit. West Ind., 100 (1864).
Rhamnidium ferreum Sargent, in Gard. and For., IV, 16 (1891).

COMMON NAME.

Black Ironwood (Fla.).

RHAMNUS Linnæus, Spec. Pl., ed. 1, I, 193 (1753).

Rhamnus crocea Nutt.   **Evergreen Bearwood.**

SYN.—*Rhamnus croceus* (Nutt.! mss.), in Torrey & Gray, Fl. N. A. I, 261 (1838).
Rhamnus ilicifolia Kellogg, in Proc. Cal. Acad. Sci., II, 37, (1863).
Rhamnus crocea var. *ilicifolia* Greene, Fl. Fran., Pt. I, 79 (1891).

Rhamnus crocea insularis (Greene) Sargent.

SYN.—*Rhamnus crocea* Lyon, in Bot. Gaz., XI, 333 (1886), not Nutt., in T. & G. (1838).
Rhamnus insularis Greene, in Bull. Cal. Acad. Sci., II, 392 (1889).
Rhamnus crocea var. *insularis* Sargent, in Gard. and For., II, 364 (1889).

Rhamnus pirifolia[1] Greene.

SYN.—*Rhamnus pirifolia* Greene, in Pittonia, III. Pt. 13, 15 (1896).

Rhamnus caroliniana Walter.   **Indian Cherry.**

SYN.—*Rhamnus Caroliniana* Walter, Fl. Caroliniana, 101 (1788).
Frangula fragilis Rafinesque, Fl. Ludovic., 97 (1817).
Sarcomphalus Carolinianus Raf., Sylva Tellur., 29 (1838).
Frangula Caroliniana Gray, Genera, II, 178, t. 167 (1849).

COMMON NAMES.

Indian Cherry (W. Va., N. C., Miss., La., Tex., Ark., Nebr.).
Buckthorn (Ark., Iowa, Nebr.).
Alder Buckthorn (Tex., Nebr.).
Yellow-wood (Ala., Fla., La.).
Stinkwood (La.).
Bog Birch (Minn.).
Stink Berry (Nebr.).
Stink Cherry (Nebr.).
Carolina Buckthorn (S. C., Pa.).
Polecat-tree (Tex.).
Polecat-wood (Ark.).
Brittlewood (Ark.).

Rhamnus purshiana[2] de C.   **Bearberry.**

SYN.—*Rhamnus alnifolia* Pursh. Fl. Am. Sept., I, 166 (1814), not L'Hér. (1788).

---

[1] This species is introduced on the authority of Prof. E. L. Greene, who describes it as distinct from his *Rhamnus insularis*, to which he had referred this plant, and states that it is a tree 20 feet high peculiar to the Santa Cruz Islands, off the coast of California. I have not been able to examine specimens of it.

[2] The following is a shrubby variety occurring in Mexico, Arizona, and southern California:

Rhamnus purshiana tomentella (Benth.) Sargent.   **Bitterbark.**
SYN.—*Rhamnus tomentella* Bentham, Pl. Hartweg., 303 (1849).
Frangula Californica var. *tomentella* Gray, Pl. Wright., Pt. II, 28 (1853).
Rhamnus Californica var. *tomentella* Brewer & Watson, Bot. Cal., I, 101 (1880).
Rhamnus Californica Hemsley, Bot. Biol. Am. Cent., I, 197 (1879).
Rhamnus Purshiana var. *tomentella* Sargent, Silva, II, 39 (1891).

Rhamnus purshiana de C.—Continued.

SYN.—*Rhamnus Purshiana* de Candolle, Prodr., II, 25 (1825).
Rhamnus Californica Eschscholtz, in Mém. Acad. St.
Pétersb., X, 281 (1826).
Rhamnus oleifolia Hooker, Fl. Bor.-Am., I, 123, t. 44 (1833).
Cardiolepis obtusa Rafinesque, Sylva Tellur., 28 (1838).
Perfonon laurifolium Raf., l. c., 29 (1838).
Endotropis oleifolia Raf., l. c., 31 (1838).
Rhamnus laurifolius (Nutt.! mss.), in Torrey & Gray, Fl.
N. A., I, 260 (1838).
Frangula Californica Gray, Genera, II, 178 (1849).
Frangula Purshiana Cooper, in Smithsonian Rep., 1858, 259
(1859).

COMMON NAMES.

Shittimwood (Oreg., Idaho, Wash.).
Cascara Sagrada (Cal., Oreg.).
Bearberry (Oreg., Idaho, Wash.).
Bearwood (Oreg.).
Yellow-wood (Oreg.).
Buckthorn (Idaho).
Pigeon-berry (Idaho).
Oregon Bearwood (Oreg. Wash.).
Coffee-berry (Cal.).
Wild Coffee-bush (Cal.).
Western Coffee (Oreg., Cal.).
Bayberry (Oreg., Cal.).
Wild Coffee (Cal.).
California Coffee (Cal.).

CEANOTHUS Linnæus, Spec. Pl., ed. 1, 195 (1753).

Ceanothus thyrsiflorus[1] Esch. **Blue Myrtle.**

SYN.—*Ceanothus thyrsiflorus* Eschscholtz, in Mém. Acad. St.
Pétersb., X, 285 (1826).
Ceanothus bicolor Rafinesque, New Fl. Bot., 3d Pt., 57 (1856),
in part.
Ceanothus elegans Lemaire, Ill. Hort., VII, t. 268 (1860).

COMMON NAMES.

Blue Myrtle.
California Lilac (Cal.).
Wild Lilac (Cal.).
Blue Blossom (Cal.).

---

[1] Katherine Brandegee (in Proc. Cal. Acad. Sci., ser. 2, IV) notes the following
hybrid of *C. thyrsiflorus* and *C. dentatus*: *Ceanothus Lobbianus* Hooker (in Bot. Mag.,
t. 4811, 1854).

Ceanothus arboreus Greene. **Tree Myrtle.**

SYN.—*Ceanothus sorediatus* Lyon, in Bot. Gaz., XI, 204, 333 (1886),
not Hook. & Arn. (1841).
Ceanothus arboreus Greene, in Bull. Cal. Acad. Sci.. II. 144
(1887).
Ceanothus velutinus var. *arboreus* Sargent, in Gard. and For.,
II, 364 (1889).

COLUBRINA [1] Brogn., in Ann. Sc. Nat., sér. 1, X, 364 (1827).

Colubrina reclinata (L'Hér.) Brongn. **Naked-wood.**

SYN.—*Ceanothus reclinatus* L'Héritier, Sert. Ang., 4 (1788).
Rhamnus elliptica Swartz, Prodr., 50 (1788).
Zizyphus Domingensis Nouveau Duhamel, III, 56 (1807).
Colubrina reclinata Brongniart, in Ann. Sc. Nat., sér. 1, x,
364 (1827).
Diplisca elliptica Rafinesque, Sylva Tellur., 31 (1838).

COMMON NAMES.

Naked-wood (Fla.).
Soldierwood (Fla.).

Family TILIACEÆ.

TILIA Linn., Spec. Pl., 514 (1753).

Tilia americana Linn. **Basswood.**

SYN.—*Tilia Americana* Linnæus, Spec. Pl., ed. 1, I, 514 (1753).
Tilia Caroliniana Miller, Gard. Dict., ed. 8, No. 4 (1768).
Tilia latifolia Salisbury, Prodr., 367 (1796).
Tilia nigra Borkhausen, Handb. Forstb., II, 1219 (1800).
Tilia pubescens Nouveau Duhamel, I., t. 51, (1801), not Ait.
(1789.)
Tilia glabra Ventenat, in Mém. Acad. Sc. Par., IV. 9, t. 2,
(1803).
Tilia Canadensis Michaux., Fl. Bor.-Am.. I, 306 (1803).

---

[1] The following are shrubs:

Colubrina texensis Gray.

SYN.—*Colubrina Texensis* Gray, in Journ. Bost. Soc. Nat. Hist., VI., 169 (1850)
Colubrina Greggii Watson, in Proc. Am. Acad. Sci., XVII, 336 (1882).

Colubrina colubrina (Linn.) Sargent.

SYN.—*Rhamnus colubrinus* Linnæus, Spec. Pl., ed. 2, I, 280 (1762).
Ceanothus colubrinus Lamarck, Ill., II, 90 (1793).
Colubrina ferruginosa Brongniart, in Ann. Sc. Nat., sér. 1, X, 369 (1827).
Colubrina colubrina Sargent, Silva, II, 47 (1891.)

Tilia americana Linn.—Continued.

Syn.—*Tilia stenopetala* Rafinesque, Fl. Ludovic., 92, (1817).
Tilia Mississippiensis Bosc, Enc. Ag., VII, 748 (1821).
Tilia longifolia Raf., Alsograph. Am., 44 (1838).
Tilia fuscata Raf., l. c., 45 (1838).
Tilia riparia Raf., l. c. (1838).
Tilia pubescens var. *macrophylla* Smith ex Raf., l. c., 47 (1838).
Tilia neglecta Spach, in Ann. Sc. Nat., sér. 2, II, 340, t. 15 (1834).
Tilia nigra laxiflora Spach, l. c. (1834).
Tilia Americana 2 laxiflora Loudon, Arb. Frut., I, 374 (1838), not *T. laxiflora* Michx. (1803).
Tilia præcox Braun, in Bot. Zeit., I, 586 (1843).—?
Tilia macrophylla in Hort. ex Koch, Dendrol., erst. Th., 480 (1869).
Tilia Ludovicia Bosc ex Koch, l. c., 481 (1869).
Tilia multiflora Hort. ex Ventenat, in Mém. Acad. Sc. Par., IV, 12 (1803).
Tilia gigantea Hort. ex Dippel, l. c. (1803).
Tilia Americana gigantea Hort. ex Dippel, l. c. (1803).
Tilia Americana multiflora Hort. ex Dippel, l. c. (1893).
Tilia Americana **b** *pubescens* Dippel, l. c. (1893), not *T. pubescens* Ait. (1789.)
Tilia Americana **c** *Moltkei* Hort. Späth ex Dippel, l. c., 66 (1893).
Tilia Americana **d** *Rosenthallii* Hort. ex Dippel, l. c. (1893).
Tilia gigantea Hort. ex Dippel, l. c., 67 (1893).
Tilia hybrida superba Hort., ex Hand-list Trees and Shr. Arb. Kew, Pt. I, 45 (1894).
Tilia longifolia dentata Hort., l. c. (1894).

COMMON NAMES.

Basswood (Me., N. H., Vt., R. I., Mass., Conn., N. Y., N. J., Del., Pa., W. Va., D. C., N. C., S. C., Ga., Ala., Miss., La., Ark., Ky., Ill., Ind., Iowa., Wis., Mich., Ohio, Ont., Nebr., Kans., Minn., N. Dak.).
American Linden (Me., N. H., R. I., N. Y., Pa., Del., N. C., Miss., Ohio, Ill., Nebr., N. Dak., Ont., Minn.).
Linn (Pa., Va., W. Va., Ala., La., Ill., Ind., Ohio, Mo., Iowa, Kans., Nebr., Wis., S. Dak.).
Linden (Vt., R. I., Pa., W. Va., Nebr., Minn.).
Limetree (R. I., N. C., S. C., Ala., Miss., La., Ill.).
Whitewood (Vt., W. Va., Ark., Minn., Ont.).
Beetree (Vt., W. Va., Wis.).
Black Limetree (Tenn.).
Smooth-leaved Limetree (Tenn.).

Tilia americana Linn.—Continued.

SYN.—White Lind (W. Va.).
Wickup (Mass.).
Yellow Basswood (Ind.).
Lein (Ind.).

Tilia pubescens Ait. **Downy Basswood.**

SYN.—*Tilia Americana* Walter, Fl. Caroliniana, 153 (1788), not
Linn. (1753).
Tilia pubescens Aiton, Hort. Kew., II, 229 (1789).
Tilia grata Salisbury, Prodr., 367 (1796).
Tilia laxiflora Michaux, Fl. Bor.-Am., I, 306 (1803).
Tilia truncata Spach, Ann. Sc. Nat., sér. 2, II, 342 (1834).
TILIA AMERICANA 3 PUBESCENS Loudon, Arb. Frut., I,
374, t. (1838).
Tilia Americana β *Walteri* Wood, Cl. Book, 272 (1855).

COMMON NAME.

Wahoo (Fla.).

Tilia pubescens leptophylla[1] Vent.

Thin leaf Downy Basswood.

SYN.—*Tilia pubescens leptophylla* Ventenat, in Mém. Acad. Sc. Par.,
IV, 11 (1803).
Tilia americana 4 *pubescens leptophylla* Loudon, Arb. Frut.,
I, 375 (1838).
Tilia mississippiensis Desf. ex Loud., l. c. (1838), not Bosc
(1821).
Tilia leptophylla Hort. ex Bayer, in Verh. Zoolog. Bot.
Gesell. Wien, XII, 58 (1862).

Tilia heterophylla Vent. **White Basswood.**

SYN.—*Tilia heterophylla* Ventenat, in Mém. Acad. Sc. Par., IV, 16,
t. 5 (1802).
Tilia alba Michaux f., Hist. Arb. Am., III, 315, t. 2 (1813),
not Ait. (1789).
Tilia americana 5 *heterophylla* Loudon, Arb. Frut., I, 375, t.
23 (1838).
Tilia glauca Rafinesque, Alsograph. Am., 45 (1838).
Tilia fulva Raf., l. c. (1838).
Tilia cinera Raf., l. c., 46 (1838).
Tilia umbellata Raf., l. c. (1838).—?
Tilia heterophylla β *alba* Wood, Cl. Book, 272 (1855).

---

[1] A variety with thinner and larger leaves found in Louisiana and fairly distinct.

Tilia heterophylla Vent.—Continued.

SYN.—*Tilia heterophylla-nigra* Bayer, in Verhandl. Zoolog. Bot.
  Gesellsch. Wien, XII (Monog. Til. Gen.), 52 (1862).
  Tilia macrophylla Hort. ex Steudel, Nom. Bot., ed. sec., II,
  687 (1841).

COMMON NAMES.

White Basswood (Ohio, Ind., Ala.).
Wahoo (Ga., Fla.).
Wild Linden.
Smooth-fruited White-leaved Limetree (Tenn.).
Large-leaved Lime Tree (Tenn.).
Silverleaf Poplar (Ky.).
Cottonwood (Ky.).
Lin (Ind.).

Family CACTACEÆ.

CEREUS[1] Miller, Gard. Dict., ed. 8 (1768).

Cereus giganteus Engelm.   **Giant Cactus.**

SYN.—*Cereus giganteus* Engelmann, in Emory's Rep., 158 (1848).
  Pilocereus Engelmanni Lemaire, Ill. Hort., IX, Misc., 97
  (1862).
  Pilocereus giganteus Förster, Handb. Cact., ed. Rümpler, 662,
  f. 88 (1886).

COMMON NAMES.

Giant Cactus (N. Mex., Ariz.).
Sahuara (Ariz.).
Saguaro (Ariz.).

Cereus thurberi Engelm.   **Thurber Cactus.**

SYN.—*Cereus Thurberi* Engelmann, in Am. Journ. Sci., ser. 2, XVII,
  234 (1854).

COMMON NAMES.

Pitahaya (Mex.).
Pitahaya dulce (Mex.).

Cereus schottii Engelm.   **Schott Cactus.**

SYN.—*Cereus Schottii* Engelmann, in Proc. Am. Acad. Sci., III
  (Syn. Cact.), 288 (1856).

---

[1] = *Cereus* Haworth, Syn. Pl. Succ., 178 (1812).

Cereus schottii Engelm.—Continued.

SYN.—Zina
Sina
Sinita } (Ariz., Mex.).
Hombre viejo
Cabeza viejo

Family RHIZOPHORACEÆ.

RHIZOPHORA Linn.. Spec. Pl., 443 (1753).

Rhizophora mangle Linn.   **Mangrove.**

SYN.—*Rhizophora Mangle* Linnæus, Spec. Pl., ed. 1. I, 443 (1753).
Rhizophora racemosa Meyer, Prim. Fl. Esseq., 185 (1818).
Rhizophora Mangle α Walker-Arnott, in Ann. Nat. Hist., I.
301 (1838).
Rhizophora Americana Nuttall, Sylva, I, 95, t. 24 (1842).
Rhizophora Mangle var. *racemosa* Eichler, in Martius. Flor.
Brasil., XII, Pt. II, 427 (1872).

COMMON NAME.

Mangrove (Fla.).

Family MYRTACEÆ.

ANAMOMIS Griseb., Fl. Brit. W. Ind., 240 (1864).

Anamomis dichotoma (Poir.) Sargent.   **Naked Stopper.**

SYN.—*Eugenia fragrans* Sims, in Bot. Mag.. XXXI, t. 1242 (1810),
not Willd. (1799).
Myrtus dichotoma Poiret, in Lamarck, Enc. Méth. Bot. Suppl.,
IV. 53 (1816).
Myrcia ? Balbisiana de Candolle. Prodr.. III. 243 (1828).
EUGENIA DICHOTOMA de C.. l. c., 278 (1828).
Anamomis punctata Grisebach. Fl. Brit. West Ind.. 240
(1864).
Anamomis dichotoma Sargent. in Gard. and For.. VI. 130
(1893).

COMMON NAMES.

Naked-wood (Fla.).
Naked Stopper.

CHYTRACULIA[1] Browne, Nat. Hist. Jam., 239 (1756).

Chytraculia chytraculia (Linn.) nom. nov.   **Stopper.**

. SYN.—*Myrtus Chytraculia* Linnæus, Amœn., V, 398 (1760).
CALYPTRANTHES CHYTRACULIA Swartz, Prodr., 79 (1788).
Eugenia pallens Poiret, in Lamarck, Enc. Méth. Bot. Suppl.,
III, 122 (1813).
Chytraculia arborea Kuntze, Rev. Gen. Pl. Par., I, 238
(1891).

Chytraculia chytraculia genuina (Berg) nom. nov.

SYN.—*Calyptranthes Chytraculia α genuina* Berg, in Linnæa,
XXVII, 27 (1854).

Chytraculia chytraculia ovalis (Berg) nom. nov.

SYN.—*Calyptranthes chytraculia β ovalis* Berg, in Linnæa, XXVII,
27 (1854).

Chytraculia chytraculia trichotoma (Berg) nom. nov.

SYN.—*Calyptranthes chytraculia γ trichotoma* Berg, in Linnæa,
XXVII, 27 (1854).

Chytraculia chytraculia pauciflora (Berg) nom. nov.

SYN.—*Calyptranthes Chytraculia δ pauciflora* Berg, in Linnæa,
XXVII, 27 (1854).
Chytraculia pauciflora (Berg) Kuntze, Rev. Gen. Pl., Par. I,
238 (1891).

Chytraculia chytraculia zuzygium (Linn.) nom. nov.

SYN.—*Myrtus Zuzygium* Linnæus, Amœn., V, 398 (1760).
Calyptranthes Zuzygium Swartz, Prodr., 79 (1788).
Calyptranthes Chytraculia ε Zuzygium Berg, in Linnæa,
XXVII, 27 (1854).
Chytraculia Suzygium Kuntze, Rev. Gen. Pl., Par. I, 238
(1891).

EUGENIA Linnæus, Spec. Pl., ed. 1, I, 470 (1753).

Eugenia buxifolia (Swartz) Willd.   **Gurgeon Stopper.**

SYN.—*Myrtus buxifolia* Swartz, Prodr., 78 (1788).
Myrtus axillaris Poiret, in Lamarck, Enc. Méth. Bot., IV.,
412 (1797), not Swartz (1788).

---

[1] SYN.—*Chytralia* Adanson, Fam. Pl., II, 80 (1763).
Calyptranthes Swartz, Prodr., 79 (1788).
Calyptranthus Jussieu, Dic. Sc. Nat., VI, 274 (1805).

Eugenia buxifolia (Swartz) Willd.—Continued.

SYN.—*Eugenia buxifolia* Willdenow, Sp. Pl., II, Par. II, 960 (1799).
Eugenia myrtoides Poiret, in Lamarck, Enc. Méth. Bot. Suppl., III, 125 (1813).
Myrtus Poireti Sprengel, Syst. Veg., II, 483 (1825).
Eugenia triplinervia γ *buxifolia* Berg, in Linnæa, XXVII, 191 (1854).
Eugenia smithii Sprengel ex Berg, l. c., 191 (1854), in part.
Eugenia unedifolia Spreng. ex Berg, l. c., 192 (1854), in part.

Eugenia monticola (Swartz) de C.   **White Stopper.**

SYN.—*Myrtus monticola* Swartz, Prodr., 78 (1788).
Eugenia axillaris Willdenow, Sp. Pl., II, 960 (1799).
Eugenia monticola de Candolle, Prodr., III, 275 (1828).
Eugenia triplinervia Berg, in Linnæa, XXVII, 191 (1854), in part.
Eugenia Smithii Sprengel ex Berg, l. c., 191 (1854), in part.
Eugenia unedifolia Spreng. ex Berg, l. c., 192 (1854), in part.

COMMON NAMES.

White Stopper (Fla.).
Stopper (Fla.).

Eugenia procera (Swartz) Poir.   **Red Stopper.**

SYN.—*Eugenia procera* Poiret, in Lamarck, Enc. Méth. Bot. Suppl., II, 129 (1813).
Myrtus procera Swartz, Prodr., 77 (1788).
Eugenia Baruensis Grisebach, in Goett. Abh., VII, 214 (1857), not Jacq. (1789).

COMMON NAMES.

Red Stopper (Fla.).
Spiceberry (Fla.).

Eugenia Garberi Sargent.   **Garber Stopper.**

SYN.—*Eugenia Garberi* Sargent, in Gard. and For., II., 28, f. 87 (1889).
EUGENIA PROCERA Sarg., in Tenth Cen. U. S., IX (Cat. For. Trees N. A.), 89 (1884), in part.

Family COMBRETACEÆ.

TERMINALIA Benth. & Hook., Gen. Pl., I, 68 (1865).

Terminalia buceras (Browne) Benth. & Hook.   **Black Olivetree.**

SYN.—*Bucida Buceras* Browne, Nat. Hist. Jam., t. 23, f. 1 (1756).
Bucida angustifolia de Candolle, Prodr., III, 10 (1828).
Terminalia Buceras Bentham & Hooker, Gen. Pl., I, 685 (1865).
Bucida Bucera var. *angustifolia* Eichler, in Martius, Fl. Brasil., XIV, Pt. II, 95 (1867).
Myrobalanus Buceras Kuntze, Rev. Gen. Pl., Par. I, 237 (1891).

CONOCARPUS Linn., Spec. Pl., 176 (1753).

Conocarpus erecta Linn.   **Florida Buttonwood.**

SYN.—*Conocarpus erecta* Linnæus, Spec. Pl., ed. 1, I, 176 (1753).
Conocarpus acutifolia Rœmer & Schultes, Syst. Veg., V, 574 (1819).

COMMON NAME.

Buttonwood (Fla.).

Conocarpus erecta arborea de C.

SYN.—*Conocarpus erecta* var. *arborea* de Candolle, Prodr., III, 16 (1828).

Conocarpus erecta procumbens (Linn.) de C.

SYN.—*Conocarpus procumbens* Linnæus, Spec. Pl., ed. 1, I, 177 (1753).
Conocarpus supina Crantz, Inst. Rei Herb., I, 355 (1766).
Conocarpus erecta var. *procumbens* de Candolle, Prodr., III, 16 (1828).

Conocarpus erecta sericea de C.

SYN.—*Conocarpus erecta* var. *sericea* de Candolle, Prodr., III, 16 (1828).
Conocarpus sericea Forst. in herb. L'Hér. ex. Don, Gen. Hist. Dichl., Pl., II, 662 (1832)

LAGUNCULARIA Gærtn. f., Frut., III, 209 (1805).

Laguncularia racemosa (Linn.) Gærtn. f.   **White Buttonwood.**

SYN.—*Conocarpus racemosa* Linnæus, Syst. Nat., ed. 10, 930 (1759).

Laguncularia racemosa (Linn.) Gærtn. f.—Continued.

Syn.—*Laguncularia racemosa* Gærtner f., Fruct., Suppl., 209, t. 217 (1805).
Schousboa commutata Sprengel, Syst. Veg., II, 332 (1825).
Bucida Buceras Vellozo, Fl. Flum., IV, t. 87 (1825), not Browne (1756).
Laguncularia glabrifolia Presl, Rel. Hænk, II, 22 (1835).

COMMON NAMES.

White Buttonwood (Fla.).
White Mangrove (Fla.).
Buttonwood (Fla.).

Family ARALIACEÆ.

ARALIA Linn., Spec. Pl., 273 (1753).

Aralia spinosa Linn.   **Angelica-tree.**

Syn.—*Chærophyllum arborescens* Linnæus, Spec. Pl., ed. 1, I, 259 (1753).—?
Aralia spinosa Linn., l. c., 273 (1753).

COMMON NAMES.

Angelica-tree.
Hercules' Club.

Family CORNACEÆ.

CORNUS Linn., Spec. Pl., 117 (1753).

Cornus florida Linn.   **(Flowering) Dogwood.**

Syn.—*Cornus florida* Linnæus, Spec. Pl., ed. 1, I, 117 (1753).
Benthamidia florida Spach, Hist. Vég., VIII, 107 (1839).

COMMON NAMES.

Flowering Dogwood (Mass., R. I., N. Y., N. J., Del., Pa., Va., N. C., S. C., Miss., La., Ark., Mo., Ill., Kans., Mich., Ont., Ohio, Ind.).
Dogwood (N. J., Pa., Del., W. Va., N. C., S. C., Ala., Fla., La., Ky., Ohio, Ind., Mich.).
Boxwood (Conn., R. I., N. Y., Miss., Mich., Ky., Ind., Ont.).
False Box Dogwood (Ky.).
New England Boxwood (Tenn.)
Flowering Cornel (R. I.).
Cornel (Tex.).

VARIETIES DISTINGUISHED IN CULTIVATION.

Cornus florida pendula Temple.   **Weeping Dogwood.**

SYN.—*Cornus florida pendula*[1] Temple, Desc. Cat. Trees, Shr. & Pl.,
26 (1889).—Dippel, Handb. Laubh., dritt. T., 244 (1896).

Cornus florida rubra Temple.   **Red-bract Dogwood.**

Syn.—*Cornus Florida Rubra* Temple, Desc. Cat. Trees, Shr. & Pl.,
6 (1889).[2]

Cornus nuttallii Aud.   **Western Dogwood.**

SYN.—*Cornus nuttallii* Audubon, Birds, T. 467 (1837).
Cornus florida Hooker, Fl. Bor.-Am., I, 277 (1833), in part.

COMMON NAMES.

Dogwood (Cal., Oreg., Wash.).
California Dogwood (Cal.).
Flowering Dogwood (Oreg., Cal.).
Western Dogwood.

Cornus alternifolia Linn. f.   **Blue Dogwood.**

SYN.—*Cornus alternifolia* Linnæus f., Syst. Veg., ed. 13, Suppl., 125
(1781).
Cornus alterna Marshall, Arb. Am., 35 (1785).
Cornus undulata Rafinesque, Alsograph. Am., 61 (1838).
Cornus rotundifolia Raf., l. c., 62 (1838).
Cornus riparia Raf., l. c. (1838).
Cornus riparia var. *rugosa* Raf., l. c. (1838).
Cornus punctata Raf., l. c. (1838).
Cornus plicata Tausch, in Flora, XXI, 733 (1838).

COMMON NAMES.

Dogwood (Vt., Mass., R. I., Conn., N. Y., N. J., Pa., Va.,
W. Va., N. C., Ga., Fla., Miss., La., Ark., Ky., Ill., Wis.,
Minn., Ohio, Ont.).

---

[1] This form was published much earlier in American nurserymen's catalogues of plants, but just how early is at present uncertain. The trade catalogue of F. L. Temple, Cambridge, Mass., for 1888-89 contains a good figure and a statement of the plant's distinguishing features under the above name and is the earliest publication of this plant known to me.

[2] Messrs. Parsons & Sons Company, proprietors of the Kissena Nurseries, Flushing, N. Y., communicate the following note on the history of this variety: "The Red Flowering Dogwood was first propagated and disseminated by this company about the year 1880. Its technical name, *Cornus florida rubra* was also given by us descriptive of its blossom, and adopted by other firms subsequently who took up its propagation." The plant is described in the trade catalogues of this firm as *Cornus florida flore rubro*, but their catalogues bear no date of issue. The name of this plant must, therefore, be rested for the present at least with Mr. Temple (l. c.) whose publication of the variety is definite.

Cornus alternifolia Linn. f.—Continued.

SYN.—Blue Dogwood (Pa.).
Purple Dogwood (Pa.).
Umbrella-tree (R. I.).
Pigeonberry (N. Y.).
Alternate-leaved Dogwood (Mich.).
Green Osier (Vt.).

NYSSA Linnæus, Spec. Pl., ed. 1, II, 1058 (1753).

Nyssa sylvatica Marsh.   **Black Gum.**

SYN.—*Nyssa sylvatica* Marshall, Arb. Am., 97 (1785).
NYSSA MULTIFLORA Wangenheim, Nordamer. Holz., 46, t.
16, f. 39 (1787).
Nyssa integrifolia Aiton, Hort. Kew., ed. 1, III, 446 (1789).
Nyssa Caroliniana Poiret, in Lamarck, Enc. Méth. Bot., IV,
507 (1797).
Nyssa Canadensis Poir., in Lam., l. c. (1797).
Nyssa villosa Michaux, Fl. Bor.-Am., II, 258 (1803).
Nyssa multiflora var. *sylvatica* Watson, Index N. A. Bot., 442
(1878).
Nyssa aquatica Coulter & Evans, in Bot. Gaz., XV, 91
(1890), not Linn. (1753), nor Marsh. (1785).

COMMON NAMES.

Black Gum (N. J., Pa., Del., Va., W. Va., N. C., S. C., Ga.,
Ala., Fla., Miss., La., Tex., Ill., Ind.).
Sour Gum (Vt., Mass., R. I., N. Y., N. J., Pa., Del., Va.,
W. Va., S. C., Fla., Tex., Ohio, Ind., Ill.).
Tupelo (Mass., R. I., N. J., Del., S. C., Ala., Fla., Miss., Tex.,
Ill., Ohio).
Pepperidge (Vt., Mass., R. I., N. Y., N. J., S. C., Tenn., Mich.,
Ohio, Ont.).
Wild Peartree (Tenn.).
Yellow Gumtree (Tenn.).
Stinkwood (W. Va.).
Tupelo Gum (Fla.).

VARIETY DISTINGUISHED IN CULTIVATION.

Nyssa sylvatica pendula (Temple) nom. nov.

SYN.—*Nyssa Multiflora Pendula* Temple, Desc. Cat. Trees, Shr. and
Pl., 6 (1888).

Nyssa biflora Walter. **Water Gum.**

SYN.—NYSSA AQUATICA Linnæus, Spec. Pl., ed. 1, II, 1058 (1753),
in part.
Nyssa biflora Walter, Fl. Caroliniana, 253 (1788).
Nyssa sylvatica var. *biflora* Sargent, Silva, V, 76 (1893).

Nyssa ogeche Marsh. **Sour Tupelo.**

SYN.—*Nyssa Ogeche* Marshall, Arb. Am., 97 (1785).
NYSSA CAPITATA Walter, Fl. Caroliniana, 253 (1788).
Nyssa coccinea Bartram, Travels, ed. 2, 17 (1791).
Nyssa tomentosa Poiret, in Lamarck, Enc. Méth. Bot., IV,
508 (1797).
Nyssa candicans Michaux, Fl. Bor.-Am., II, 259 (1803).
Nyssa Montana Gærtner, Fruct., III, 201, t. 216 (1805).

COMMON NAMES.

Sour Tupelo (S. C., Fla.).
Ogeechee Lime (S. C., Fla.).
Gopher Plum (Fla.).
Tupelo.
Wild Limetree.
Limetree.

Nyssa aquatica Linn. **Tupelo Gum.**

SYN.—*Nyssa aquatica* Linnæus, Spec. Pl., ed. 1, II, 1058 (1753), in
part; Marshall (1785).
NYSSA UNIFLORA Wangenheim, Nordamer. Holz., 83, t. 27,
f. 57 (1787).
Nyssa denticulata Aiton, Hort. Kew., ed. 1, III, 446 (1789).
Nyssa palustris Salisbury, Prodr., 175 (1796).
Nyssa angulosa Poiret, in Lamarck, Enc. Méth. Bot., IV, 507
(1797).
Nyssa tomentosa Michaux, Fl. Bor.-Am., II, 259 (1803).
Nyssa angulisans Michx., l. c. (1803).
Nyssa grandidentata Michx. f., Hist. Arb. Am., II, 252, t. 19
(1812).
Nyssa capitata grandidentata Browne, Trees Am., 426 (1846).

COMMON NAMES.

Large Tupelo (Ala., La., Tex.).
Tupelo Gum (Ala., Miss., La.).
Sour Gum (Ark., Mo.).
Swamp Tupelo (S. C., La.).
Cotton Gum (N. C., S. C., Fla.).

Nyssa aquatica Linn.—Continued.

Syn.—Tupelo (N. C., S. C.).
　　Wild Olivetree (La.).
　　Olivier à grandes feuilles (La.).
　　Olivetree (Miss.).

Family ERICACEÆ.

VACCINIUM Linn., Spec. Pl., ed. 1, I, 349 (1753).

Vaccinium arboreum Marsh.　　　　　　　**Tree Huckleberry.**

Syn.—*Vaccinium arboreum* Marshall. Arb. Am., 157 (1785).
　　Vaccinium mucronatum Walter, Fl. Caroliniana, 139 (1788),
　　　　not Linn. (1764).
　　Vaccinium diffusum Aiton, Hort. Kew., ed. 1, II, 11 (1789).
　　Arbutus obtusifolius Rafinesque, Fl. Ludovic., 55 (1817).
　　Batodendron arboreum Nuttall, in Trans. Am. Phil. Soc., 2d
　　　　ser., VIII, 261 (1843).

COMMON NAMES.

Farkleberry (N. C., S. C.. Fla.. Miss., La., Mo.).
Sparkleberry (N. C., S. C., Ala., Fla.).
Myrtle Berries (La.).
Bluet (La.).
Tree Huckleberry (S. C.).
Gooseberry (N. C.).

ARBUTUS Linn., Spec. Pl., 395 (1753).

Arbutus menziesii Pursh.　　　　　　　**Madroña.**

Syn.—*Arbutus Menziesii* Pursh. Fl. Am. Sept., I, 282 (1814).
　　Arbutus procera Douglas in Lindley, in Bot. Reg., XXI,
　　　　t. 1753 (1836).
　　Arbutus laurifolia Hooker, Fl. Bor.-Am., II, 36 (1840), not
　　　　Linn. f. (1781).

COMMON NAMES.

Madroña (Cal.. Oreg.).
Madrove (Cal.).
Laurelwood (Oreg.)
Madrone-tree.
Laurel (Oreg.).
Manzanita (Oreg.).

Arbutus xalapensis H., B. & K.   **Mexican Madroña.**

SYN.—*Arbutus Xalapensis* Humboldt, Bonpland & Kunth, Nov.
Gen. Spec., III, 279 (1818).
Arbutus mollis H., B. & K., l. c., 280 (1818).
Arbutus laurifolia Lindley, in Bot. Reg., XXV, t. 67 (1839),
not Linn. f. (1781).
Arbutus varians Bentham, Pl. Hartweg., 77 (1849).
Arbutus macrophylla Martens & Galeotti, in Bull. Acad. Brux.,
IX, Pt. I, 534 (1842).
Arbutus prunifolia Klotzsch, in Linnæa, XXIV, 73 (1851).
Arbutus Menziesii Torrey, in Bot. Mex. Bound. Surv., 108
(1859), not Pursh (1814).
Arbutus Texana Buckley, in Proc. Acad. Sci., Phila. 1861,
460 (1861).
ARBUTUS XALAPENSIS var. TEXANA Gray, Syn. Fl. N. A.,
2nd ed., II, Pt. I, Suppl., 397 (1886).

COMMON NAMES.
Manzanita.
Madroña.
Madrone-tree.
Laurel.

Arbutus arizonica (Gray) Sargent.   **Arizona Madroña.**

SYN.—*Arbutus Menziesii* Rothrock, in Wheeler's Rep., VI., 25, 183
(1878), not Pursh (1814).
Arbutus Xalapensis Sargent, in Tenth Cen. U. S. (Cat. For.
Trees N. A.), IX., 97 (1884), not H., B. and K. (1818).
ARBUTUS XALAPENSIS var. ARIZONICA Gray, Syn. Fl. N. A.,
2nd ed., II, Pt. I, Suppl., 396 (1886).
Arbutus Arizonica Sargent, in Gard. and For., IV, 317, f. 54
(1891).

ANDROMEDA Linn., Spec. Pl., 393 (1753).

Andromeda ferruginea Walter.   **Andromeda.**

SYN.—*Andromeda ferruginea* Walter, Fl. Caroliniana, 138 (1788).
Lyonia ferruginea Nuttall, Genera, I, 266 (1818).
Lyonia squamulosa Martens & Galeotti, in Bull. Acad. Brux.,
IX, 542 (1842).

COMMON NAME.
Titi (Fla.).

Andromeda ferruginea arborescens Michx.

SYN.—*Andromeda ferruginea* var. *arborescens* Michaux, Fl. Bor.-
Am., I, 252 (1803).

Andromeda ferruginea arborescens Michx.—Continued.

SYN.—*Andromeda rigida* Pursh, Fl. Am. Sept., I, 292 (1814).
Lyonia rigida Nutt., l. c. (1818).

Andromeda ferruginea fruticosa Michx.

SYN.—*Andromeda ferruginea* var. *fruticosa* Michaux, Fl. Bor.-Am.,
I, 252 (1803).
Lyonia ? rhomboidalis Don, Gen. Hist. Dichl. Pl., III, 831
(1834).
Andromeda rhomboidalis Sargent, Silva, V, 132 (1893), not
Veill., in Nouv. Duham. (1801).

OXYDENDRUM de C., Prodr., VII, 601 (1839).

Oxydendrum arboreum (Linn.) de C.   **Sourwood.**

SYN.—*Andromeda arborea* Linnæus, Spec. Pl., ed. 1, I, 394 (1753).
Andromeda arborescens Persoon, Syn. Pl., I, 480 (1805).
Lyonia arborea Don, in Edinburgh New Phil. Journ., XVII,
159 (1834).
Oxydendrum arboreum de Candolle. Prodr., VII, 601 (1839).

COMMON NAMES.

Sourwood (W. Va., Del., N. C., S. C., Ga., Fla., Ala., Miss.,
La., Ky., Ohio).
Sorrel-tree (Pa., Del., N. C., S. C., Miss., La., Ohio).
Sour Gum Bush (Ohio).
Sour Gum (W. Va.).
Arrow-wood (W. Va.).
Titi (S. C.).
Lily of the Valley-tree.

KALMIA Linn., Spec. Pl., 391 (1753).

Kalmia latifolia Linn.   **Mountain Laurel.**

SYN.—*Kalmia latifolia* Linnæus, Spec. Pl., ed. 1, I, 391 (1753).
Kalmia serotina Hoffmannsegg, Verz. Pflanz., 70 (1824).—?
Kalmia nitida Forbes, Hort. Wob., 93 (1838).—?
Kalmia myrtifolia André, in Rev. Hort., 10, f. 1 (1883).
Chamædaphne latifolia Kuntze, Rev. Gen. Pl., Par. I, 388
(1891).

COMMON NAMES.

Kalmia latifolia Linn.—Continued.

SYN.—Calico Bush (Vt., R. I., N. Y., Pa., Del., N. C., S. C., Ala.,
Miss., La.).
Spoonwood (N. H., Mass., R. I., Pa., Miss.).
Ivy (Conn., Md., Va., N. C., S. C., Miss.).
Poison Ivy (Tenn., Ala.).
Poison Laurel (Ala.).
Mountain Laurel (Vt., Mass., W. Va., Ky., Tenn.).
Sheep Laurel (Pa., Ohio).
Wood Laurel (Pa.).
Small Laurel (W. Va.).
Kalmia (Pa., S. C.).
Calico-tree (Tenn.).
Calico Flower (Tenn.).
Mountain Ivy (Va.).
Big-leaved Ivy (lit. domestic medicine).
Ivywood (S. C.).

RHODODENDRON Linn., Spec. Pl., 392 (1753).

Rhododendron maximum Linn.   **Rhododendron.**

SYN.—*Rhododendron maximum* Linnæus, Spec. Pl., ed. 1, I, 392
(1753).
Rhododendron procerum Salisbury, Prodr., 287 (1796).

COMMON NAMES.

Great Laurel (N. H., Mass., R. I., N. Y., N. J., Pa., N. C.,
Minn.).
Rose Bay (R. I., Pa., N. C., S. C., Miss.).
Bigleaf Laurel (Pa.).
Big Laurel (W. Va.).
Laurel (R. I., Va., N. C., S. C.).
Mountain Laurel (Pa., S. C.).
Rhododendron (R. I., N. Y., Pa., Va., S. C.).
Dwarf Rose Bay-tree (Tenn.).
Spoon Hutch (N. H.).
Wild Rose Bay,
Deertongue Laurel. } (Lit. of domestic medicine.)

VARIETIES DISTINGUISHED IN CULTIVATION.

Rhododendron maximum roseum Pursh.

SYN.—*Rhododendron maximum* var. *roseum* Pursh, Fl. Am. Sept.,
I, 297 (1814).

Rhododendron maximum purpureum Pursh.

SYN.—*Rhododendron maximum* var. *purpureum* Pursh, Fl. Am.
Sept., I, 297 (1814).
Rhododendron purpureum Don, Gen. Hist. Dichl. Pl., III,
843 (1834).

Rhododendron maximum album Pursh.

SYN.—*Rhododendron maximum* var. *album* Pursh, Fl. Am. Sept., I,
297 (1814).
Rhododendron Purshii Don, Gen. Hist. Dichl. Pl., III, 843
(1834).

Rhododendron catawbiense[1] Michx.   **Catawba Rhododendron.**

SYN.—*Rhododendron catawbiense* Michaux, Fl. Bor.-Am., I, 258
(1803).

Family MYRSINACEÆ.

ICACOREA[2] Aublet, Pl. Guian., II, Suppl., 1 (1775).

Icacorea paniculata (Nutt.) Sudworth.   **Marlberry.**

SYN.—*Cyrilla paniculata* Nuttall, in Am. Journ. Sci., V, 290 (1822).
Pickeringia paniculata Nutt., in Journ. Acad. Sci. Phila.,
VII, Pt. I, 95 (1834).
ARDISIA PICKERINGIA Nutt., Sylva, III, 69, t. 102 (1849).
Bladhia paniculata (Nutt.) Sudworth, in Gard. and For., IV,
239 (1891).
Icacorea paniculata (Nutt.) Sudw., l. c., VI, 324 (1893).

COMMON NAMES.

Marlberry, (Fla.).
Cherry (Fla.).

JAQUINIA[3] Linnæus, Diss. Fl. Jam., Append. (1759); Amœn., V,
388 (1760).

Jaquinia armillaris Jacq.   **Joewood.**

SYN.—*Jaquinia armillaris* Jacquin., Enum. Pl. Carib., 15 (1760).

[1] Heretofore this species has not been included among our arborescent species,
commonly occurring as a low or straggling shrub. On an excursion through the
southern Alleghany Mountain region in 1889 I discovered a dozen or more individuals
of this species 3 to 4 inches in diameter and with single trunks 12 to 15 feet high,
with the habit of a small tree. They were standing in a small, rich mountain creek
bottom about 1 mile southeast of Blowing Rock, Watauga County, North Carolina.

[2] =*Ardisia* Swartz, Prodr., 48 (1788).

[3] Usually written *Jacquinia*, but Linnæus' original spelling is retained as adopted
subsequently by Jacquin.

Jaquinia armillaris Jacq.—Continued.

SYN.—*Jaquinia arborea* Vahl., Eclog., I, 26 (1796).
Jacquinia amarillis β *arborea* Grisebach, Fl. Brit. West.
Ind., 397 (1864.)

COMMON NAME.

Joewood (Fla.).

Family SAPOTACEÆ.

CHRYSOPHYLLUM Linnæus, Spec. Pl., ed. 1, I, 192 (1753).

Chrysophyllum monopyrenum Swartz. **Satinleaf.**

SYN.—*Chrysophyllum Cainito* Miller, Gard. Dict., ed. 8, No. 1 (1768),
not Linn. (1753.)
CHRYSOPHYLLUM OLIVIFORME Lamarck, Enc. Méth. Bot.,
I, 552 (1783), not Linn. (1759).
Chrysophyllum monopyrenum Swartz, Prodr., 49 (1788).
Chrysophyllum ferrugineum Gœrtner f., Fruct., III, 122, t. 202,
(1805).
Chrysophyllum microphyllum A. de Candolle, in de C.,
Prodr., VIII, 158 (1844), not Jacq. (1763).

COMMON NAME.

Satinleaf (Fla.).

SIDEROXYLON Linnæus, Spec. Pl., ed. 1, I, 192 (1753).

Sideroxylon mastichodendron Jacq. **Mastic.**

SYN.—*Sideroxylum mastichodendron* Jacquin, Coll., II, 253, t. 17,
f. 5 (1788).
Bumelia pallida Swartz, Prodr., 49 (1788).
Achras pallida Poiret, in Lamarck, Enc. Méth. Bot., VI,
533 (1804).
Bumelia salicifolia Willdenow, Sp. Pl., I, Pt. II, 1086 (1797),
in part.
Bumelia Mastichodendron Roemer & Schultes, Syst. Veg.,
IV, 493 (1819).
Sideroxylum pallidum Sprengel, Syst. Veg., I, 666 (1825).
Bumelia fœtidissima Nuttall, Sylva, III, 39, t. 94 (1849), not
Willd. (1797).

COMMON NAMES.

Mastic (Fla.).
Wild Olive (Fla.).

BUMELIA[1] Swartz, Prodr., 49 (1788).

Bumelia tenax (Linn.) Willd.   **Tough Buckthorn.**

Syn.—*Sideroxylon tenax* Linnæus, Mantiss., 48 (1767).
  Chrysophyllum Carolinense Jacquin, Obs., III, 3, t. 54 (1768).
  Sideroxylon sericeum Walter, Fl. Caroliniana, 100 (1788).
  Bumelia tenax Willdenow, Sp. Pl., I, Pt. II, 1085 (1797).
  Sideroxylon chrysophylloides Michaux, Fl. Bor.-Am., I, 123
    (1803).
  Bumelia chrysophylloides Pursh, Fl. Am. Sept., I, 155·(1814).
  Sclerocladus tenax Rafinesque, Sylva Tellur., 35 (1838).
  Sclerozus tenax Raf., Autikon Bot., 73 (1840).
  Lyciodes tenax [L.] (W.) Kuntze, Rev. Gen. Pl., Par. II, 407
    (1891).

COMMON NAMES.

  Black Haw (Fla.).
  Tough Buckthorn (S. C.).
  Ironwood (S. C., Fla.).

Bumelia lanuginosa (Michx.) Pers.   **Shittimwood.**

Syn.—*Sideroxylon tenax* Walter, Fl. Caroliniana, 100 (1788), not
    Linn. (1767).
  Sideroxylon lanuginosum Michaux, Fl. Bor.-Am., I, 122 (1803).
  Bumelia lanuginosa Persoon, Syn. Pl., I, 237 (1805).
  Chrysophyllum Ludovicianum Rafinesque, Fl. Ludovic., 53
    (1817).
  Bumelia oblongifolia Nuttall, Genera, I, 135 (1818).
  Bumelia arachnoidea Rafinesque, New Fl. Bot., 3d Pt., 28
    (1836).
  Bumelia ferruginea Nuttall, Sylva, III, 34 (1849).
  Bumelia tomentosa A. de Candolle, in de C., Prodr., VIII,
    190 (1844).
  Bumelia arborea Buckley, in Proc. Acad. Sci. Phila. 1861,
    461 (1861).
  Bumelia pauciflora Engelmann, ex Gray, Syn. Fl., N. A.,
    ed. 1, II, 68 (1878).
  Lyciodes lanuginosum [Michx.] (Pers.) Kuntze, Rev. Gen.
    Pl., Par. II, 406 (1891).

[1] A shrub found in the south Atlantic region is—

Bumelia reclinata Vent.

  Syn.—*Bumelia reclinata* Ventenat, Choix Pl., t. 22 (1803).
    Sideroxylon reclinatum Michaux, Fl. Bor.-Am., I, 122 (1803).
    Bumelia lycioides var. *reclinata* Gray, Syn. Fl. N. A., ed. 1, II, 68 (1878).

Bumelia lanuginosa (Michx.) Pers.—Continued.

SYN.—Gum Elastic.
Shittimwood (Tex.).
Black Haw (Fla.).

Bumelia lanuginosa rigida Gray.

SYN.—*Bumelia lanuginosa* var. *rigida* Gray, Syn. Fl. N. A., 2nd ed.,
II, Pt. I, 68 (1886).
Bumelia spinosa Watson, in Proc. Am. Acad. Sci., XVIII, 112
(1883), not de C. (1844).

Bumelia lycioides (Linn.) Gærtn. f. **Bumelia.**

SYN.—*Sideroxylon lycioides* Linnæus, Spec. Pl., ed. 2, I, 279 (1762).
Sideroxylon decandrum Linn., Mantiss., 48 (1767).
Sideroxylon læve Walter, Fl. Caroliniana, 100 (1788).
Bumelia lycioides Gærtner f., Fruct., III, 127, t. 202 (1805).
Bumelia pubescens Tenore, Sem. Hort. Neap. (1827).
Lyciodes spinosum Kuntze, Rev. Gen. Pl., Par. II, 406 (1891).

Ironwood (Va., S. C., Ga., Fla., Miss., Ky.).
Southern Buckthorn (Miss., La., Tex., Ill.).
Carolina Buckthorn (N. C.).
Buckthorn (S. C.).
Chittimwood (Tex.).
Mock Orange (Fla.).

Bumelia[1] **angustifola** Nutt. **Saffron Plum.**

SYN.—*Bumelia angustifolia* Nuttall, Sylva, III, 38, t. 93 (1849).
Bumelia reclinata Torrey, in Bot. Mex. Bound. Surv., 109
(1859), not Vent. (1803).
Bumelia parvifolia Chapman, Fl. S. States, ed. 1, 275 (1860),
not A. de C. (1844).
Bumelia cuneata Gray, Syn. Fl. N. A., ed. 1, II, 68 (1878),
not Swartz (1797).

---

[1] Recently a new arborescent species of this genus has been described from the vicinity of Culiacan, Mexico, as *Bumelia Palmeri* Rose (in Gard. and For., VII, 195, 1894). Specimens were collected by Dr. Edward Palmer in 1891. It appears to have been long known to the Mexicans and much prized for its prune-like, edible fruit, but until recently entirely overlooked by botanists. It attains a height of 45 to 50 feet and a diameter of 2½ to 4 feet.

Bumelia arborescens Rose (in Contr. U. S. Nat. Herb., I, No. 29, 339, 1895) is an arborescent species detected by Dr. Edward Palmer in the region of Colima, Mexico, in 1891. It is reported to have a diameter of 1 foot, no mention being made of its height growth.

Bumelia angustifola Nutt.—Continued.

SYN.—*Lyciodes angustifolium* Kuntze, Rev. Gen. Pl., Par. II, 406 (1891).

COMMON NAMES.

Saffron Plum (Fla.).
Downward Plum (Fla.).
Ants-wood (Fla.).

DIPHOLIS A. de C., Prodr., VIII, 188 (1844).

Dipholis salicifolia (Linn.) A. de C. **Bustic.**

SYN.—*Achras silicifolia* Linnæus, Spec. Pl., ed. 2, I, 470 (1762).
Bumelia silicifolia Swartz, Prodr., 50 (1788).
Sideroxylum salicifolium Lamarck, Ill., II, 42 (1793).
Dipholis salicifolia A. de Candolle, in de C., Prodr., VIII, 188 (1844).

COMMON NAMES.

Bustic (Fla.).
Cassada (Fla.). ·

MIMUSOPS Linnæus, Spec. Pl., 349 (1753).

Mimusops sieberi A. de C. **Wild Sapodilla.**

SYN.—*Mimusops Sieberi* A. de Candolle, in de C., Prodr., VIII, 204 (1844).
Achras Zapotilla var. *parviflora* Nuttall, Sylva, III, 28, t. 90 (1849).
Mimusops dissecta Grisebach, Fl. Brit. West Ind., 400 (1864), in part.
Achras Bahamensis Baker, in Hooker, Icon., XVIII, t. 1795 (1888).
Mimusops Floridana Engler, in Bot. Jahrb., zwölf. B., 524 (1890).

COMMON NAMES.

Wild Dilly (Fla.).
Wild Sapodilla (Fla.).

Family EBENACEÆ.

DIOSPYROS Linn., Spec. Pl., 1057 (1753).

Diospyros virginiana Linn. **Persimmon.**

SYN.—*Diospyros Virginiana* Linnæus, Spec. Pl., ed. 1, II, 1057 (1753).

Diospyros virginiana Linn.—Continued.

SYN.—*Diospyros Guajacana* Romans, Nat. Hist. Fla., 20 (1775).
Diospyros concolor Moench, Meth., 471 (1794).
Diospyros guaiacana Robin, Voy. Louis, III, 417 (1807).
Diospyros pubescens Pursh, Fl. Am. Sept., I, 265 (1814), not
Pers. (1807).
Diospyros Caroliniana Rafinesque, Fl. Ludovic., 139 (1817).
Diospyros virginiana β *pubescens* Nuttall, Genera, II, 240
(1818).
Diospyros Virginiana var. *Macrocarpa* Rafinesque, Med. Flor.,
I, 155 (1828).
Diospyros Virginiana var. *Concolor* Raf., l. c. (1828).
Diospyros Virginiana var. *Microcarpa* Raf., l. c. (1828).
Diospyros fertilis Loddiges ex Loudon, Arb. Frut., II, 1197
(1838).
Diospyros Persimon Wikström, in Jahr. Schwed. 1830, 92
(1834).
Diospyros ciliata Rafinesque, New Fl. Bot., 3d Pt., 25 (1836),
not A. de C. (1844).
Diospyros lucida Hort. ex Loudon, in Gard. Mag., 394 (1841),
Diospyros angustifolia Audibert ex Spach, Hist. Vég., IX.
405 (1840).
Diospyros intermedia Hort. ex Loudon, Enc. Trees, 627 (1842).
Diospyros Virginiana *dulcis* Prince ex Browne, Trees Am.,
369 (1846).
Diospyros distyla ex Koch, Dendrol., erst. Th., zw. Ab., 205
(1872).
Diospyros Mexicana Scheele ex Hiern., in Trans. Camb. Phil.
Soc., XII, 238 (1873).

COMMON NAMES.

Persimmon (Conn., N. Y., N. J., Pa., Del., Va., W. Va., N. C.,
S. C., Ga., Fla., Miss., La., Ky., Mo., Tex., Ark., Ill., Ind.,
Iowa, Ohio).
Date Plum (N. J., Tenn.).
Plaqueminier (La.).
Simmon (Fla.).
Possumwood (Fla.).

Diospyros texana Scheele. **Mexican Persimmon.**

SYN.—*Diospyros texana* Scheele, in Linnæa, XXII, 145 (1849).

COMMON NAMES.

Mexican Persimmon (Tex.).
Black Persimmon (Tex.)
Chapote (Tex.).

Family SYMPLOCACEÆ.[1]

SYMPLOCOS Jacquin, Enum. Pl. Carib., 5 (1760).

Symplocos tinctoria (Linn.) L'Hér.　　　　　　**Sweetleaf.**

SYN.—*Hopea tinctoria* Linnæus, Mantiss., 105 (1767).
Symplocos tinctoria L'Héritier, in Trans. Linnæan Soc., I, 176 (1791).
Protohopea tinctoria Miers, in Journ. Linn. Soc., XVII, 290 (1879).
Eugenioides tinctorium (L'Hér.) Otto Kuntze, Rev. Gen. Pl., Par. II, 976 (1891).

COMMON NAMES.

Sweetleaf (Del., N. C., S. C., Ala., Fla.).
Yellow-wood (N. C., S. C., Ala.).
Horse Sugar (Del., Ala., La.).
Florida Laurel (Fla.).

Family STYRACACEÆ.

MOHRODENDRON[2] Britton, in Gard. and For., VI, 463 (1893).

Mohrodendron carolinum (Linn.) Britton.　　**Silverbell-tree.**

SYN.—*Halesia Carolina* Linnæus, Syst. Nat., ed. 10, 1044 (1759).
HALESIA TETRAPTERA Ellis, in Phil. Trans., LI, 932, t. 22, f. A. (1761).

---

[1] Usually included in Styracaceæ.
[2] SYN.—*Halesia* Ellis, in Linnæus, Syst. Nat., ed. 10, 1044 (1859), not P. Browne (1756).
Mohria Britton, in Gard. and For., VI, 434 (Oct. 18, 1893), not Swartz (1806).
Carlomohria Greene, in Erythea, I, 236 (Nov. 3, 1893) (*genus nudum*—?).
Mohrodendron Britton, in Gard. and For., VI, 463 (Nov. 8, 1893).
According to Article V of the emendations of the Paris botanical code (Appendix, p. 249), Professor Greene's *Carlomohria* must be considered a *genus nudum*, since the author did not formally cite any previously published species in connection with his new genus, nor accompany the genus with characters. It is otherwise plainly evident, however, that he proposed *Carlomohria* as an equivalent name for the preoccupied *Halesia*. The connection can certainly be taken in no other light than that of valid publication, which, but for adherence to the above rule, must give *Carlomohria* precedence over Dr. Britton's *Mohrodendron*, a genus published five days later and in accordance with the provisions of Article V. *Mohrodendron* is here retained in a desire to support the principle underlying the American rule, but with the belief that the apparent inadequacy of the above rule will be remedied by future consent of American botanists so as to give standing to several genera of which *Carlomohria* is an example.

Mohrodendron carolinum (Linn.) Britton—Continued.

SYN.—*Halesia stenocarpa* Koch, in Wochenschr. Gärtn. Pflanz., I, 190 (1858).

Mohria Carolina (L.) Britton, in Gard. and For., VI, 434 (1893).

Mohrodendron Carolinum (L.) Britt., l. c., 463 (November, 1893).

Carlomohria Carolina Greene, in Erythea, I, 246 (December, 1893).

COMMON NAMES.

Snowdrop-tree (R. I., Pa. (cult.), N. C., S. C., Fla., La.).
Silverbell-tree (R. I. (cult.), Ala., Fla., Miss.).
Silverbell (Pa., cult.).
Wild Olive Tree (Tenn.).
Bell-tree (Tenn.).
Four-winged Halesia (Ala.).
Opossum-wood (Ala.).
Rattlebox (Tex.).
Calicowood (Tex., Ill.).
Tisswood (Tenn.).

Mohrodendron meehani (Sarg.) nom. nov. **Meehan Silverbell-tree.**

SYN.—*Halesia tetraptera Meehani* Sargent, in Gard. and For., V, 534, f. 91 (1892).

Halesia Meehani Meehan,[1] in Hort. et in lit. in Sarg., l. c., 534 (1892).

Mohrodendron[2] **dipterum** (Ellis) Britton.   **Snowdrop-tree.**

SYN.—HALESIA DIPTERA Ellis, in Phil. Trans., LI, 932, t. 22, f. B. (1761).

Halesia reticulata Buckley, in Proc. Acad. Sci. Phila. 1860, 444 (1860).

Mohria diptera (L.) Britton, in Gard. and For., VI, 434(1893).

Mohrodendron dipterum (L.) Britt.,l. c., 463 (November, 1893).

---

[1] The necessary form of this citation makes it appear that Mr. Thomas Meehan named this plant after himself, which was not the case, the plant being so named by his nurserymen. Mr. Meehan is responsible only for the communication of the plant's history to Garden and Forest long after it was named.

[2] The following shrubby species is found in southern Georgia and the adjacent region of Florida:

Mohrodendron parviflorum (Michx.) Britton.   **Small-flower Snowdrop-tree.**

SYN.—HALESIA PARVIFLORA Michaux, Fl., Bor.-Am., II, 40 (1803).

Mohria parviflora (Michx.) Britton, in Gard. and For., VI, 434 (1893).

Mohrodendron parviflorum (Michx.) Britt., l. c., 463 (November, 1893).

Carlomohria parviflora Greene, in Erythea, I, 246 (December, 1893).

Mohrodendron dipterum (Ellis) Britton—Continued.

SYN.—*Carlomohria diptera* Greene, in Erythea, I, 246 (December, 1893).

COMMON NAMES.

Snowdrop-tree (R. I., Del. (cult.), S. C., Ala., La., Tex.).
Silverbell-tree (R. I., Del. (cult.), Miss., Tex.).
Cow Licks (La.).

Family OLEACEÆ.

FRAXINUS Linn., Spec. Pl., 1057 (1753).

Fraxinus cuspidata Torr.   **Fringe Ash.**

SYN.—*Fraxinus cuspidata* Torrey, in Bot. Mex. Bound. Surv., 166 (1859).

Fraxinus dipetala[1] Hook & Arn.   **Shrubby Fringe Ash.**

SYN.—*Fraxinus dipetala* Hooker & Arnott, in Bot. Beechey's Voy., 362, t. 87 (1840).
Ornus dipetala Nuttall, Sylva, III, 66, t. 101 (1849).
Chionanthus fraxinifolia Kellogg, in Proc. Cal. Acad. Sci., V, 18 (1873).

Fraxinus greggii Gray.   **Gregg Ash.**

SYN.—*Fraxinus Schiedeana* var. *parvifolia* Torrey, in Bot. Mex. Bound. Surv., 166 (1859), not *F. parvifolia* Lam. (1786), nor Willd. (1805).
Fraxinus Greggii Gray, in Proc. Am. Acad. Sci., XII, 63 (1876).

Fraxinus quadrangulata Michx.   **Blue Ash.**

SYN.—*Fraxinus quadrangulata* Michaux, Fl. Bor.-Am., II, 255 (1803).
Fraxinus tetragona, Du Mont de Courset, Bot. Cult., ed. 2, II, 583 (1811).
Fraxinus quadrangularis Loddiges, Cat. ed. (1836).

---

[1]The following American varieties are not known to be arborescent:

Fraxinus dipetala brachyptera Gr.   **Short-wing Fringe Ash.**

SYN.—*Fraxinus dipetala* var. *brachyptera* Gray, Syn. Fl. N. A., II, 74 (1878).

Fraxinus dipetala trifoliolata Torr.   **Trifoliate Fringe Ash.**

SYN.—*Fraxinus dipetala* var. ? *trifoliolata* Torrey, in Bot. Mex. Bound. Surv., 167 (1859).
Fraxinus dipetala var. *trifoliata* Torr. ex Sargent, Silva, VI, 31 (1894).

Fraxinus quadralaganta Michx.—Continued.

SYN.—*Fraxinus quadrangulata* var. *nervosa* Loudon, Arb. Frut., II, 1235 (1838).
Fraxinus americana quadrangulata Browne, Trees Am., 397 (1846).
Fraxinus americana quadrangulata nervosa Browne, l. c. (1846).
Fraxinus quadrangulata var. *subpubescens* Wesmael, in Bull. Soc. Belg., XXX, 114 (1892), not *F. americana subpubescens* Browne (1846).

COMMON NAME.

Blue Ash (Pa. (cult.), Ala., Ky., Mo., Ill., Mich.).

Fraxinus nigra Marsh.   **Black Ash.**

SYN.—*Fraxinus Novæ-Angliæ* Du Roi, Harbk. Baumz., ed. 1, I, 290 (1771), not Mill. (1768).
Fraxinus Nigra Marshall, Arb. Am., 51 (1785).
FRAXINUS SAMBUCIFOLIA Lamarck, Euc. Méth. Bot., II, 549 (1786).
Fraxinus nigra crispa Loddiges, Cat. ed. (1836).
Fraxinus americana sambucifolia Browne, Trees Am., 396 (1846).
Fraxinus americana sambucifolia crispa Browne, l. c. (1846).
Fraxinus coarctata Hort. ex Dippel, Handb. Laubh., erst. T., 101 (1889).
Fraxinus imbricata Hort. ex Dippel, l. c. (1889).
Fraxinus nigra cucullata Dippel, l. c., 102 (1889).
Fraxinus nigra subsp. *nigra* Wesmael, in Bull. Bot. Belg., XXX, 112 (1892).

COMMON NAMES.

Black Ash (Me., N. H., Vt., Mass., R. I., N. Y., N. J., Pa., Del., W. Va., Mo., Ohio, Ont., Mich., Ill., Minn.).
Hoop Ash (Vt., N. Y., Del., Ohio, Ill., Ind.).
Basket Ash (Mass.).
Brown Ash (N. H., Tenn.).
Swamp Ash (Vt., R. I., N. Y.).
Water Ash (W. Va., Tenn., Ind.).

Fraxinus anomala Watson.   **Dwarf Ash.**

SYN.—*Fraxinus anomala* Watson, in King's Rep., V, 283 (1871).

COMMON NAMES.

Dwarf Ash (Ariz., Utah).
Ash (Utah).

Fraxinus anomala triphylla[1] Jones.   **Trifoliate Dwarf Ash.**

SYN.—*Fraxinus anomala* var. *triphylla* Jones, in Proc. Cal. Acad.
Sci., sec. ser., V, Pt. I, 707 (1895).

Fraxinus velutina Torr.   **Leatherleaf Ash.**

SYN.—*Fraxinus velutina* Torrey, in Emory's Rep., 149 (1848).
FRAXINUS PISTACIÆFOLIA Torrey, in Pac. R. R. Rep., IV,
128 (1856).
Fraxinus coriacea Watson, in Am. Nat., VII, 302 (1873), in
part.
Fraxinus pistaciæfolia var. *coriacea* Gray, Syn. Fl. N. A., ed.
i, II, Pt. I, 74 (1878).
Fraxinus Americana var. *pistaciæfolia* Wenzig, in Bot.
Jahrb., viert. B., 182 (1883).
Fraxinus pistaciæfolia var. *velutina* (Torr.) Sudworth, in Rep.
Sec. Agric., 1892, 326 (1893).

COMMON NAME.

Ash (Tex., Ariz., Nev.).

Fraxinus americana Linn.   **White Ash.**

SYN.—*Fraxinus Americana* Linnæus, Spec. Pl., ed. 1, II, 1057
(1753).
Fraxinus Nova Anglia Miller, Gard. Dic., ed. 8, No. 5 (1768).
Fraxinus alba Marshall, Arb. Am., 51 (1785).
Fraxinus acuminata Lamarck, Enc. Méth. Bot., II, 547 (1786).
Fraxinus Caroliniensis Wangenheim, Nordamer. Holz., 81
(1787).
Fraxinus Canadensis Gærtner, Fruct., I. 222, t. 49 (1788).
Fraxinus epiptera Michaux, Fl. Bor.-Am., II. 256 (1803).
Fraxinus mixta Bosc, in Mém. Inst. Fr., IX, 1808, 209 (1811).
Fraxinus pannosa Bosc, l. c. (1811).
Fraxinus discolor Muhlenberg, Cat. Pl. Am. Sep., 111 (1813),
not Raf. (1817).
Fraxinus americana 2 *latifolia* London, Arb. Frut., II, 1232
1232 (1838).
Fraxinus Novæ-Angliæ Wangenheim, ex de Candolle, Prodr.,
VIII., 277 (1844), not Du Roi (1772).
Fraxinus americana epiptera Browne, Trees Am., 399 (1846).
Fraxinus glauca Hort. ex Koch, Dendrol., zw. Th. erst. Ab.,
253 (1872).

---

[1] This form is maintained as a distinct variety on Mr. Jones's authority, although
a common character of the species is to bear leaves on the same individual with one
to three leaflets. Mr. Jones does not state (l. c.) whether the plant is a tree or
shrub.

Fraxinus americana Linn.—Continued.

SYN.—*Fraxinus americana juglandifolia serrata* Dippel, Handb.
Laubh., erst. T., 75 (1889).
Fraxinus juglandifolia subserrata Dippel, l. c. (1889).
Fraxinus elliptica Dippel, l. c., (1889).
Fraxinus Am. foliis albo-marginatis Hort. ex Dippel, l. c., 76
(1889).
Fraxinus Americana macrophylla Hort. ex Dippel, l. c.,
(1889).
Fraxinus salicifolia Hort. ex Dippel, l. c. (1889).
Fraxinus Americana var. *normale* Wesmael, in Bull. Soc.
Bot. Belg., XXX, 107 (1892).
Fraxinus Americana var. *acuminata* Wesm., l. c. (1892).

COMMON NAMES.

White Ash (Me., N. H., Vt., Mass., R. I., Conn., N. Y., N. J.,
Del., Pa., Va., W. Va., N. C., S. C., Ga., Fla., Ala., Miss.,
La., Tex., Ky., Mo., Ill., Ind., Iowa, Kans., Nebr., Mich.,
Ohio, Ont., Minn., N. Dak., Wis.).
Ash (Ark., Iowa, Wis., Ill., Mo., Minn.).
American Ash (Iowa).
Franc-Frene (Quebec).
Cane Ash [1] (Ala., Miss., La.).

Fraxinus americana curtissii (Vasey) nom. nov.
Small-fruit White Ash.

SYN.—*Fraxinus albicans* Buckley, in Proc. Phila. Acad. Sci. 1862, 4
(1862), in part.—?
Fraxinus Curtissii Vasey, in Rep. in Com. Agric. 1875, 168
(Cat. Forest Trees, 20) (1876).
FRAXINUS AMERICANA MICROCARPA var. Gray, Syn. Fl.
N. A., ed. 1, II, 75 (1878).

Fraxinus texensis (Gray) Sargent. **Texas Ash.**

SYN.—*Fraxinus albicans* Buckley, in Proc. Acad. Sci. Phila. 1862, 4
(1862), in part.—?
Fraxinus coriacea Watson, in Am. Nat., VII, 302 (1873), in
part.
FRAXINUS AMERICANA var. TEXENSIS Gray, Syn. Fl. N.
A., ed. 1, II, Pt. I, 75 (1878).
Fraxinus Texensis Sargent, Silva, VI, 47 (1894).

COMMON NAME.

Mountain Ash (Tex.).

---

[1] So called from its growing in cane-brakes of the forest (teste Dr. C. Mohr).

328

Fraxinus pennsylvanica Marsh. **Red Ash.**

SYN.—*Fraxinus pennsylvanica* Marshall, Arb. Am., 51 (1785).
FRAXINUS PUBESCENS Lamarck, Enc. Méth. Bot., II, 548
(1786).
Fraxinus pubescens β longifolia Vahl, Enum., I, 52 (1804).
Fraxinus pubescens γ latifolia Vahl, l. c. (1804).
Fraxinus pubescens var. *subpubescens* Persoon, Syn. Pl., II,
605 (1807).
Fraxinus longifolia Bosc, in Mém. Inst. Fr., IX, 209 (1811).
Fraxinus subvillosa Bosc, l. c. (1811).
Fraxinus lancea Bosc, l. c. (1811).—?
Fraxinus rubicunda Bosc, l. c. (1811).—?
Fraxinus fusca Bosc, l. c. (1811).—?
Fraxinus rufa Bosc, l. c. (1811).—?
Fraxinus tomentosa Michaux f., Hist. Arb. Am., III, 112, t. 9
(1813).
Fraxinus discolor Rafinesque, Fl. Ludovic., 37 (1817), not
Muhl. (1813).
Fraxinus americana pubescens Browne, Trees Am., 395 (1846).
Fraxinus americana subpubescens Browne, l. c. (1846).
Fraxinus oblongocarpa Buckley, in Proc. Acad. Sci. Phila.
1862, 4 (1862).
Fraxinus epiptera Hort. ex Dippel, Handb. Laubh., erst. T.,
76 (1889), not Michx., (1803), nor Dippel, l. c., 74 (1889).
Fraxinus pubescens **a** *ovata* Dippel, l. c., 76 (1889).
Fraxinus pubescens **b** *coriacea* Dippel, l. c. (1889).
Fraxinus coriacea Bosc ex Dippel, l. c. (1889), not Watson
(1873).—?
Fraxinus arbutifolia Hort. ex Dippel, l. c. (1889).
Fraxinus pubescens **c** *nana* Dippel, l. c., 77 (1889).
Fraxinus pubescens subpubescens Hort. ex Dippel, l. c. (1889).
Fraxinus pubescens **d** *longifolia* Dippel, l. c., (1889).
Fraxinus pubescens **e** *Boscii* Dippel, l. c., (1889).
Fraxinus pubescens aucubæfolia Dippel, l. c. (1889).
Fraxinus aucubæfolia Hort. ex Dippel, l. c. (1889).
Fraxinus aucubæfolia nova Hort. ex Dippel, l. c. (1889).
Fraxinus viridis var. *pubescens* Hitchcock, in Trans. Acad.
Sci. St. Louis, V, 507 (1891).

COMMON NAMES.

Red Ash (Me., N. H., Vt., Mass., R. I., N. Y., N. J., Pa., Del.,
W. Va., N. C., S. C., Fla., Ga., Ala., Miss., La., Ky., Mo.,
Ill., Kans., Nebr., Mich., Minn., Ont.).
Brown Ash (Me.).
Black Ash (N. J.).
River Ash (R. I., Ont.).

Fraxinus pennsylvanica Marsh.—Continued.

SYN.—Bastard Ash (Vt.).
Ash (Nebr.).
Piss Ash (Vt.).

Fraxinus pennsylvanica profunda[1] (Bush) nom. nov.

SYN.—*Fraxinus Americana profunda* Bush, in Fifth Ann. Rep. Mo. Bot. Gard., 147 (1894).

Fraxinus lanceolata Borkh.   **Green Ash.**

SYN.—*Fraxinus juglandifolia* Willdenow, Berl. Baumz., 117 (1796), not Lam. (1786).
Fraxinus Caroliniana Willd., l. c., 119 (1796), not Mill. (1768), nor Lam. (1786).
Fraxinus lanceolata Borkhausen, Handb. Forst. Bot., I, 826 (1800).
Fraxinus juglandifolia β *subintegerrima* Vahl, Enum., I, 50 (1804).
Fraxinus Caroliniana β *latifolia* Willdenow, Sp. Pl., IV, Par. II, 1103 (1805).
Fraxinus expansa Willd., Berl. Baumz., zw. Ausg., 150 (1811).
Fraxinus concolor Muhlenberg, Cat. Pl. Am. Sep., 101 (1813).
FRAXINUS VIRIDIS Michaux f., Hist. Arb. Am., III, 115, t. 10, not fruit (1813).
Fraxinus Americana Hooker, Fl. Bor.-Am., II, 51 (1838), in part.
Fraxinus pubescens Torrey, Fl. N. Y., II, 126, t. 90 (1843), not Lam. (1786).
Fraxinus americana juglandifolia Browne, Trees Am., 398 (1846).
Fraxinus Novæ-Angliæ Koch, Dendrol., erst. Th., 251 (1872) not Mill. (1768), nor Wang. ex de C. (1844).
Fraxinus Americana subsp. *Novæ-Angliæ* Wesmael, in Bull. Bot. Soc. Belg., XXX, 108 (1892).
Fraxinus Pennsylvanica var. *lanceolata* Sargent, Silva, VI, 5, t. CCLXXII (1894).

COMMON NAMES.

Green Ash (Mass., R. I., Conn., N. Y., N. J., Pa., Del., N. C., S. C., Ala., Miss., La., Tex., Mo., Ill., Kans., Nebr., Mich., Minn., S. Dak., Ohio, Ont., Iowa).

---

[1] It is likely that this form will prove to be a distinct species. The very large size of the fruit appears to be constant in at least the few specimens examined from Missouri and the Gulf region. The structure and character of the wood suggest a nearer relationship to the Red Ash than to the White Ash, but further investigation seems necessary before the true status of the plant can be definitely determined.

Fraxinus lanceolata Borkh.—Continued.

Syn.—Blue Ash (Ark., Iowa).
White Ash (Kans., Nebr.).
Swamp Ash (Fla., Ala., Tex.).
Ash (Ark., Iowa, Nebr.).
Water Ash (Iowa).

Fraxinus berlandieriana A. de C.   **Berlandier Ash.**

Syn.—*Fraxinus Berlandieriana* A. de Candolle, in de C., Prodr., VIII, 278 (1844).
Fraxinus viridis var. *Berlandieriana* Torrey, in Bot. Mex. Bound. Surv., 166 (1859).
Fraxinus trialata Buckley, in Proc. Acad. Sci. Phila. 1862, 5 (1862).
Fraxinus pubescens var. *Berlandieriana* Wenzig, in Bot. Jahrb., viert. B., 183 (1883).
Fraxinus pubescens var. *Lindheimeri* Wenz., l. c., 184 (1883).
Fraxinus Americana var. *Berlandieriana* Wesmael, in Bull. Soc. Bot. Belg., XXX, 108 (1892).

Fraxinus caroliniana Mill.   **Water Ash.**

Syn.—*Fraxinus Caroliniana* Miller, Gard. Dict., ed. 8, No. 6 (1768).
Fraxinus americana Marshall, Arb. Am., 50 (1785), not Linnæus (1753).
Fraxinus juglandifolia Larmarck, Enc. Méth. Bot., II, 548 (1786). —?
Fraxinus excelsior ? Walter, Fl. Caroliniana, 254 (1788), not Linn. (1753).
FRAXINUS PLATYCARPA Michaux, Fl. Bor.-Am., II, 256 (1803).
Fraxinus pallida Bosc, in Mém. Inst. Fr., IX, 201 (1811).
Fraxinus pubescens Bosc, l. c., 210 (1811), not Lam. (1786).
Fraxinus triptera Nuttall, Genera, II, 232 (1818).
Fraxinus curvidens Hoffmannsegg, Verz. Pflanzenkult., 29 (1824).
Samarpses triptera Rafinesque, New Fl. and Bot., 3d Pt., 93 (1836).
Fraxinus americana caroliniana Browne, Trees Am., 398 (1846).
Fraxinus americana platycarpa Browne, l. c. (1846).
Fraxinus americana triptera Browne, l. c., 399 (1846).
Fraxinus pauciflora Nuttall, Sylva, III, 61, t. 100 (1849).
Fraxinus Nuttallii Buckley, in Proc. Acad. Sci. Phila. 1860, 444 (1860).
Fraxinus nigrescens Buckley, l. c., 1862. 5 (1862).
Fraxinus Cubensis Grisebach, Cat. Pl. Cub., 170 (1866).

Fraxinus caroliniana Mill.—Continued.

SYN.—*Fraxinus Platycarpa* var. *Floridana* Wenzig, in Bot. Jahrb.,
viert. B., 185 (1883).
Fraxinus nigra subsp. Caroliniana Wesmael, in Bull. Soc.
Bot. Belg., XXX, 113 (1892).

COMMON NAMES.

Water Ash (N. C., S. C., Fla., Ala., Miss., La., Tex.).
Carolina Ash (Pa., cult.).
Poppy Ash (Ala.).
Pop Ash (Fla.).

Fraxinus oregona Nutt.   **Oregon Ash.**

SYN.—*Fraxinus pubescens β* Hooker, Fl. Bor.-Am., II, 51 (1838).
Fraxinus latifolia Bentham, Bot. Voyage Sulph., 33 (1844).
Fraxinus Oregona Nuttall, Sylva, III, 59, t. 99 (1849).
Fraxinus Oregona β Nutt., l. c., 59 (1849).
Fraxinus Californica Koch, Dendrol., zw. Th., erst. Ab., 260
(1872).
Fraxinus Oregona var. *riparia* Wenzig, in Bot. Jahrb., viert.
B., 187 (1883).
Fraxinus Americana subsp. Oregona Wesmael, in Bull. Soc.
Bot. Belg., XXX, 110 (1892).

COMMON NAME.

Oregon Ash (Cal., Wash., Oreg.).

CHIONANTHUS Linn., Spec. Pl., 8 (1753).

Chionanthus virginica Linn.   **Fringetree.**

SYN.—*Chionanthus Virginica* Linnæus, Spec. Pl., ed. 1, I, 8 (1753).
Chionanthus Zeylonica Linn., l. c. (1753).
Chionanthus trifida Moench, Meth., 478 (1794).
Chionanthus Virginica var. *latifolia* Aiton, Hort. Kew., ed.
1, I, 14 (1789).
Chionanthus Virginica var. *angustifolia* Ait., l. c. (1789).
Chionanthus vernalis Salisbury, Prodr., 14 (1796).
Chionanthus cotinifolia Willdenow, Sp. Pl., I, 47 (1797).
Chionanthus triflora Stokes, Bot. Mat. Med., I, 19 (1812).
Chionanthus Virginica var. *montana* Pursh, Fl. Am. Sept., I,
8 (1814).
Chionanthus Virginica var. *maritima* Pursh, l. c. (1814).
Chionanthus maritima Rafinesque, New Fl. Bot., 3d Pt., 86
(1836).
Chionanthus maritima var. *rhombifolia* Raf., l. c. (1836).
Chionanthus obovata Raf., l. c., 87 (1836).

Chionanthus virginica Linn.—Continued.

Syn.—*Chionanthus heterophylla* Raf., 1. c. (1836).
Chionanthus montana Raf., 1. c. (1836).
Chionanthus longifolia Raf., 1. c., 88 (1836).
Chionanthus angustifolia Raf., 1. c. (1836).
Chionanthus latifolia Aiton ex Steudel, Nom. Bot., ed. sec.,
I, 350 (1840).
Chionanthus fragrans Edwards ex Steudel, 1. c., 351 (1840).
Chionanthus verna Baillon, Hist. Pl., I, 295 (1869).

COMMON NAMES.

Fringetree (R. I. (cult.), N. Y. (cult.), N. J., Pa., Del., D. C.,
N. C., S. C., Fla., Miss., La., Tex., Mo.).
White Fringe (Mass., R. I., Pa.).
American Fringe (W. Va.).
White Ash (W. Va.).
Old Man's Beard (N. C., S. C., Ala., Fla., Miss., La.).
Flowering Ash (S. C.).
Snowflower-tree (Tenn.).

OSMANTHUS Laureiro, Fl. Coch., 28 (1790).

Osmanthus americanus (Linn.) Benth. & Hook.   **Devilwood.**

Syn.—*Olea Americana* Linnæus, Mantiss., 24 (1767).
Osmanthus Americanus Bentham & Hooker, Genera Pl., II,
677, (1876).
COMMON NAMES.
Devilwood (Ala., Fla.).
Wild Olive (Fla.).

Family BORRAGINACEÆ.

CORDIA Linn., Spec. Pl., 190 (1753).

Cordia sebestena Linn.   **Geigertree.**

Syn.—*Cordia Sebestena* Linnæus, Spec. Pl., ed. 1, I, 190 (1753).
Cordia juglandifolia Jacquin, Enum. Pl. Carib., 14 (1760).
Cordia speciosa Willdenow, in Roemer & Schultes, Syst.
Veg., IV, 799 (1819).
Sebestena scabra Rafinesque, Sylva Tellur., 38 (1838).
Cordia Sebestena var. *rubra* Eggers, Vidensk. Medd. For.,
Kjöbenh., 1876, 132 (1876).
COMMON NAME.
Geigertree (Fla.).

Cordia boissieri A. de C.   Anacahuita.

SYN.—*Cordia Boissieri* A. de Candolle, in de C., Prodr., IX, 478 (1845).

BOURRERIA Browne, Nat. Hist. Jam., 168 (1756).

Bourreria havanensis (Roem. & Sch.) Miers.   Strongback.

SYN.—*Ehretia Bourreria* Linnæus, Spec. Pl., ed. 2, I, 275 (1762), in part.
Ehretia Havanensis Roemer & Schultes, Syst. Veg., IV, 805 (1819).
Bourreria tomentosa γ *Havanensis* Grisebach, Fl. Brit. West Ind., 482 (1864).
Bourreria recurva Miers, in Ann. & Mag. Nat. Hist., ser. 4, III, 203 (1869).
Bourreria ovata Miers, l. c. (1869).
Bourreria Havanensis Miers, l. c., 207 (1869).

COMMON NAMES.

Strongbark (Fla.).
Strongback (Bahama Islands).

Bourreria havanensis radula (Poir.) Gr.   Bristle-leaf Strongback.

SYN.—*Ehretia radula* Poiret, in Larmarck, Enc. Méth. Bot., Suppl., II, 2 (1811).
Bourreria radula Don, Gen. Hist. Dichl. Pl., IV, 390 (1836).
Cordia Floridana Nuttall, Sylva, III, 83, t. 107 (1849).
Bourreria virgata Grisebach, in Mem. Am. Acad., new ser., VIII, 528 (1862).
Bourreria Havenensis var. *radula* Gray, Syn. Fl. N. A., ed. 1, II, Pt. I, 181 (1878).

EHRETIA Browne, Nat. Hist. Jam., 168 (1756).

Ehretia elliptica de C.   Anaqua.

SYN.—*Ehretia elliptica* de Candolle, Prodr., IX, 503 (1845).
Ehretia ciliata Miers, in Ann. and Mag. Nat. Hist., ser. 4, III, 111 (1869).
Ehretia exasperata Miers, l. c., 112 (1869).

COMMON NAME.

Knackaway (Tex.).
Anaqua (Tex.).

Family VERBENACEÆ.

CITHAREXYLUM[1] Linn., Spec. Pl., 625 (1753).

Citharexylum villosum Jacq.   **Fiddlewood.**

Syn.—*Citharexylum villosum* Jacquin, Icon. Pl. Rar., I, 12, t. 118
(1786).
Citharexylum molle Salisbury, Prodr., 108 (1796).

COMMON NAME.

Fiddlewood (Fla.).

AVICENNIA Linn., Spec. Pl., 110 (1753).

Avicennia nitida Jacq.   **Blackwood.**

Syn.—*Avicennia nitida* Jacquin, Enum. Pl. Carib., 25 (1760).
Avicennia tomentosa Meyer, Prim. Fl. Esseq., 221 (1818),
not Jacq. (1760).
Avicennia Floridana Rafinesque, in Atlantic Journ., 148
(1832).
Avicennia Meyeri Miquel, in Linnæa, XVIII, 262 (1844).
Avicennia oblongifolia Nutt. ? ex Chapman, Fl. S. States,
ed. 1, 310 (1860).

COMMON NAMES.

Blackwood (Fla.).
Blacktree (Fla.).
Black Mangrove (Fla.).

Family BIGNONIACEÆ.[2]

CATALPA Scopoli, Introd. Hist. Nat., 170 (1777).

Catalpa catalpa (Linn.) Karsten.   **Common Catalpa.**

Syn.—*Bignonia Catalpa* Linnæus, Spec. Pl., ed. 1, II, 622 (1753).

---

[1]Linnæus formerly wrote *Citharexylon* (Amoen. I, 406, 1749).

[2]*Tabebuia Donnell-Smithii* Rose, (in Bot. Gaz., XVII, 418, t. 26, 1892; in Contr.
U. S. Nat. Herb., I. No. 9, 346, 1895) belongs to this family. Although as yet not
found to occur within our borders, this species is of considerable interest commer-
cially in the United States, where the lumber is known as "Prima vera" or "White
Mahogany." It is a tree 50 to 75 feet in height and 2 to 4 feet in diameter. At present it is known to be abundant in the lower part of the department of Escuintla,
Mexico, where it is locally well known, but doubtless overlooked by botanists until
collected there by Capt. John Donnell Smith in 1890, and at Colima by Dr. Edward
Palmer in 1891. Large quantities of the logs are shipped from Manzanillo, State of
Colima, Mexico, to Cincinnati, Ohio, and San Francisco, Cal., where they are cut
into veneers and cabinet lumber.

Catalpa catalpa (L.) Kartsen—Continued.

SYN.—CATALPA BIGNONIOIDES Walter, Fl. Caroliniana, 64 (1788).
Catalpa cordifolia Moench, Meth., 464 (1794).
Catalpa ternifolia Cavanilles, Desc. Pl., 26 (1802).
Catalpa syringæfolia Sims, in Bot. Mag., XXVII, t. 1094 (1808).
Catalpa communis Du Mont de Courset, Bot. Cult., ed. 2, III, 242 (1811).
Catalpa catalpa Karsten, Deutsch. Fl., 927 (1882).

COMMON NAMES.

Catalpa[1] (Mass., R. I., Conn., N. Y., N. J., Pa., Del., W. Va., N. C., S. C., Ala., Ga., Fla., Miss., La., Ark., Ky., Mo., Ill., Kans., Nebr., Iowa, Mich., Wis., Ohio, Minn.).
Indian Bean (Mass., R. I., N. Y., N. J., Pa., N. C., Ill.).
Beantree (N. J., Del., Pa., Va., La., Nebr.).
Catawba (W. Va., Ala., Fla., Kans.).
Cigartree (R. I., N. J., Pa., W. Va., Mo., Ill., Wis., Iowa).
Catawba-tree (Del.).
Indian Cigar-tree (Pa.).
Smoking Bean (R. I.).

Catalpa speciosa Warder.   **Hardy Catalpa.**

SYN.—*Catalpa cordifolia* Jaume, in Nouveau Duhamel, II, t. 5 (1802), not Moench (1794).
Catalpa bignonioides Lesquereux, in Owen's 2d Rep. Arkan., 375 (1860), not Walt. (1788).
Catalpa speciosa (Warder in Hort.) Engelmann, in Bot. Gaz., V, 1 (1880).

COMMON NAMES.

Hardy Catalpa (Ill., Iowa, Kans., Mich. (cult.).
Western Catalpa (Pa. (cult.), Ohio, Kans., Nebr. (cult.), Ill.).
Catalpa (R. I., N. Y. (cult.), La., Ill., Ind., Mo., Wis., Iowa, Nebr., Minn. (cult.).
Cigartree (Mo., Iowa (cult.).
Bois Puant (La.).
Indian Bean (Ind.).
Shawneewood (Ind.).

CHILOPSIS Don, in Edinb. New Phil. Journ., IX, 261 (1823).

Chilopsis linearis (Cav.) Sweet.   **Desert Willow.**

SYN.—*Bignonia linearis* Cavanilles, Icon., III, 35, t. 269 (1794).

[1]This species is not supposed to be indigenous except in parts of Georgia, Florida, Alabama, and Mississippi, but through its hardiness and special attractiveness as an ornamental tree it has been widely introduced in cultivation.

Chilopsis linearis (Cav.) Sweet—Continued.

SYN.—*Chilopsis saligna* Don, in Edinburgh New Phil. Journ., IX, 261 (1823).

Chilopsis linearis Sweet, Hort. Brit., ed. 1, 283 (1827).

Chilopsis glutinosa Engelmann, in Wislizenus, Mem. Tour North. Mex., 10 (1848).

COMMON NAMES.

Desert Willow (Cal., Tex., N. Mex., Ariz., Utah, Nev.).

Texas Flowering Willow (Tex.).

Flowering Willow (Tex.).

CRESCENTIA Linn., Spec. Pl., 626 (1753).

rescentia ovata Burmann.   **Black Calabash-tree.**

SYN.—*Crescentia ovata* Burmann f., Fl. Ind., 132 (1768).

Crescentia latifolia Miller, Gard. Dict., ed. 8, No. 2 (1768).

CRESCENTIA CUCURBITINA Linnæus, Mantiss., 250 (1767).

Crescentia latifolia Lamarck, Enc. Méth. Bot., I, 558 (1783).

Crescentia lethifera Tussac, Fl. Antill., IV, 50, t. 17 (1827).

Crescentia obovata Bentham, Bot. Voyage Sulphur, 130, t. 46 (1844).

Crescentia cujete Billb. ex Beurling, in Vet. Akad. Handl. Stockh. 1854, 138 (1856), not Linn (1753), nor Velloso (1825).

Crescentia palustris Forsyth ex Seeman, in Trans. Linn. Soc., XXIII, 20 (1860).

Crescentia species Cooper, in Smithsonian Rep. 1860, 439 (1861).

COMMON NAMES.

Black Calabash Tree (Fla.).

Black Calabash (Fla.).

Family RUBIACEÆ.

EXOSTEMA[1] Richard, ex Humb. and Bonp. Pl. Æquin., 1, 131 (1808).

Exostema caribæum (Jacq.) Roem. & Schult.   **Princewood.**

SYN.—*Cinchona Caribæa* Jacquin, Enum. Pl. Carib., 16 (1760).

Cinchona Jamaicensis Wright, in Trans. Royal Phil. Soc., LXVII, 504, t. 10 (1778.)

Cinchona Caribbeana Wright, l. c., 506 (1778).

Cinchona myrtifolia Stokes, Bot. Mat. Med., I, 359 (1812.)

---

[1] The original spelling is preserved; otherwise written *Exostemma*.

Exostema caribæum (Jacq.) Roem. & Schult.—Continued.

SYN.—*Exostema Caribæum* Roemer & Schultes, Syst. Veg., V, 18 (1819).

COMMON NAME.

Princewood (Fla.).

PINCKNEYA Michx., Fl. Bor.-Am., I, 103 (1803).

Pinckneya pubens Michx.   **Fevertree.**

SYN.—*Pinckneya pubens* Michaux, Fl. Bor.-Am., I, 103, t. 13 (1803).
Cinchona caroliniana Poiret, in Lamarck, Enc. Méth. Bot., VI, 40 (1804).
Pinckneya pubescens Persoon, Syn. Pl., I, 197 (1805).

COMMON NAMES.

Georgia Bark (S. C., Fla.).
Fevertree (Ala.[1]).
Florida Quinine Bark (Fla.).

GUETTARDA Ventenat, Choix Pl., 1 (1803).

Guettarda elliptica Swartz.   **Guettarda.**

SYN.—*Guettarda elliptica* Swartz, Prodr., 59 (1788).
Guettarda Blodgetti (Shuttleworth Mss. in Herb.) Chapman, Fl. S. States, ed. 1, 178 (1860).

COMMON NAME.

Nakedwood (Fla.).

Family CAPRIFOLIACEÆ.

SAMBUCUS Linn., Spec. Pl., 269 (1753).

Sambucus mexicana Presl.   **Mexican Elder.**

SYN.—*Sambucus Mexicana* de Candolle, Prodr., IV, 322 (1830).
Sambucus glauca Bentham, Pl. Hartweg., 313 (1849), not Nutt. (1841).
Sambucus velutina Durand & Hilgard, in Journ. Acad. Sci. Phila., new ser., III, 39 (1854).
Sambucus Canadensis var. *Mexicana* Sargent, Silva, V, 88, t. CCXXI (1893).

---

[1] Teste Dr. Chas. Mohr.

Sambucus mexicana Presl.—Continued.

SYN.—Elder (N. Mex., Tex.).
Elderberry-tree.

Sambucus callicarpa[1] Greene.   **Redberry Elder.**

SYN.—*Sambucus callicarpa* Greene, Fl. Fran., Pt. III, 342 (1892).
Sambucus racemosa Gray, in Brewer & Watson, Bot. Cal.,
I, ed. 1, 278 (1876), not Linn. (1753).

Sambucus glauca Nutt.   **Pale Elder.**

SYN.—*Sambucus Ebulus cerulea*[2] Rafinesque, Alsograph. Am., 48
(1838). — ?
Sambucus glauca (Nutt.!), in Torrey & Gray, Fl. N. A., II,
13 (1841).
Sambucus Mexicana Newberry, in Pac. R. R. Rep., VI, 75
(1857), not de C. (1830).
Sambucus Californica Koch, Dendrol., zw. Th., 72 (1872).

Elder (Cal., Utah, Oreg.).
Elderberry (Cal.).
Black Elderberry (Utah).
Mountain Elder (Western States).

VIBURNUM Linn., Spec. Pl., 267 (1753).

Viburnum lentago Linn.   **Sheepberry.**

SYN.— *Viburnum lentago* Linnæus, Spec. Pl., ed. 1, I, 268 (1753).
Viburnum pyrifolium Bigelow, Fl. Bost., ed. 2, 116 (1824),
not Poir. (1808).

---

[1] I have not seen this plant, and it is included in the present catalogue on Professor Greene's authority. Professor Sargent (Silva, V, 91, 1892) doubtfully refers it to *S. glauca*, but the characters described for Professor Greene's plant— "clustered" "fruit, bright red"—seem difficult to reconcile with the solitary stem and black berries with blue bloom of *S. glauca*. It is not unlikely, however, that *S. callicarpa* may prove to be a form of the northern red-berried elder, which is believed to be distinct from the Old World *S. racemosa*.

[2] The great uncertainty as to the identity of Rafinesque's plant makes it unsafe to take up his name, which if clearly established should replace Nuttall's *S. glauca*. Presumably Rafinesque never saw specimens of his *S. cerulea*, but founded it on a statement of Captains Lewis and Clark—berries "of a pale sky blue" (Hist. Exp. Rky. Mts. and Pac. Oc., II, 160, 1814). By exclusion, Rafinesque's *S. cerulea* is the only plant in the region with fruit of this character.

Virburnum lentago Linn.—Continued.

COMMON NAMES.

SYN.—Sheepberry (Vt., N. H., Mass., R. I., Conn., N. Y., N. J.,
Pa., Del., S.C.,Ky., Ill., Iowa, Mich., Nebr.,Minn., N. Dak.,
Ohio).
Naunyberry (Vt., N. Y., Mich., Ohio, Ont., Iowa, Minn.,
N. Dak.).
Nanny Plum (Vt.).
Black Haw (Ill., Mo., Minn., N. Dak.)
Wild Raisin (Me.).
Sweetberry (Minn.).
Sweet Viburnum (R. I., Tenn., Nebr.).
Viburnum (R. I.).

Viburnum prunifolium[1] Linn. **Stagbush.**

SYN.—*Viburnum prunifolium* Linnæus, Spec. Pl.,ed. 1,1, 268 (1753).
Viburnum pyrifolium Poiret, in Lamarck, Enc. Méth. Bot.,
VIII, 653 (1808).
Viburnum amblodes Rafinesque, Alsograph. Am., 55 (1838).

COMMON NAMES.

Black Haw (R. I., N. Y., N. J., Pa., Del., Va., W. Va., N. C.,
S. C., Ala., Ga., Fla., Miss., La., Tex., Ky., Mo., Kans.,
Ill., Ind., Ohio).
Sloe (Tenn.).
Sheepberry (N. J.).
Nannyberry (N. J.).
Alisier (La.).
Stagbush.
Haw (Md., Va.).

Viburnum ferrugineum (Torr. and Gr.) Small. **Rusty Stagbush.**
SYN.—VIBURNUM PRUNIFOLIUM FERRUGINEUM β Torrey & Gray,
Fl. N. A., II, 15 (1841).
Viburnum ferrugineum Small, in Mem. Torr. Bot. Club, IV,
123, Pl., 78 (1894).

---

[1] The following is a well-marked variety recently detected in New Jersey. It is
not yet known, however, whether this plant attains arborescent size:
Viburnum prunifolium globosum Nash.
SYN.—*Viburnum prunifolium globosum* Nash, in Bull. Torr. Bot. Club, XX, 70
(1893).

APPENDIX.

LAWS OF BOTANICAL NOMENCLATURE ADOPTED BY THE CONGRESS OF BOTANISTS AT PARIS, AUGUST, 1867.

CHAPTER I.—GENERAL CONSIDERATIONS AND LEADING PRINCIPLES.

Article 1. Natural history can make no real progress without a regular system of nomenclature, acknowledged and used by a large majority of naturalists of all countries.

Art. 2. The rules of nomenclature should neither be arbitrary nor imposed by authority. They must be founded on considerations clear and forcible enough for every one to comprehend and be disposed to accept.

Art. 3. The essential point in nomenclature is to avoid or to reject the use of forms or names that may create error or ambiguity or throw confusion into science. Next in importance is the avoidance of any useless introduction of new names. Other considerations, such as absolute grammatical correctness, regularity, or euphony of names, a more or less prevailing custom, respect for persons, etc., notwithstanding their undeniable importance, are relatively accessory.

Art. 4. No custom contrary to rule can be maintained if it leads to confusion or error. When a custom offers no serious inconvenience of this kind, it may be a motive for exceptions, which we must, however, abstain from extending or imitating. In the absence of rule, or where the consequences of rules are questionable established custom becomes law.

Art. 5. The principles and forms of nomenclature should be as similar as possible in botany and in zoology.

Art. 6. Scientific names should be in Latin. When taken from another language a Latin termination is given to them, except in cases sanctioned by custom. If translated into a modern language it is desirable that they should preserve as great a resemblance as possible to the original Latin names.

Art. 7. Nomenclature comprises two categories of names: 1. Names, or rather terms, expressing the nature of the groups comprehended one within another. 2. Names particular to each of the groups of plants or animals that observation has made known to us.

CHAPTER II.—ON THE MANNER OF DESIGNATING THE NATURE AND SUBORDINATION OF THE GROUPS THAT CONSTITUTE THE VEGETABLE KINGDOM.

Art. 8. Every individual plant belongs to a species (species), every species to a genus (genus), every genus to an order (ordo familia), every order to a cohort (cohors), every cohort to a class (classis), every class to a division (divisio).

Art. 9. In many species we distinguish likewise varieties and variations, and in some cultivated species modifications still more numerous; in many genera, sections; in many orders, tribes.

341

Art. 10. Finally, if circumstances require us to distinguish a greater number of intermediate groups, it is easy, by putting the syllable "sub" before the name of the group, to form subdivisions of that group; in this manner suborder (subordo) designates a group between an order and a tribe; subtribe (subtribus), a group between a tribe and a genus, etc. The ensemble of subordinate groups may thus be carried, for uncultivated or spontaneous plants only, to twenty degrees, in the following order:

Regnum vegetabile.
  Divisio.
    Subdivisio.
      Classis.
        Subclassis.
          Cohors.
            Subcohors.
              Ordo.
                Subordo.
                  Tribus.
                    Subtribus.
                      Genus.
                        Subgenus.
                          Sectio.
                            Subsectio.
                            Species.
                              Subspecies (vel Proles, Angl. Race).
                              Varietas.
                                Subvarietas.
                                Variatio.
                                  Subvariatio.
                                  Planta.

Art. 11. The definition of each of these names of groups may vary in a certain degree, according to individual opinion and the state of science, but their relative rank, sanctioned by custom, must not be inverted. Any classification containing inversions, such as the division of genera into orders or of species into genera, is inadmissible.

Art. 12. The fertilization of one species by another gives rise to a hybrid (hybridus); that of a modification or subdivision of a species by another modification of the same species produces a half-breed (mistus, mule of florists).

Art. 13. The arrangement of species in a genus or in a subdivision is made by means of typographical signs, letters, or figures. Hybrids are classed after one of the species from which they originate, with the sign X prefixed to the generic name.

The rank of subspecies under species is marked by letters or figures; that of varieties by the series of Greek letters α, β, γ, etc. Groups below varieties and half-breeds (mule of florists) are indicated by letters, figures, or typographical signs, according to the will of the author.

Art. 14. Modifications of cultivated species should, where possible, be classed under the wild or spontaneous species from which they are derived.

For this purpose the most striking are treated as subspecies, and when constant from seed they are called races (proles).

Modifications of a secondary order take the name of varieties, and if there be no doubt as to their almost constant heredity by seed they are termed subraces (subproles).

Modifications of minor importance, more or less comparable to subvarieties, variations, or subvariations of uncultivated species, are indicated according to their origin in the following manner: 1. Satus (seedling; Gall. semis; Germ. Samling), for a form obtained from seed. 2. Mistus (blending; Gall. métis; Germ. Blendlinge), for a form arising from cross-fertilization in a species. 3. Lusus (sport; Germ. Spielart), for a form originating from a leaf bud or from any other organ, and propagated by division.

CHAPTER III.—ON THE MANNER OF DESIGNATING EACH GROUP OR ASSOCIATION OF PLANTS.

SECTION 1.—GENERAL PRINCIPLES.

ART. 15. Each natural group of plants can bear in science but one valid designation, namely, the most ancient, whether adopted or given by Linnæus, or since Linnæus, provided it be consistent with the essential rules of nomenclature.

ART. 16. No one ought to change a name or a combination of names without serious motives, derived from a more profound knowledge of facts or from the necessity of relinquishing a nomenclature that is in opposition to essential rules. (Art. 3, first paragraph, 4, 11, 15, etc.; see sec. 6.)

ART. 17. The form, the number, and the arrangement of names depend upon the nature of each group, according to the following rules:

SECTION 2.—NOMENCLATURE OF THE DIFFERENT KINDS OF GROUPS.

ART. 18. The names of divisions and subdivisions, of classes and subclasses, are drawn from their principal characters. They are expressed by words of Greek or Latin origin, some similarity of form and termination being given to those that designate groups of the same nature (Phanerogams, Cryptogams; Monocotyledons, Dicotyledons, etc.).

ART. 19. Among Cryptogams, the old family names, such as Filices, Musci, Fungi, Lichenes, Algæ, may be used for names of classes and subclasses.

ART. 20. Cohorts are designated preferably by the name of one of their principal orders, and as far as possible with a uniform termination.

Subcohorts (rarely used) may be designated in the same manner.

ART. 21. Orders (Ordines, Familiæ) are designated by the name of one of their genera, with the final "aceæ" (Rosaceæ, from Rosa; Rauunculaceæ, from Rauunculus, etc.).

ART. 22. Custom warrants the following exceptions:

(1) When the Latin name of the genus from which is taken that of the order ends in ix or is (genitive icis or idis), the termination "iceæ" or "ideæ" is admitted (Salicineæ, from Salix; Tamariscineæ, from Tamarix; Berberideæ, from Berberis).

(2) When the genus from which the name is derived has an unusually long name, no tribe in the order taking its appellation after the same genus, the termination in "eæ" is admitted (Dipterocarpeæ, from Dipterocarpus).

(3) Some large orders, named long since, have retained the exceptional names under which they are generally known (Cruciferæ, Leguminosæ, Guttiferæ, Umbelliferæ, Compositæ, Labiatæ, Cupuliferæ, Coniferæ, Palmeæ, Gramineæ, etc.).

(4) An old generic name, which has become that of a section or of a species, may be preserved as the foundation of that of the order (Lentibularieæ, from Lentibularia; Hippocastaneæ, from Æsculus hippocastanum; Caryophylleæ, from Dianthus caryophyllus, etc.).

ART. 23. The names of suborders (subordines, subfamiliæ) are derived from the name of one of the genera that form part of them, with the final "eæ."

ART. 24. The names of tribes and subtribes are taken from that of one of the genera included in the group, with the final "eæ" or "ineæ."

ART. 25. Genera, subgenera, and sections receive names, commonly substantive, which may be compared to our own proper family names. These names may be derived from any source whatsoever, and may even be arbitrarily imposed, under the restrictions mentioned further on.

ART. 26. A name may be given to subsections as well as to inferior generic sub-divisions, or these may simply be indicated by a number or by a letter.

ART. 27. When the name of a genus, subgenus, or section is taken from the name of a person it is composed in the following manner:

The name, cleared of titles or of any accessory particle, takes the final "a" or "ia."

The spelling of the syllables unaffected by this final is preserved without altera-tion, even with letters or diphthongs now employed in certain languages, but not in Latin. Nevertheless, ä, ö, and ü of the German language become ae, oe, and ue, while é and è of the French language become e.

ART. 28. Botanists who have generic names to publish show judgment and taste by attending to the following recommendations:

(1) Not to make names too long or difficult to pronounce.

(2) To give the etymology of each name.

(3) If they have formerly made a name that has not been accepted, not to estab-lish another genus under the same name, particularly in the same order or in a neighboring one.

(4) Not to dedicate genera to persons in all respects strangers to botany, or at least to natural history, nor to persons quite unknown.

(5) Not to draw names from barbarous tongues unless those names be frequently quoted in books of travel and have an agreeable form that adapts itself readily to the Latin tongue and to the tongues of civilized countries.

(6) If possible, by the composition or the termination of the word to call to mind the affinities or the analogies of the genus.

(7) To avoid adjective nouns.

(8) Not to give to a genus a name whose form is more properly that of a section (Eusideroxylon, for example).

(9) To avoid taking up names that have already been used, but have not been approved, and applying them to genera different from the former, unless it be wished again to dedicate a genus to a botanist; but even in this case it is desirable, 1, that the nullity of the first genus should be unquestionable; 2, that the order in which it is proposed to reestablish the name be quite distinct from the former one.

(10) To avoid making choice of names used in zoology.

ART. 29. Botanists constructing names for subgenera or for sections will do well to attend to the recommendations of the foregoing article, as well as to these:

(1) Give, where possible, to the principal division of a genus a name that by some modification or addition may call the name of the genus to mind. (For instance, eu at the beginning of the name when it is of Greek origin; astrum, ella, at the end of a name when Latin, or any other modification consistent with the rules of grammar and the usages of the Latin language.)

(2) Avoid calling a section by the name of the genus it belongs to with the final "oides" or "opsis." Give, on the contrary, the preference to this final for a section having some resemblance to another genus by adding in that case "oides" or "opsis" to the name of that other genus, if it be of Greek derivation, so as to form the name of the section.

(3) Avoid taking as a sectional name one already in use as such in another genus or which is that of a genus.

ART. 30. When it is required to express the name of a section together with a generic name and that of a species, the name of the section is put between the two others in a parenthesis.

ART. 31. All species, even those that singly constitute a genus, are designated by the name of the genus to which they belong followed by a name termed specific, more commonly of the adjective kind.

ART. 32. The specific name ought in general to indicate something of the appear-ance, the characters, the origin, the history, or the properties of the species. If derived from the name of a person, it usually calls to mind the name of him who discovered or described it, or who may have been otherwise concerned with it.

ART. 33. Names of persons used as specific names have a genitive or an adjective form (Clusii or Clusiana). The first is used when the species has been described or distinguished by the botanist whose name it takes; in other cases the second form is preferred. Whatever be the form chosen, every specific name derived from the name of a person should begin with a capital letter.

ART. 34. A specific name may be an old generic name or a substantive proper name. It then takes a capital and does not agree with the generic name (Digitalis Sceptrum, Coronilla Emerus).

ART. 35. No two species of the same genus can bear the same specific name, but the same specific name may be given in several genera.

ART. 36. In constructing specific names, botanists will do well to give attention to the following recommendations:

(1) Avoid very long names, as well as those that are difficult to articulate.

(2) Avoid names that express a character common to all, or to almost all, the species of a genus.

(3) Avoid names designating little-known or very limited localities, unless the species be very local.

(4) Avoid, in the same genus, names too similar in form—above all, those that only differ in their last letters.

(5) Readily adopt unpublished names found in traveler's notes or in herbaria, unless they be more or less defective (see art. 17).

(6) Avoid names that have been already used in the genus, or in some nearly allied genus, and have become synonyms.

(7) Name no species after anyone who has neither discovered, nor described, nor figured, nor studied it in any way.

(8) Avoid specific names composed of two words.

(9) Avoid specific names having, etymologically, the same meaning as the generic name.

ART. 37. Hybrids whose origin has been experimentally demonstrated are designated by the generic name, to which is added a combination of the specific names of the two species from which they are derived, the name of the species that has supplied the pollen being placed first with the final i or o, and that of the species that has supplied the ovulum coming next, with a hyphen between (Amaryllis vittato-reginæ, for the Amaryllis proceeding from A. reginæ, fertilized by A. vittata).

Hybrids of doubtful origin are named in the same manner as species. They are distinguished by the absence of a number and by the sign × being prefixed to the generic name (× Salix capreola, Kern.).

ART. 38. Names of subspecies and varieties are formed in the same way as specific names, and are added to them according to relative value, beginning with those of the highest rank. Half-breeds (mules of florists) of doubtful origin are named and ranked in the same manner.

Subvarieties, variations, and subvariations of uncultivated plants may receive names analogous to the foregoing, or merely numbers or letters, for facilitating their arrangement.

ART. 39. Half-breeds (mules of florists) of undoubted origin are designated by a combination of the two names of the subspecies, varieties, subvarieties, etc., that have given birth to them, the same rules being observed as in the case of hybrids.

ART. 40. Seedlings, half-breeds of uncertain origin, and sports should receive from horticulturists fancy names in common language, as distinct as possible from the Latin names of species or varieties. When they can be traced back to a botanical species, subspecies, or variety, this is indicated by a succession of names (Pelargonium zonale, Mrs. Pollock).

SECTION 3.—ON THE PUBLICATION OF NAMES AND ON THE DATE OF EACH NAME OR COMBINATION OF NAMES.

ART. 41. The date of a name or of a combination of names is that of its actual and irrevocable publication.

ART. 42. Publication consists in the sale or the distribution among the public of printed matter, plates, or autographs. It consists, likewise, in the sale or the distribution, among the leading collections, of numbered specimens, accompanied by printed or autograph tickets bearing the date of the sale or distribution.

ART. 43. The communication of new names in a public meeting and the placing of names in collections or in gardens open to the public do not constitute publication.

ART. 44. The date put to a work is presumed to be correct till there is evidence to the contrary.

ART. 45. A species is not looked upon as named unless it has a generic name as well as a specific one.

ART. 46. A species announced in a work under generic and specific names, but without any information as to its characters, can not be considered as being published. The same may be said of a genus announced without being characterized.

ART. 47. Botanists will do well to conform to the following recommendations:

(1) To give accurately the date of publication of their works or portions of works, and that of the sale or the distribution of named and numbered plants.

(2) To publish no name without clearly indicating whether it is that of an order or of a tribe, of a genus or of a section, of a species or of a variety—in short, without giving an opinion as to the nature of the group to which the name is given.

(3) To avoid publishing or mentioning in their works unpublished names which they themselves do not accept, especially if the authors of such names have not expressly authorized them to do so. (See art. 36, 5.)

SECTION 4.—ON THE PRECISION TO BE GIVEN TO NAMES BY THE QUOTATION OF THE AUTHOR WHO FIRST PUBLISHED THEM.

ART. 48. For the indication of the name or names of any group to be accurate and complete it is necessary to quote the author who first published the name or combination of names in question.

ART. 49. An alteration of the constituent characters or of the circumscription of a group does not warrant the quotation of another author than the one who first published the name or combination of names.

When the alteration is considerable the words mutatis charact., or pro parte, or excl. syn., excl. sp., excl. var., or any other abridged indication are added to the quotation of the original author, according to the nature of the changes that have been made and of that of the group that is dealt with.

ART. 50. Names published from a private document, such as an herbarium, a non-distributed collection, etc., are individualized by the addition of the name of the author who publishes them, notwithstanding the contrary indication that he may have given. In like manner, names used in gardens are individualized by the mention of the author who first publishes them.

The herbarium, the collection, or the garden should be fully quoted in the text. (Lam. ex Commers. Mss. in Herb. Par.; Lindl. ex Hort. Lodd.)

ART. 51. When a group is moved without alteration of name to a higher or lower rank than that which it held before, the change is considered equivalent to the creation of an entirely new group, and the author who has effected the change is the one to be quoted.

ART. 52. Authors' names put after those of plants are abbreviated unless they are very short.

For this purpose preliminary particles or letters that do not strictly speaking form part of the name are suppressed, and the first letters are given without any omission whatsoever. If a name of one syllable is long enough to make it worth while to abridge it, the first consonants only are given (Br. for Brown); if the name has two or more syllables, the first syllable and the first letter of the following one are taken; or, the two first, if they are both consonants (Juss. for De Jussieu; Rich. for Richard).

When it is found necessary to give more of a name for the sake of avoiding confusion between names beginning with the same syllables, the same system is to be followed. For instance, two syllables are given, together with the one or two first consonants of the third; or else one of the last characteristic consonants of the name is added (Bertol. for Bertoloni, so that it may be distinguished from Bertero; or Michx. for Michaux, to prevent confusion with Micheli. Christian names or accessory designations serving to distinguish two botanists of the same name are abridged in the same way (Adr. Juss. for Adrien de Jussieu, Gærtn. fil. or Gærtn. f. for Gærtner son).

When it is a settled custom to abridge a name in another manner it is best to conform to it (L. for Linnæus, St.-Hil. for Saint-Hilaire).

Section 5.—On Names that are to be Retained where a Group is Divided, Remodeled, Transferred, or Moved from One Rank to Another, or when Two Groups of the Same Rank are United.

Art. 53. An alteration of characters or a revision carrying with it the exclusion of certain elements of a group or the addition of fresh ones does not warrant a change in the name or names of a group.

Art. 54. When a genus is divided into two or more genera its name must be retained and given to one of the chief divisions. If the genus contains a section or some other division which, judging by its name or by its species, is the type or the origin of the group, the name is reserved for that part of it. If there is no such section or subdivision, but one of the parts detached contains, however, a great many more species than the others, it is to that that the original name is to be applied.

Art. 55. In case two or more groups of the same nature are united into one the name of the oldest is preserved. If the names are of the same date, the author chooses.

Art. 56. When a species is divided into two or more species, if one of the forms happens to have been distinguished earlier than the others the name is retained for that form.

Art. 57. When a section or a species is moved into another genus, when a variety or some other division of a species is given as such to another species, the name of the section, the specific name, or that of the division of the species is maintained unless there arise one of the obstacles mentioned in articles 62 and 63.

Art. 58. When a tribe is made into an order, when a subgenus or a section becomes a genus, or a division of a species becomes a species, or vice versa, the old names are maintained, provided the result be not the existence of two genera of the same name in the vegetable kingdom, two divisions of a genus, or two species of the same name in the same genus, or two divisions of the same name in the same species.

Section 6.—On Names that are to be Rejected, Changed, or Altered.

Art. 59. Nobody is authorized to change a name because it is badly chosen or disagreeable, or another is preferable or better known, or for any other motive, either contestable or of little import.

Art. 60. Every one is bound to reject a name in the following cases:

(1) When the name is applied, in the vegetable kingdom, to a group that has before received a name in due form.

(2) When it is already in use for a class or for a genus, or is applied to a division or to a species of the same genus, or to a subdivision of the same species.

(3) When it expresses a character or an attribute that is positively wanting in the whole of the group in question, or at least in the greater part of the elements it is composed of.

(4) When it is formed by the combination of two languages.

(5) When it is in opposition to the rules laid down in section 5.

ART. 61. The name of a cohort, subcohort, order, suborder, tribe, or subtribe must be changed if taken from a genus found not to belong to the group in question.

ART. 62. When a subgenus, a section, or a subsection passes as such into another genus the name must be changed if there is already in that genus a group of the same rank under the same name.

When a species is moved from one genus into another its specific name must be changed if it is already borne by one of the species of that genus. So, likewise, when a subspecies, a variety, or some other subdivision of a species is placed under another species its name must be changed if borne already by a form of like rank of that species.

ART. 63. When a group is transferred to another, keeping there the same rank, its name will have to be changed if it leads to misconception.

ART. 64. In the cases foreseen in articles 60, 61, 62, and 63 the name to be rejected or changed is replaced by the oldest admissible one existing for the group in question; in the absence of this, a new one is to be made.

ART. 65. The name of a class, of a tribe, or of any other group above the genus may have its termination altered so as to suit rule or custom.

ART. 66. When a name derived from Latin or Greek has been badly written or badly constructed, when a name derived from that of a person has not been written consistently with the true spelling of that name, or when a fault of gender has carried with it incorrect terminations of the names of species or of their modifications every botanist is authorized to rectify the faulty names or terminations, unless it be a question of a very ancient name current under its incorrect form. This right must be used reservedly, especially if the change is to bear upon the first syllable and, above all, upon the first letter of the name.

When a name is drawn from a modern language it is to be maintained just as it was made, even in the case of the spelling having been misunderstood by the author and justly deserving to be criticized.

SECTION 7.—ON NAMES OF PLANTS IN MODERN LANGUAGES.

ART. 67. Latin scientific names, or those that are immediately derived from them, are used by botanists preferably to names of another kind or having another origin unless these are very intelligible and in common use.

ART. 68. Every friend of science ought to be opposed to the introduction into a modern language of names of plants that are not already there unless they are derived from a Latin botanical name that has undergone but a slight alteration.

LAWS OF BOTANICAL NOMENCLATURE ADOPTED BY THE BOTANICAL CLUB OF THE AMERICAN ASSOCIATION FOR THE ADVANCEMENT OF SCIENCE. AT ROCHESTER, N. Y., AUGUST, 1892.

Resolved, That the Paris code of 1867 be adopted except where it conflicts with the following:

I. *The law of priority.*—Priority of publication is to be regarded as the fundamental principle of botanical nomenclature.

II. *Beginning of botanical nomenclature.*—The botanical nomenclature of both genera and species is to begin with the publication of the first edition of Linnæus's Species Plantarum, in 1753.

III. *Stability of specific names.*—In the transfer of a species to a genus other than the one under which it was first published the original specific name is to be retained unless it is identical with the generic name or with a specific name previously used in that genus.

IV. *Homonyms.*—The publication of a generic name or a binomial invalidates the use of the same name for any subsequently published genus or species, respectively.

V. Publication of genera.—Publication of a genus consists only (1) in the distribution of a printed description of the genus named; (2) in the publication of the name of the genus and the citation of one or more previously published species as examples or types of the genus, with or without a diagnosis.

VI. Publication of species.—Publication of a species consists only (1) in the distribution of a printed description of the species named; (2) in the publishing of a binomial with reference to a previously published species as a type.

VII. Similar generic names.—Similar generic names are not to be rejected on account of slight differences, except in the spelling of the same word; for example, Apios and Apium are to be retained, but of Epidendrum and Epidendron, Asterocarpus and Astrocarpus, the latter is to be rejected.

VIII. Citation of authorities.—In the case of a species which has been transferred from one genus to another the original author must always be cited in parenthesis, followed by the author of the new binomial.

The following amendments were accepted by the same association, at Madison, Wis., August, 1893:

1. The amendment of Section III of the Rochester Code of Nomenclature by striking out all after the word "retained."

2. That the general sequence of natural orders as taken up in Engler and Prantl's Naturliche Pflanzenfamilien be adopted.

3. That in determining the name of a genus or species to which two or more names have been given by an author in the same volume or on the same page of a volume, precedence shall decide.

A committee was authorized to proceed with the publication of a revised list, matters concerning it not determined by the club being referred to the committee with power, which led to the following additional rules promulgated by the committee:

I. That the original name is to be maintained, whether published as species, subspecies, or variety.

II. That varieties are to be written as trinomials.

III. That specific or varietal names derived from persons or places, or used as the genitive of generic names or substantives, are to be printed with a capital initial letter.

IV. That no comma is to be inserted between the specific or varietal name and the name of the author cited.

AMERICAN ORNITHOLOGISTS' UNION CODE OF NOMENCLATURE, ADOPTED IN 1886.

A.—GENERAL PRINCIPLES.

I. Zoological nomenclature is a means, not an end, of zoological science.

II. Zoological nomenclature is the scientific language of systematic zoology, and vernacular names are not properly within its scope.

III. Scientific names are of the Latin nor language, and when derived from another language are to be latinized in form; but names which have been used in zoological nomenclature as if they were Latin words can not be changed or rejected if they are otherwise unobjectionable.

IV. Zoological nomenclature has no necessary connection with botanical nomenclature, and names given in one of these two systems can not conflict with those of the other system; use of a name in botany, therefore, does not prevent its subsequent use in zoology.

V. A name is only a name, having no meaning until invested with one by being using as the handle of a fact; and the meaning of a name so used in zoological nomenclature does not depend upon its signification in any other connection.

B.—CANONS OF ZOOLOGICAL NOMENCLATURE.

OF THE KINDS OF NAMES IN ZOOLOGY.

I. Zoological nomenclature includes two kinds of names: (1) Common names, definitive of the relative rank of groups in the scale of classification; (2) proper names, appellative of each group of organisms.

II. All members of any one group in zoology are included in and compose the next higher group, and no inversion of the relative rank of groups is admissible.

III. Proper names of groups above genera consist preferably of a single word, taken as a noun and in the nominative plural.

IV. Proper names of families uniformly consist of a single word ending in idæ; of subfamilies, of a single word ending in inæ; of other groups, of one word or more of no fixed termination.

V. Proper names of families and subfamilies take the tenable name of some genus, preferably the leading one, which these groups respectively contain, with change of termination into -idæ or -inæ. When a generic name becomes a synonym, a current family or subfamily name based upon such generic name becomes untenable.

VI. Proper names of genera and subgenera are single words, preferably nouns, or to be taken as such, in the nominative singular, of no definite construction and no necessary signification.

VII. Proper names of all groups in zoology, from kingdom to subgenus, both inclusive, are written and printed with a capital initial letter.

VIII. Proper names of species and of subspecies or "varieties" are single words, simple or compound, preferably adjectival or genitival, or taken as such when practicable, agreeing in gender and number with any generic name with which they are associated in binomial nomenclature, and written with a small initial letter.

IX. Proper names do not attach to individual organisms nor to groups of lower grades than subspecies, names which may be applied to hybrids, to monstrosities, or other individual peculiarities, or to artificial varieties, such as domestic breeds of animals, having no status in zoological nomenclature.

OF THE BINOMIAL SYSTEM AS A PHASE OF ZOOLOGICAL NOMENCLATURE.

X. Binomial nomenclature consists in applying to every individual organism, and to the aggregate of such organisms not known now to intergrade in physical characters with other organisms, two names, one of which expresses the specific distinctness of the organism from all others, the other its superspecific indistinctness from or generic identity with certain other organisms, actual or implied, the former name being the specific, the latter the generic designation, the two together constituting the technical name of any specifically distinct organism.

OF THE TRINOMIAL SYSTEM AS A PHASE OF ZOOLOGICAL NOMENCLATURE.

XI. Trinomial nomenclature consists in applying to every individual organism, and to the aggregate of such organisms known now to intergrade in physical characters, three names, one of which expresses the subspecific distinctness of the organism from all other organisms, and the other two of which express, respectively, its specific indistinctness from or generic identity with certain other organisms, the first of these names being the subspecific, the second the specific, and the third the generic designation, the three written consecutively, without the intervention of any other word, term, or sign, constituting the technical name of any subspecifically distinct organism.

OF THE BEGINNING OF ZOOLOGICAL NOMENCLATURE PROPER AND OF THE OPERATION OF THE LAW OF PRIORITY.

XII. The law of priority begins to be operative at the beginning of zoological nomenclature.

XIII. Zoological nomenclature begins at 1758, the date of the tenth edition of the Systema Naturæ of Linnæus.

XIV. The adoption of a "statute of limitation" in modification of the lex prioritatis is impracticable and inadmissable.

XV. The law of priority is to be rigidly enforced in respect to all generic, specific, and subspecific names.

XVI. The law of priority is only partially operative in relation to names of groups higher than genera, and only where names are strictly synonymous.

OF NAMES PUBLISHED SIMULTANEOUSLY.

XVII. Preference between competitive specific names published simultaneously in the same work, or in two works of the same actual or ostensible date (no exact date being ascertainable), is to be decided as follows:

1. Of names, the equal pertinency of which may be in question, preference shall be given to that which is open to least doubt.

2. Of names of undoubtedly equal pertinency (a) that founded upon the male is to be preferred to that founded upon the female; (b) that founded upon the adult to that on the young, and (c) that founded on the nuptial condition to that of the pre⁻nuptial or postnuptial conditions.

3. Of names of undoubtedly equal pertinency, and founded upon the same condition of sex, age, or season, that is to be preferred which stands first in the book.

· XVIII. Preference between competitive generic names published simultaneously in the same work, or in two works of the same actual or ostensible date (no exact date being ascertainable), is to be decided as follows:

1. A name accompanied by the specification of a type takes precedence over a name unaccompanied by such specification.

2. If all or none of the genera have types indicated, that generic name takes precedence the diagnosis of which is most pertinent.

OF THE RETENTION OF NAMES.

XIX. A generic name when once established is never to be canceled in any subsequent subdivision of the group, but retained in a restricted sense for one of the constituent portions.

XX. When a genus is subdivided the original name of the genus is to be retained for that portion of it which contained the original type of the genus when this can be ascertained.

XXI. When no type is clearly indicated, the author who first subdivides a genus may restrict the original name to such part of it as he may judge advisable, and such assignment shall not be subject to subsequent modification.

XXII. In no case should the name be transferred to a group containing none of the species originally included in the genus.

XXIII. If, however, the genus contains both exotic and nonexotic species—from the standpoint of the original author—and the generic term is one originally applied by the ancient Greeks or Romans, the process of elimination is to be restricted to the nonexotic species.

XXIV. When no type is specified, the only available method of fixing the original name to some part of the genus to which it was originally applied is by the process of elimination, subject to the single modification provided for by Canon XXIII.

XXV. A genus formed by the combination of two or more genera takes the name first given in a generic or subgeneric sense to either or any of its components. If both or all are of the same date, that one selected by the reviser is to be retained.

XXVI. When the same genus has been defined and named by two authors, both giving it the same limits, the later name becomes a synonym of the earlier one; but in case these authors have specified types from different sections of the genus, and these sections be raised afterwards to the rank of genera, then both names are to be retained in a restricted sense for the new genera.

XXVII. When a subgenus is raised to full generic rank its name is to be retained as that of the group thus raised. In like manner, names first proposed or used in a subspecific sense are tenable in case the subspecies be raised to full specific standing, and are to have priority over a new name for the subspecies so elevated.

XXVIII. When it become necessary to divide a composite species or subspecies, the old specific or subspecific name is to be retained for that form or portion of the group to which it was first applied or to which it primarily related. If this can not be positively ascertained, the name as fixed by the first reviser is to be retained.

XXIX. When a species is separated into subspecies, or when species previously sup_ posed to be distinct are found to intergrade, the earliest name applied to any form of the group shall be the specific name of the whole group, and shall also be retained as the subspecific designation of the particular form to which it was originally applied. In other words, the rule of priority is to be strictly enforced in respect to subspecific names.

XXX. Specific names when adopted as generic are not to be changed.

XXXI. Neither generic nor specific names are to be rejected because of barbarous origin, for faulty construction, for inapplicability of meaning, or for erroneous signification.

XXXII. A nomen nudum, generic or specific, may be adopted by a subsequent author, but the name takes both its date and authority from the time when and the author by whom the name becomes clothed with significance by being properly defined and published.

OF THE REJECTION OF NAMES.

XXXIII. A generic name is to be changed which has been previously used for some other genus in the same kingdom; a specific or subspecific name is to be changed when it has been applied to some other species of the same genus or used previously in combination with the same generic name.

XXXIV. A nomen nudum is to be rejected as having no status in nomenclature.

XXXV. An author has no right to change or reject names of his own proposing, except in accordance with rules of nomenclature governing all naturalists, he having only the same right as other naturalists over the names he has himself proposed.

XXXVI. A name resting solely on an inadequate diagnosis is to be rejected on the ground that it is indeterminable and therefore not properly defined.

XXXVII. If an author describes a genus and does not refer to it any species, either then or previously described, the genus can not be taken as established or properly defined unless the characters given have an unmistakable significance.

XXXVIII. A species can not be considered as named unless both generic and specific names have been applied to it simultaneously—i. e., unless the species has been definitely referred to some genus.

XXXIX. A name which has never been clearly defined in some published work is to be changed for the earliest name by which the object shall have been so defined, if such name exist; otherwise, a new name is to be provided, or the old name may be properly defined and retained, its priority and authority to date from the time and author so defining it.

OF THE EMENDATION OF NAMES.

XL. The original orthography of a name is to be rigidly preserved unless a typographical error is evident.

OF THE DEFINITION OF NAMES.

XLI. A name to be tenable must have been defined and published.

XLII. The basis of a generic or subgeneric name is either (1) a designated recognizably described species, or (2) a designated recognizable plate or figure, or (3) a published diagnosis.

XLIII. The basis of a specific or subspecific name is either (1) an identifiable published description, or (2) a recognizable published figure or plate, or (3) the original type specimen or specimens absolutely identified as the type or types of the species or subspecies in question; but in no case is a type specimen to be accepted as the basis of a specific or subspecific name when it radically disagrees with or is contradictory to the characters given in the diagnosis or description based upon it.

XLIV. In determining the pertinence of a description or figure on which a genus, species, or subspecies may respectively rest, the consideration of pertinency is to be restricted to the species scientifically known at the time of publication of the description or figure in question or to contemporaneous literature.

XLV. Absolute identification is requisite in order to displace a modern current name by an older obscure one.

XLVI. In describing an organism which is considered to represent a new genus as well as a new species, it is not necessary to formally separate the characters into two categories, generic and specific, in order to render tenable the names given to the organism in question, although such a distinction is desirable.

OF THE PUBLICATION OF NAMES.

XLVII. Publication consists in the public sale or distribution of printed matter—books, pamphlets, or plates.

XLVIII. The reading of a paper before a scientific society or a public assembly does not constitute publication, and new genera and species first announced in this way date only from the time of their subsequent and irrevocable publication.

XLIX. The date borne by a publication is presumed to be correct till proved otherwise; although it is well known that in many instances, as in the proceedings or transactions of societies and in works issued in parts, the date given is not that of actual publication; and when this fact can be substantiated, the actual date of publication, if it can be ascertained, is to be taken.

Remarks.—It is notorious that the dates on the title-page of the completed volume of works issued in parts often antedate, sometime postdate, the actual publication of the different parts or are otherwise erroneous. Also, that the volumes of proceedings of learned societies not infrequently bear simply the date of the period or year to which they relate, even when not published till months, and sometimes years, after the ostensible date; and that serial publications, when not issued promptly, as not infrequently happens, are sometimes antedated by several months. This state of things is happily less prevalent now than formerly, and is more frequently the result of inattention or failure to appreciate the importance of precision in such matters than from any motive of unfairness. At the present time authors in good standing are careful to make permanent record of the date of publication of each part of a work issued in successive brochures or printer's signatures; and societies not infrequently give the exact date of the appearance of each signature or part of their various publications. This, it is needless to urge, is a practice which should become general.

Where doubt arises as to the priority of publication between a properly dated work and one improperly or dishonestly dated, it would hardly be unfair to throw the *onus probandi* on the publishers of the latter, or to favor the work the date of which is not open to question.

Finally, respecting the matter of publication, your committee would submit the following:

Naturalists would do well (a) to indicate exactly the date of publication of their works, parts of works, or papers; (b) to avoid publishing a name without indicating the nature of the group (whether generic, subgeneric, or supergeneric) it is intended to distinguish; (c) to avoid including in their publications any unaccepted manuscript names, since such names only needlessly increase synonymy; (d) societies, Government or other surveys, or other publishing boards should indicate the date of issue of each part of works published serially or in installments, as well as of all volumes and completed works.

Furthermore, the custodians of libraries, public or private, would do well to indicate, either in the work itself or in the proper book of record, the date of reception of all publications received, particularly in the case of those of a serial character or which are issued in parts. (This, it may be observed, is a practice carefully adhered to in well-regulated libraries of the present time.)

OF THE AUTHORITY FOR NAMES.

L. The authority for a specific or subspecific name is the describer of the species or subspecies. When the first describer of a species or subspecies is not also the authority it is to be inclosed in parentheses; e. g., Turdus migratorius L., or Merula migratoria (L).

Remarks.—Ordinarily the use of authorities may be omitted, as in incidental reference to species of a well-known fauna in faunal lists, etc.; but, on the other hand, the use of authorities may be of the greatest importance in giving exact indication of the sense in which a name is used—for instance, in check lists or monographic and revisionary works.

In writing the names of subspecies the authority for the specific or second element of the name may nearly always be omitted.

The relation of authorities may be otherwise indicated; as, e. g., Merula migratoria L. sp., or Merula migratoria Sw. & Rich. ex L., or Merula migratoria Sw. & Rich. (L. sub. Turdus), etc.; but the method first above mentioned has the merit of the greater simplicity and brevity.

Two very different practices have prevailed among naturalists in respect to authorities for names. The B. A. Code gave preference to the authority for the specific name, for the following reasons: "Of the three persons concerned with the construction of a binomial title * * * we conceive that the author who first describes and names a species which forms the groundwork of later generalizations possesses a higher claim to have his name recorded than he who afterwards defines a genus which is found to embrace that species, or who may be the mere accidental means of bringing the generic and specific names into contact By giving the authority for the specific name in preference to all others, the inquirer is referred directly to the original description, habitat, etc., of the species, and is at the same time reminded of the date of its discovery." Agassiz and others opposed this practice and gave preference to the referrer of the species to its proper genus, on the ground that it required greater knowledge of the structure and relationship of species to properly classify them than to simply name and describe them. By this school the authority is considered as constituting part of the name. This method is also in accordance with the usage of the older zoologists and botanists, from Linnæus down. But it often happens that the authority for the combination of names used is not that of the classifier, but of the author, who has merely "shuffled names," or worked out the synonymy in accordance with nomenclatural rules, and has had nothing to do with the correct allocation of the species.

LI. The authority for a name is not to be separated from it by any mark of punctuation (except as provided for under Canon L).

Remarks.—In respect to punctuation and typography in relation to names and their authorities usage varies, but it is quite generally conceded that no comma need be used between the name and its authority; "the authority," as Verrill has suggested, "being understood to be a noun in the genitive case, though written in the nominative form or more frequently abbreviated." In printing, the authority is usually and advisably distinguished by use of type differing from that of the name; if the latter be in italic type the authority may be in roman, or if in small capitals or in antique the authority may be in italic type, etc.

LII. The name of the authority, unless short, is to be abbreviated, and the abbreviation is to be made in accordance with commonly recognized rules, and irregularly formed and nondistinctive abbreviations are to be avoided.

C.—RECOMMENDATIONS FOR ZOOLOGICAL NOMENCLATURE IN THE FUTURE.

OF THE CONSTRUCTION AND SELECTION OF NAMES.

I. As already provided under Canon II, the rules of Latin orthography are to be adhered to in the construction of scientific names.

II. In latinizing personal names only the termination should be changed, except as in cases provided for under Recommendation IV.

III. The best zoological names are those which are derived from the Latin or Greek and express some distinguishing characteristic of the object to which they are applied.

OF THE TRANSLITERATION OF NAMES.

IV. Names adopted from languages written in other than Roman characters, as the Greek, Russian, Arabic, Japanese, etc., or from languages containing characters not represented in the Roman alphabet, as the Spanish, French, German, Scandinavian, Western Slavonian, etc., should be rendered by the corresponding Roman letters or combinations of letters.

OF THE DESCRIPTION OF ZOOLOGICAL OBJECTS.

V. When naming a new species or subspecies always give a diagnosis, as short as possible, but still containing all the essential features by which the species or subspecies may be distinguished from the other known members of the genus to which it is referred. Base the diagnosis on the type specimen and indicate the museum where the type is deposited and the catalogue number by which it may be identified. Give a comparison with the nearest allied forms and tabulate, if possible, the characters of the new form in a "key" to the genus, or a section of it.

VI. When establishing a new genus always mention at least the family to which it is considered to belong and a single typical species; give, then, the diagnostic characters by which the members of the genus may be distinguished from those of the allied genera.

OF THE BIBLIOGRAPHY OF NAMES.

VII. In preparing tables of bibliographical references in works of a revisionary or monographic character, all published works which throw light upon the history of the organisms in question are subject to citation.

VIII. Citations are to be made in chronological order, the earliest name given to the organism standing first and the other designations following in due sequence; then, under each designation are to be arranged, also in chronological order, the several works or papers which treat of the organism under such designation. The date of publication is always to be made a part of the citation.

IX. When the diagnostic characters or the limits of a group have been changed, such change should be shown by an abridged indication of the character of the change, as "mut. char.," "pro parte," to follow the citation.

OF THE SELECTION OF VERNACULAR NAMES.

X. Vernacular names, though having no standing in scientific nomenclature and being not strictly subject to the law of priority, have still an importance that demands the due exercise of care in their selection, especially with reference to their fitness and desirability.

ADDITIONS AND EMENDATIONS.

Page 97, 31st line: Read **Juniperus sabinoides** (H. B K.) Sarg., and add *Cupressus sabinoides* Humb., Bonpl., and Kunth, Nov. Gen. Sp., II, 3 (1817), and *Juniperus sabinoides* Sargent, Silva, X, 91 (1896), not Griseb. (1844), nor Endl. (1847), to synonymy.

Page 106, 26th line: Read **Yucca constricta** Buckl., and add *Yucca constricta* Buckley, in Proc. Acad. Sci. Phila., 1862, 8 (1862), to synonymy.

Page 107, 25th line: Add following species now known to be arborescent:

Yucca aloifolia Linn. **Spanish Bayonet.**

SYN.—*Yucca aloifolia* Linnæus, Spec. Pl., ed. 1, I, 319 (1753).

Yucca gloriosa Linn. **Spanish Dagger.**

SYN.—*Yucca gloriosa* Linnæus, Spec. Pl., ed. 1, I, 319 (1753).

Page 111, 5th line: Read *Hicorius.*
Page 140, 8th line (et seq.): Read *Betula.*
Page 142, 32d line: Read *Betula pumila × lenta.*
Page 163, 9th line: Read QUERCUS.
Page 200, 28th line: Read *Laurus.*
Page 225, 9th line (et seq.): Read *oxyacantha.*
Page 270, 2d line: Read *Azedarach.*
Page 297, 21st line: Read *Scutia.*
Page 339, 29th line: A revision of this species has just been published, the following being the synonymy as now understood:

Viburnum rufotomentosum Small. **Rusty Stagbush.**

SYN.—*Viburnum prunifolium β ferrugineum* Torrey and Gray, Fl. N. A., II, 15 (1841), not *V. Lentago ferrugineum* Rafinesque (1838).
Viburnum ferrugineum Small, in Mem. Torr. Bot. Club, IV, 123, pl. 78 (1894).
Viburnum rufotomentosum Small, in Bull. Torr. Bot. Club, vol. 23, p. 410 (1896).

The following new species have been described since going to press:

Abies arizonica Merriam. **Arizona Cork Fir.**

SYN.—*Abies arizonica* Merriam, in Proc. Biolog. Soc., Washington, X, 115, (1896).

Dr. Merriam first discovered this fir on San Francisco Mountain, Arizona, in September, 1890, and believed it to be distinct from the allied *A. subalpina*, but for the lack of specimens it has since remained undescribed. During the present season Drs. Fernow and Merriam visited the same locality and secured specimens which appear to amply distinguish it from the Alpine fir. It is a tree 30 to 40 feet in height, very characteristic in its fine-grained, whitish, elastic, and corky bark. At present known only from San Francisco Mountain and the neighboring peaks.

Cratægus saligna Greene.

SYN.—*Cratægus saligna* Greene, in Pittonia, III, 99 (1896).

Professor Greene has recently described this species as a "slender tree, 10 to 18 feet high, with rather few branches and long, willowy, more or less drooping branchlets." "Plentiful along the lower Cimarron River, Colorado," and "allied to *C. rivularis.*"

INDEX OF LATIN NAMES.

[Accepted names in heavy-face type; synonyms in *italics*; names commonly used hitherto. in SMALL CAPITALS.]

ABIES Juss , 50.
Abies acutissima Hort. ex Beissn., 37.
Abies alba Chapm , 33.
Abies alba Michx., 36.
Abies alba Torr., 38.
Abies alba var. *arctica* Parl., 36
Abies alba argentea Hort. ex Gord., 37.
Abies alba aurea Beissn., 38.
Abies alba cærulea Hort. ex Carr., 37.
Abies alba compacta pyramidalis Hort. ex Beissn , 38
Abies alba echinoformis Hort. ex Carr , 38.
Abies alba glauca Plumbly ex Gord. 37.
Abies alba nana Loud. 38.
Abies alba nana glauca Hort. ex Beissn., 38.
Abies alba pendula Hort. ex Carr., 38.
Abies alba prostrata Hort. ex Carr., 38.
Abies Albertiana Murr., 44.
Abies annabilis Murr., 53.
Abies amabilis (Loud.) Forb., 56.
Abies amabilis Hort. ex Beissn., 56.
Abies amabilis Hort. ex Carr., 58.
Abies Americana Koch, 34.
Abies Americana Mill., 41.
Abies Americana alba Hort. ex Beissn., 37.
Abies Americana cærulea Beissn., 37.
Abies Americana rubra Hort. ex Beissn., 35.
Abies arctica Murr., 33.
Abies arctica Cunn. ex Gord., 35.
Abies aromatica Raf., 53.
Abies balsamea (Linn.) Mill., 50.
Abies balsamea Bigel., 54.
Abies balsamea argentea Beissn., 52.
Abies balsamea argentifolia nom. nov., 52.
Abies balsamea brachylepis Willk., 51.
Abies balsamea cærulea Carr., 52.
Abies balsamea denudata Carr.,52.
Abies balsamea β Fraseri Nutt., 50.
Abies balsamea β Fraseri Spach., 50.
Abies balsamea globosa Hort. ex Beissn., 52.
Abies balsamea hemisphærica nom. nov., 52.
Abies balsamea hudsonia (Knight) Veitch, 51.
Abies balsamea hudsonica Beissn., 51.
Abies balsamea longifolia (Loud.) Endl., 51.
Abies balsamea nana Carr., 52
Abies balsamea nudicaulis Carr., 51.
Abies balsamea paucifolia nom. nov., 52.
Abies balsamea prostrata (Knight) Carr., 51.
Abies balsamea tenuifolia Hort. ex Carr., 52.
Abies balsamea variegata Carr., 52.

Abies balsamea versicolor nom nov., 52.
Abies balsamifera Michx. 50
Abies bifolia Murr., 53.
Abies bracteata Hook. & Arn., 56
Abies Bridgei Kell. 44
Abies cærulea Forb., 37.
Abies cærulea Booth ex Loud., 36.
Abies cærulescens Hort. ex Koch, 36.
Abies californica Hort., ex Stend., 46.
Abies californica vera Hort. ex Beissn., 56.
Abies campylocarpa Murr., 58.
Abies Canadensis Mill., 34.
Abies Canadensis Michx., 41.
Abies Canadensis Coop., 44.
Abies canadensis alba-spica Barron ex Gord., 43.
Abies canadensis aurea Beissn., 43.
Abies canadensis compacta nana Hort. ex Beissn., 42.
Abies canadensis fastigiata Hort. ex Beissn., 43.
Abies canadensis globosa Beissn., 42.
Abies canadensis globularis Hort. ex Beissn., 42.
Abies Canadensis gracilis Wat. ex Gord., 42.
Abies canadensis macrophylla Hort. ex Beissn., 43.
Abies canadensis macrophylla Hort. Beissn., 45.
Abies canadensis var. microphylla Lindl. ex Hoop., 43.
Abies canadensis nana Hort. ex Carr., 42.
Abies canadensis parvifolia Veitch, 43.
Abies canadensis pendula Beissn., 43.
Abies Caroliniana Chapm., 44.
Abies commutata Gord., 39.
Abies concolor (Gord.) Parry, 54.
Abies concolor Lindl. & Gord., 54.
Abies concolor angustata nom. nov., 55.
Abies concolor fastigiata Charg., 55.
Abies concolor var. *lasiocarpa* Beissn., 56.
Abies concolor var. *lasiocarpa pendula* Hort. ex Beissn., 55.
Abies concolor var. *lasiocarpa variegata* Hort. ex Beissn., 55.
Abies concolor lowiana (Murr.) Lem., 56.
Abies concolor varia nom. nov., 55.
Abies concolor pendens (Beissn.) nom. nov., 55.
Abies concolor purpurea nom. nov., 55.
Abies concolor purpurea compressa nom. nov., 55.
Abies concolor violacea Hort. ex Beissn., 55.
Abies concolor violacea compacta Hort. ex Beissn., 55.
Abies curvifolia Salisb., 36.
Abies denticulata Michx., 33.

357

369

CORDIA Linn., 332.
Cordia bolssieri A. de C., 333.
Cordia Floridana Nutt., 333.
Cordia juglandifolia Jacq., 332.
Cordia sebestena Linn., 332.
Cordia Sebestena var. *rubra* Egg., 332.
Cordia speciosa Willd., 332.
CORNACEÆ, 308.
CORNUS Linn., 308.
Cornus alterna Marsh., 309.
Cornus alternifolia Linn. f., 309.
Cornus florida Linn., 308.
Cornus florida Hook., 309.
Cornus florida pendula Temple, 309.
Cornus florida rubra Temple, 309.
Cornus nuttallii Aud., 309.
Cornus plicata Tausch, 309.
Cornus punctata Raf., 309.
Cornus riparia Raf., 309.
Cornus riparia var. *rugosa* Raf., 309.
Cornus rotundifolia Raf., 309.
Cornus undulata Raf., 309.
Corypha Palmetto Walt., 104.
COTINUS Adans., 273.
Cotinus Americana Nutt., 273.
Cotinus Coggygria Engl., 273.
Cotinus cotinoides (Nutt.) Britt. 273.
Cotoneaster spathulata Wenz., 231.
CRATÆGUS Linn., 214.
Cratægus acerifolia Hort., 219.
Cratægus acerifolia Moench, 231.
Cratægus æmula Hort., 234.
Cratægus æstivalis (Walt.) Torr. & Gr., 235.
Cratægus alpestris ex Koch, 216.
Cratægus amœna Salisb., 212.
Cratægus angustifolia Borckh., 217.
Cratægus apiifolia Med., 223.
Cratægus apiifolia (Marsh.) Michx., 232.
Cratægus apiifolia minor Loud., 232.
Cratægus arborescens Ell., 232.
Cratægus arbutifolia Ait., 235.
Cratægus atropurpurea Stev. ex Nym., 228.
Cratægus aurea Hort., 231.
Cratægus Azarella Griseb., 224.
Cratægus berberifolia Torr. & Gr., 218.
Cratægus betulæfolia Lodd., 234.
Cratægus Bosciana Roem., 218.
Cratægus brachyacantha Sarg. & Engelm., 215.
Cratægus brachyacantha var. *maxima* Kuntze, 215.
Cratægus brevispina Dougl., 214.
Cratægus brevispina Hort., 225.
Cratægus Bruantii Hort., 228.
Cratægus calycina Peterm. ex Nym., 222.
Cratægus caroliniana Lodd., 217.
Cratægus Caroliniana Pers., 233.
Cratægus Carrierei Carr., 216.
Cratægus chloracarpa Koch, 219.
Cratægus coccinea Linn., 218.
Cratægus coccinea Brand., 220.
Cratægus coccinea Kelmanni Hort., 221.
Cratægus coccinea corallina Loud., 218.
Cratægus coccinea c flabellata Dipp., 219.
Cratægus coccinea glandulosa Hort., 221.
Cratægus coccinea indentata Loud., 218.
Cratægus coccinea var. *macracantha* Dudl., 220.

Cratægus coccinea 4 maxima Lodd., 218.
Cratægus coccinea var. *mollis* Torr. & Gr., 220.
Cratægus coccinea δ oligandra Torr. & Gr., 218.
Cratægus coccinea populifolia Torr. & Gr , 219.
Cratægus coccinea d pruinosa Dipp., 220.
Cratægus coccinea spinosa Godef., 218.
Cratægus coccinea var. *typica* Reg., 219.
Cratægus coccinea β viridis Torr. & Gr., 218.
Cratægus coccinea var. *viridis* Torr., 220.
Cratægus cordata (Mill.) Ait., 231.
Cratægus coronaria Salisb., 206.
Cratægus Coursetiana Roem., 216.
Cratægus crus-galli Linn., 215.
Cratægus Crus-galli Wang., 230.
Cratægus crus-galli angustifolia (Ehr.) nom. nov., 217.
Cratægus Crus-galli var. *arbutifolia* Arb. Kew, 216.
Cratægus crus-galli berberifolia (Torr. & Gr.) Sarg., 218.
Cratægus crus-galli fontanesiana (Spach) Wenz., 217.
Cratægus Crus-galli δ linearis de C., 217.
Cratægus Crus-galli var. *ovalifolia* Lindl., 217.
Cratægus Crus-galli var. *pruinosa* Hort. Kew, 216.
Cratægus crus-galli prunellifolia (Poir.) nom. nov., 216.
Cratægus crus-galli prunifolia (Marsh.) Torr. & Gr., 217.
Cratægus Crus-galli var. *pyracanthifolia* Ait., 216.
Cratægus Crus-galli var. *pyracanthifolia* Reg., 232.
Cratægus crus-galli salicifolia (Med.) Ait., 216.
Cratægus Crus-galli b *salicifolia* a *linearis* Dipp. 217.
Cratægus Crus-galli var. *splendens* Ait., 215.
Cratægus cuneifolia Borkh., 230.
Cratægus digyna Pall., 222.
Cratægus dissecta Borkh., 223.
Cratægus diversifolia Steud., 224.
Cratægus douglasii Lindl., 214.
Cratægus Douglasii Mac., 220.
Cratægus Douglasii var. *rivularis* Sarg., 214.
Cratægus Downingii Hort., 220.
Cratægus elegans Poir., 223.
Cratægus elliptica Soland., 233.
Cratægus Elliptica Ell., 235.
Cratægus eriocarpa Lodd., 227.
Cratægus fastigiata Hort., 226.
Cratægus fissa Bosc ex de C., 223.
Cratægus fissa Lee ex Loud., 227.
Cratægus flava Hort. ex Loud., 226.
Cratægus flava Darl., 230.
Cratægus flava Ait., 232.
Cratægus flava Ell., 234.
Cratægus flava elliptica (Ait.) Sarg., 233.
Cratægus flava var. *lobata* Lindl., 233.
Cratægus flava var. *pubescens* Gr., 234.
Cratægus flexispina Hort., 231.
Cratægus flexispina Sarg., 233.
Cratægus flexispina var. *pubescens* (Gr.) Millsp., 234.
Cratægus flexuosa Schw., 230.
Cratægus flexuosa de C., 234.
Cratægus florida Lodd., 235.
Cratægus Fontanesiana Steud., 217.
Cratægus glandulosa Willd., 218.

18158—No. 14——24

Thuja occidentalis rosenthali Beissn., 67.
Thuja occidentalis silver-queen Beissn., 70.
Thuja occidentalis spaethi Beissn., 66.
Thuja occidentalis spihlmanui Beissn., 67.
Thuja occidentalis tatarica Beissn., 67.
Thuja occidentalis theodonensis Beissn., 67.
Thuja occidentalis varia nom. nov., 68.
Thuja occidentalis vervœneana Gord , 67.
Thuja occidentalis viridis Beissn., 67.
Thuja occidentalis walthamensis Gord., 66.
Thuja occidentalis wareana Gord., 66.
Thuja occidentalis wareana globosa Beissn., 66.
Thuja occidentalis wareana lutescens Beissn., 66.
Thuja pigmœa Hort. ex. Gord., 73.
Thuja plicata Don, 70.
Thuja plicata argenteo-versicolor nom. nov., 72.
Thuja plicata atrovirens (Gord.) nom. nov , 71.
Thuja plicata aureo-variegata Hort. ex Beissn , 72.
Thuja plicata aurescens (Beissn) nom. nov , 71.
Thuja plicata compacta (Carr.) Beissn., 72.
Thuja plicata cristatiformis nom. nov., 73.
Thuja plicata erecta (Gord.) nom. nov., 73.
Thuja plicata flava nom nov., 72.
Thuja plicata gracillima (Beissn.) nom. nov. 71.
Thuja plicata llaveana Gord., 72.
Thuja plicata Llaveana Hort. ex Gord., 72.
Thuja plicata minima Gord., 73.
Thuja plicata penduliformis nom. nov., 73.
Thuja plicata pumila (Gord.) nom. nov., 73.
Thuja plicata variegata Carr., 72.
Thuja prostrata Hort. ex Gord., 73.
Thuja recurva nana Hort. ex Gord. 73.
Thuja Sibirica Hort. ex Gord., 66.
Thuja sphœroidea nana Hort. ex Gord., 78.
Thuja Vervœneana Van-Geert ex Gord., 67.
Thuya Bodmeri Hort. ex Beissn., 68.
Thuya Craigiana Murr., 64.
Thuya craigiana Hort., 71.
Thuya Craigiana glauca Hort. ex Beissn., 65.
Thuya cristata Hort. ex Beissn., 69.
Thuya excelsa Bong., 79.
Thuya gigantea Carr., 64.
Thuya gigantea Nutt., 70.
Thuya gigantea atrovirens Hort. ex Gord. 71.
Thuya gigantea aurea Hort. ex Beissn., 72.
Thuya gigantea aureo-variegata Hort. ex Beissn., 72.
Thuya gigantea aurescens Beissn., 71.
Thuya gigantea compacta Hort. Kew., 72.
Thuya gigantea gracilis Hort. ex Beissn., 72.
Thuya gigantea gracillima Hort. ex Beissn., 67.
Thuya gigantea lutescens Hort. ex Beissn., 71.
Thuya gigantea var. pendula Hort. Kew., 73.
Thuya gigantea var. plicata Don, 71.
Thuya gigantea var. plicata cristata Hort. Kew., 73.
Thuya gigantea var. plicata lutea Hort. Kew., 72.
Thuya gigantea semperaurea Hort. ex Beissn., 71.
Thuya globosa Hort. ex Beissn., 69.
Thuya Lobbi atrovirens Smith ex Gord., 71.
Thuya Lobbi aurea Hort. ex Beissn., 72.
Thuya Lobbi aureo-variegata Hort. ex Beissn., 72.
Thuya Lobbi gracilis Hort. ex Beissn., 71.
Thuya Lobbi lutescens Hort. ex Beissn., 72.
Thuya Lobbi semperaurea Hort. ex Beissn., 71.
Thuya Menziesii Dougl., 71.
Thuya obtusa Moench, 65.
Thuya occidentalis albo-spica Hort. ex Beissn., 70.

Thuya occidentalis atrovirens Hort. ex Beissn., 67.
Thuya occidentalis aurescens Hort. ex Beissn., 67.
Thuya occidentalis columnaris Hort. ex Beissn., 66.
Thuya Occidentalis conpocta Carr., 72.
Thuya Occidentalis dumosa Hort. ex Gord., 72.
Thuya occidentalis erecta viridis Hort. ex Beissn., 67.
Thuya occidentalis globosa compacta Hort. ex Beissn., 69.
Thuya occidentalis globosa viridis Hort. ex Beissn. 69.
Thuya occidentalis lutea nana Hort. ex Beissn., 68.
Thuya occidentalis magnifica Hort ex Beissn., 67.
Thuya occidentalis nana Carr. 72.
Thuya occidentalis Ohlendorffi Hort. ex Beissn., 66.
Thuya occidentalis pendula glauca Hort. ex Beissn. 68.
Thuya occidentalis var. plicata Loud. ex Gord , 71.
Thuya occidentalis pumila Hort. ex Beissn 70.
Thuya occidentalis pyramidalis Hort. *x Beissn., 66.
Thuya occidentalis stricta Hort. ex Beissn., 66
Thuya occidentalis Tom Thumb Hort. ex Beissn., 65.
Thuya occidentalis Victoria Hort. ex Beissn., 70.
Thuya odorata Marsh., 65.
Thuya plicata argenteo-variegata Hort. ex Beissn., 72.
Thuya plicata aurea Hort. ex Beissn , 68.
Thuya plicata aurea Beissn., 72.
Thuya plicata dumosa Hort. ex Gord., 72.
Thuya plicata pygmœa Beissn , 73.
Thuya plicata Wareana Hort. ex Beissn., 66.
Thuya recurvata Hort. ex Beissn. 68.
Thuya sphœroidalis Rich., 76.
Thuya sphœroidea Spreng., 76.
Thuya tetragona Hort. ex Beissn.. 66.
Thuyœcarpus juniperinus Trantv., 100.
Thuyopsis borealis argenteo-variegata Hort. ex Beissn., 80.
Thuyopsis borealis aureo-variegata Hort. ex Beissn., 80.
Thuyopsis borealis compacta Hort ex Beissn., 81.
Thuyopsis borealis compressa Hort. ex Beissn., 81.
Thuyopsis borealis glauca Hort. ex Beissn., 80.
Thuyopsis borealis gracilis Hort. ex Beissn , 81.
Thuyopsis borealis pendula Hort. ex Beissn., 81.
Thuyopsis borealis viridis Hort. ex Beissn., 80.
TILIA Linn.; 300.
Tilia alba Michx. f., 302.
Tilia americana Linn., 300.
Tilia Americana Walt., 302.
Tilia Americana gigantea Hort. ex Dipp., 301.
Tilia americana 5 heterophylla Loud., 302.
Tilia Americana 2 laxiflora Loud., 301.
Tilia Americana c Moltkei Hort. Späth, 301.
Tilia Americana multiflora Hort. ex Dipp., 301.
Tilia Americana b pubescens Dipp., 301.
TILIA AMERICANA 3 PUBESCENS Lond., 302.
Tilia americana 4 pubescens leptophylla Lond., 302.
Tilia Americana d Rosenthallii Hort. ex Dipp., 301.
*Tilia Americana β Walteri-*Wood, 302.
Tilia Canadensis Michx., 300.
Tilia Caroliniana Mill., 300.
Tilia cinera Raf., 302.
Tilia fulva Raf., 302.
Tilia fuscata Raf., 301.

INDEX OF COMMON NAMES.